AF789073

# AEROSPACE SENSORS

# AEROSPACE SENSORS

ALEXANDER V. NEBYLOV

MOMENTUM PRESS

MOMENTUM PRESS, LLC, NEW YORK

*Aerospace Sensors*
Copyright © Momentum Press®, LLC, 2013.

All rights reserved. No part of this publication may be reproduced, stored in a retrieval system, or transmitted in any form or by any means—electronic, mechanical, photocopy, recording, or any other—except for brief quotations, not to exceed 400 words, without the prior permission of the publisher.

First published by Momentum Press®, LLC
222 East 46th Street, New York, NY 10017
www.momentumpress.net

ISBN-13: 978-1-60650-059-0 (paperback)
ISBN-10: 1-60650-059-7 (paperback)
ISBN-13: 978-1-60650-061-3 (e-book)
ISBN-10: 1-60650-061-9 (e-book)

DOI: 10.5643/9781606500613

Cover design by Jonathan Pennell
Interior design by Exeter Premedia Services Private Ltd.,
Chennai, India

10 9 8 7 6 5 4 3 2 1

Printed in the United States of America

# CONTENTS

# Series Preface

The present series is concerned with sensors *per se*, and because the subject matter is so wide-ranging in both scope and maturity, this must be reflected within the individual volumes. So, whereas care has been taken to include a considerable amount of practical material, the proportion of such leavening is inevitably variable. The present volume will be found to include material on the basic processes that are addressed by the sensors used in most aspects of aerospace technology, plus considerable detail on the relevant sensors themselves and their applications. In the context of aerospace engineering, however, there are many items of complex equipment—mostly radio and navigationally oriented—that can be considered as sensors in their own right. This situation has been addressed in a companion volume that is in production at the time of writing, and will act as an adjunct to the present work.

Sensors cannot, of course, be divorced from their associated instrumentation, and problems that arise in atmospheric flight (aircraft) are to some considerable extent different from those relevant to space flight (spacecraft). Nevertheless, many aeronautically-oriented sensors do qualify for use in spacecraft, especially those like NASA's now-retired shuttle, that need terrestrial and landing instrumentation. Hence, there is still common ground between the two, and this has been addressed by the inclusion of some basic material on both atmospheric and space flight.

Much of aerospace engineering is currently—and happily—an international endeavor, and the present volume clearly recognizes this by the inclusion of material from both Eastern and Western countries. In particular, it has benefited greatly from the expertise of the Volume Editor and contributor, Alexander V. Nebylov, one of Russia's most eminent authorities in the field.

J. Watson,
Series Editor
Swansea, UK, 2012

# PREFACE

This book is devoted to modern sensors and their applications in control systems relevant to aerospace vehicles.

Two centuries ago, a person who wished to move faster than walking or running would mount a good horse and enjoy riding using all his or her sensory perceptions. Since that time, many kinds of vehicles have been created for satisfying the desire for fast and elaborate transportation. In the air, and more recently in space, considerable increases in the speed-ranges of various forms of transport have taken place, but have demanded the creation of a huge variety of sensing elements for detecting the state and behavior of the relevant vehicles.

A horse-rider does not need to supervise the function of a horse's legs, and to turn, it is enough to give a clear command via the harness. However, all aerospace vehicles demand the monitoring of many parameters and it is only recently that control systems have become sufficiently "intelligent" to relieve the pilot of constantly monitoring the behavior of those parameters and interpreting their aggregate meaning. Thus, a modern aerospace vehicle may be compared to a good horse, which itself knows how to operate each leg so as to orient its body in the direction desired by its rider.

Another example is afforded by birds, which may be considered excellent examples of "intelligent" flight control. They too can be compared with modern aircraft, which some believe are actually winning the competition with nature, and which demand near-perfect sensors and control algorithms for the realization of such phenomenal performance.

The automatic control of aerospace systems with huge numbers of operating parameters is one of the highest technological achievements of modern civilization, and does indeed compete functionally with those inherent in natural life-forms, including human beings. However, the operating principles of the various necessary sensors and automatic systems are often essentially different from those utilized in nature and form the knowledge base of leading designers and firms in the field of aerospace instrumentation. It should also be noted that the majority of aerospace sensors differ considerably from those designed for applications in automobile, ship, railway, and other forms of transportation, or those used in industrial, chemical, medicinal, and other areas. The topic of aerospace sensors therefore merits special treatment, and it is hoped that this book will to some extent fulfill this requirement.

The intent of the volume is to present the fundamentals of design, construction and application of numerous aerospace sensors, a concept born in the International Federation of Automatic Control (IFAC), especially in its Aerospace Technical Committee. An international team of twelve authors represents four countries from Eastern and Western Europe and North America, and all of whom have considerable experience in aerospace sensor and systems design.

The nine chapters in this volume cover the majority of sensors for aircraft and missiles, and many for spacecraft, satellite, and space probes. Principles of operation, design, and achievable performance for different sensors are presented along with particulars of their construction. The introductory Chapter 1 briefly reviews the characteristics of atmospheric and space environments, this knowledge being essential for understanding the operation of aerospace sensors. Material on aerospace vehicle classification, specific design criteria, and the requirements of onboard systems and sensors is presented.

Chapter 2 is devoted to modern achievements in the development of the oldest group of aircraft sensors—membranous aneroid and other atmospherically-based instruments. Flight altitude and components of speed, attack, and slip angle measurements are also considered.

Chapters 3 and 4 introduce radio-altimeters and other autonomous radio sensors for motional parameters such as ground speed and crab angle. Airborne weather sensors and collision avoidance devices are also reviewed.

Chapters 5 and 6 cover accelerometers and gyroscopes of various kinds which are broadly used as basic sensors in the construction of gimbaled and strapdown inertial navigation systems (INS) and for direct applications in aerospace vehicles.

Chapter 6 was written by six co-authors and became the largest in the book. It recognizes the particularly important role of INS and separate gyroscopic sensors for aerospace vehicular navigation and motion control.

Chapter 7 presents the different aspects of magnetic, gyro-magnetic, and electronic compass design and their application to flight.

In Chapter 8, engine parameter information collection systems are considered. Fuel quantity and consumption sensors, pressure pick-ups, tachometers, vibration control, and temperature sensors are all described.

Finally, Chapter 9 is devoted to principles and examples of sensor integration. The most important facets of sensor system choice, integrated measuring system optimization, and the simulation of sensor integration by appropriate algorithms are described. The examples of sensor integration considered include the noninvariant and robust integrated metering of speed, radio-inertial measurements, airborne gravimetry, and orbital verticality.

The book is written for practicing engineers, designers, and researchers in the area of control systems for various aerospace vehicles including aircraft, UAVs, missiles, spacecraft, satellites, and space probes. It may also be used as a study guide for both undergraduate and graduate students and for postgraduates in aerospace engineering, aeronautics, astronautics, and various related areas. Moreover, it will be found useful by other people wishing to satisfy their general interest in the modern aerospace technologies that are so important in shaping our twenty-first century life-styles.

Alexander V. Nebylov
St.-Petersburg, Russia

# ACKNOWLEDGMENTS

I would like to thank the contributing authors of this book, all of whom kindly responded positively to my invitations to participate in an international team tasked with writing reviews of various aspects of modern aerospace sensor technology. All these authors are distinguished specialists in their own technological fields, and already had heavy design and research workloads. It was not easy for them to find time to write their chapters, and especially to conform to the required expository style. I am very grateful to all these authors for their responsible approaches to these constraints, and deeply respect their abiding commitments to their chosen subjects in science and engineering.

I also wish to express my sincere gratitude to the editor of the series, Dr. Joe Watson, for his consistent support during the compilation of the book, especially during a difficult period when a change of publisher occurred. His recommendations and helpful critiques promoted improvements in the structure and quality of the content.

Especial gratitude is due to my colleague from the IFAC Aerospace Technical Committee (ATC), Prof. Klaus Schilling, who appointed me editor of the book after mooting the idea of its inclusion within a series at IFAC. Prof. Schilling kindly discussed with me the structure of the chapters at the initial preparatory stage. I am also grateful to the members of the ATC and particularly to its present chairman, Prof. Houria Siguerdidjane.

My numerous colleagues, professors, and researchers at the State University of Aerospace Instrumentation in St. Petersburg also merit my profound gratitude, as do several generations of its students, for the fruitful dialogues that have allowed me to realize my own vision of the problems inherent in developing and perfecting aerospace sensors.

I am also indebted to the editorial and production staff of Momentum Press, especially Mr. Joel Stein, for their valued suggestions and constructive advice during the preparation of the manuscript.

Finally, I would like to thank my family, because my work on the book was performed at the expense of time otherwise available for companionship with my wife Elena and my son Vladimir, both of whom who suffered this with fortitude over several years.

Any criticism regarding the contents and material in the book and the quality of its presentation will be accepted with gratitude.

A.V. Nebylov,
Editor

# About the Series Editor

Dr. Joseph Watson is an Electrical Engineering graduate of the University of Nottingham, England, and the Massachusetts Institute of Technology, where he held a King George VI Memorial Fellowship. A Fellow of the IET, Senior Member of the IEEE, and a member of Sigma Xi (the Scientific Research Society of America), he has published books and papers in various areas including electronic circuit design, nucleonics, biomedical electronics and gas sensors; and is a former Associate Editor for the IEEE Sensors Journal. He has also been visiting professor at the University of Calgary, Canada, and the University of California, Davis and Santa Barbara; and has held various consultancies with firms in the USA, Canada, and Japan. Since retirement from Swansea University, UK, he has continued as chairman of the UK-based Gas Analysis and Sensing Group, and as series editor for the present volumes.

# About the Editor

Alexander Nebylov graduated with honors as an Engineer in Missile Guidance from the Leningrad Institute of Aircraft Instrumentation in 1971. He led many R&D projects in aerospace instrumentation, motion control systems and avionics, and is a scientific consultant for various Russian design bureaus and research institutes. He was awarded the Candidate of Science degree in 1974 and the Doctor of Science degree in 1985, both in Information Processing and Control Systems, from the State Academy of Aerospace Instrumentation. He achieved the academic rank of full professor in the Higher Attestation Commission of the USSR in 1986.

For the last two decades, Dr. Nebylov has been with the State University of Aerospace Instrumentation in St. Petersburg as professor and chairman of Aerospace Devices and Measuring Complexes, and director of the International Institute for Advanced Aerospace Technologies. He is an author of fourteen books and numerous scientific papers and has also been a member of the leadership of the IFAC Aerospace Technical Committee since 2002. He has also been a member of the Presidium of the International Academy of Navigation and Motion Control since 1995. In 2006 the title of Honored Scientist of the Russian Federation was bestowed on Prof. Nebylov.

# CHAPTER 1

# INTRODUCTION

Alexander V. Nebylov
*State University of Aerospace Instrumentation*
*St. Petersburg, Russia*

## 1.1 GENERAL CONSIDERATIONS

### *1.1.1 TYPES OF AEROSPACE VEHICLES AND MISSIONS*

Aircraft are designed to fly within the atmosphere, which makes possible the creation of motion, lift, and attitude control. Spacecraft are designed for travel in open space; the presence of air is not necessary because the thrust is created by rocket engines. There are three main ways for an aircraft to gain lift within the atmosphere: the fixed wing for normal airplanes, the rotor (or rotary wing) for helicopters, and an envelope filled with any gas less dense than the surrounding air for aerostats such as balloons or airships (dirigibles).

There are also some special aircraft such as surface-effect vehicles that fly at low levels above land or water. In the case of cushioncrafts (hovercraft), air that is impelled by a propeller and retained by a special skirt forms an air cushion between the hull and the surface. In the case of wing-in-ground-effect vehicles (WIGs), or ekranoplans intended for use over water, a dynamic air cushion is produced between the specially configured wing and the surface. In both cases the close proximity of the ground or water surface requires appropriate approaches in the measurement of the relevant flight parameters.

Taking into account the problems encountered in the provision of appropriate sensors, aerospace vehicles can be classified on the basis of several main characteristics. The aircraft or spacecraft may be unmanned, have an obligatory pilot and optional or mandatory other members, or a crew and passengers. The flight of manned aerospace vehicles has to be smooth and without significant linear or angular accelerations in order to not injure the crew and passengers. The maximum tolerance overload for a trained pilot is roughly between four and eight gravitational units (4g and 8g). This limitation makes the range of some flight parameters narrower and simplifies the designed functioning of sensors.

Stringent safety requirements for manned vehicles make it necessary that all equipment, especially sensors, should be highly reliable, and in some cases it is important that the crew should be able to replace or repair a failed sensor during flight. The acceptable level of failure probability for navigation and motion control sensors in manned aerospace vehicles should be no more than $10^{-5}$ to $10^{-6}$. For some unmanned vehicles—for example, guided missiles—such requirements can be relaxed, making the sensors cheaper and lighter.

For manned aerospace vehicles, life support systems must be controlled, for which special sensors may be required, in particular those for the measurement of relative oxygen and carbon dioxide content, air temperature, and humidity, and for emergency rescue systems. However, in all cases the mass of the vehicle must be kept in mind when sensors and sensor systems are being designed. This mass can vary from a few kilograms for microsatellites and space probes, some short-range guided missiles and pilotless air scouts, to several hundred tons for space stations, heavy launchers, and great airliners. It is possible to install any number of sensors on a heavy vehicle without restraint, but each additional sensor can appreciably decrease the payload of a light vehicle, making it necessary to limit their number and weight. Micromechanical sensors, microelectronics, and a mechatronic approach can often provide solutions to the weight/reliability problem.

The nature of the vehicle's mission can also affect the requirements for onboard sensors and such missions can include

- transport operations with the aim of delivering passengers or cargo to a given area or port;
- travel with no fixed destination (space station, navigation satellite, patrol aircraft, etc.);
- the launch of payloads into certain orbits or trajectories;
- the rendezvous and docking of two vehicles;
- flights to test vehicles' performance and characteristics;
- sport or tourist flights for pilot or passenger pleasure;
- combat missions; and
- the exploration of deep space.

Each kind of mission requires its own variety of onboard sensors and criteria for control algorithm optimization.

Aerospace vehicles also travel at different speeds. The greatest speeds ever achieved by man-made objects have been by aerospace vehicles, especially spacecraft. As a rule, increasing speed makes the task of accurately positioning the vehicle more difficult. This is especially obvious when satellite navigation systems are used as the basic means for positioning. Furthermore, not only a vehicle's speed, but also its motion dynamics, influences the accuracy of instruments that measure flight parameters.

Even a single aerospace vehicle can have different modes of flight with different requirements for motion measurement. At some stages of flight such demands may be more rigorous, and the quality and complexity of measurement may vary over time. For example, spacecraft docking maneuvers have to be very precise, and accurate orientation when correcting orbit and descent into the atmosphere is mandatory. For guided missiles, the terminal phase of trajectory control is of front-rank importance; for passenger planes, safe landing requires the greatest accuracy from the navigation system; and for warplanes, attack maneuvers may put the maximum load on the relevant sensors. So, the requirements for sensors and sensor systems and the criteria for system optimization may not be invariant, but may depend on the phase of flight.

### *1.1.2 THE ROLE OF SENSORS AND CONTROL SYSTEMS IN AEROSPACE*

Aerospace vehicles can be considered as representing a class of transportation with unique and specific features. No other transport vehicles can perform such complex modes of motion along required trajectories with such great speed and carry such responsibility for mission success. The specific cost (per unit mass) of aerospace vehicles exceeds that of cars and other means of transport many times over, and it is well known that some defense tasks may be solved only by using aerospace technologies. All these features of aerospace vehicles define the specific role and design of sensors and their control systems. Aerospace vehicles have four other characteristics, as follows, that are not shared by other vehicles.

- Aerospace crafts are the only vehicles permitting motion in 3D space with extreme dynamics (as compared with submarines, for example). Hence it is always necessary to control the vehicle's 6D motion (linear and angular), for which complex means of automation plus the direct application of automatic control systems with perfect sensors are required.
- The specific payload cost aboard aerospace vehicles is so great that for economic reasons it is advisable to decrease the number of crew members. In fact, the creation of pilotless vehicles may prove economically expedient. Furthermore, for some military applications, only pilotless vehicles can be used, as is the case of deep space probes which in principle cannot return to earth. Naturally, pilotless crafts must have full sets of sensors that provide all the necessary data for automatic control.
- Many kinds of aerospace vehicles cannot principally be serviced or repaired during a long flight, and consequently it is very important that all systems and sensors be extremely reliable. The survivability of automatic systems aboard such vehicles can be achieved only by the application of structural redundancy including backup and interchangeability of sensors. All this must be considered at the design stage.
- Many aerospace vehicles must be able to fly far from, and without communicating with, any navigational infrastructure, which requires them to be autonomous and self-maintaining. The role of onboard sensors and systems in this case is therefore more important than for external data or devices. Thus, criteria for the selection of these sensors must be defined early and the sensors carefully selected.

### *1.1.3 SPECIFIC DESIGN CRITERIA FOR AEROSPACE VEHICLES AND THEIR SENSORS*

Aerospace vehicles are the most sophisticated, highly developed, and knowledge-intensive form of transportation permitting people to travel at greater velocities than in any other mode of transport. Their design and production are the privileges of technologically advanced societies, requiring great scientific, engineering, industrial, and financial resources and the joint effort of different enterprises and even countries (Bradshaw and Counsell 1992; Etkin 1982; Fleeman 2001; Grewal, Lawrence, and Andrews 2007; Kayton and Fried 1997; Lawrence 2001; McLeon 1990; Siouris 2004; Wertz and Larson 2000).

Several criteria are especially important in the design of aerospace vehicles and their sensors.

1. Engines must be reliable, and their operating conditions must be known—any engine failure in flight greatly endangers any aerospace vehicle. No other means of transportation places such importance on power plant reliability, which is truly of vital concern. Accomplishing a mission, and often even surviving, depends totally on engines being in full working order. Hence, engine reliability has to be verified by employing specialized sensors.
2. Any aerospace structure and equipment has to be extremely light, because given a fixed lifting force or engine thrust, each additional kilogram of weight decreases the payload and the capacity for acceleration. This requirement of light construction does not permit great rigidity and makes any aerospace vehicle rather flexible. However, it is essential to prevent as much flexing and oscillation as possible, and this can be accomplished both by optimizing the construction and by applying appropriate flight control. Special sensors for flexing and oscillation distributed along the vehicle body are necessary, and the design of all onboard sensors must be performed in the light of mass saving. Hence micromechanical and mechatronic techniques become very important.
3. All aerospace vehicles move in 3D space and require accurate and reliable 3D navigation and means of motion control. For aircraft, altitude control is of the same importance as position (latitude and longitude) and airspeed and become critical at landing. In deep space, altitude is not defined at all and each of the three axes, $x$, $y$, and $z$, is of equal importance.
4. Accurate attitude control and orientation with reference to relative points in space is very important for aerospace vehicles. It is necessary to measure the attitude angles with special sensors of different designs and to have special actuators that are able to influence these angles.
5. Aerospace vehicles have specific serviceability, reparability, and scheduled and unscheduled maintenance requirements. For reasons of safety, and because such vehicles are very expensive, it is inexpedient to scrimp on service—this has to be accomplished using the best equipment and by the best technicians. For airplanes and helicopters the planned service and deep testing of main power plants and units must be unfailingly regular and frequent. On the other hand, this service has to be rather quick, again requiring the best equipment and expertise. For spacecraft, frequent service in stationary service areas is difficult and often impossible, and this makes careful prelaunch inspection and service especially important. The main sensors on space vehicles, and particularly on probes for investigating deep space, have to be backed up. This redundancy makes sure that the mission does not fail because only one sensor malfunctions, for example.
6. The requirement for clean aerodynamics in aircraft and space carriers places many limits on body design and sensor distribution, especially for modes of flight where fairings have to withstand great thermodynamic changes. Fairings and cowls must not prevent sensors from functioning properly, and those constructed from special materials along with specially designed sensors can mitigate the problem.
7. Aerospace sensors must be temperature-resistant and able to withstand considerable falls in temperature. Even if a sensor is installed inside a temperature-controlled module, any power failure or abrupt depressurization in the module can lead to a precipitous

temperature drop. On some space probes the sensors must remain functional over temperature changes of a thousand degrees or more. The radiation resistance and pressure resistance of sensors are also important, especially on space vehicles and probes designed to land on other planets.

### 1.1.4 PHYSICAL PRINCIPLES INFLUENCING PRIMARY AEROSPACE SENSOR DESIGN

Aerospace sensors involve a large collection of different units for measuring many physical parameters that together characterize the state of an aerospace vehicle. Some of them, such as sensors for temperature or engine operational parameters, use ordinary principles of metrology and do not differ much from similar sensors on automobiles or ships (see Chapter 8). However, the most important sensors for aerospace vehicle parameters are essentially different from sensors for other kinds of vehicles, this being due largely to the great range of speeds and altitudes in addition to the more rapidly changing dynamics faced by aerospace vehicles (see Chapters 2–4).

The main parameters relevant to aerospace vehicle position and movement are geographical coordinates, altitude, velocity (three components), attitude (three angles), three angular rates, probably linear and angular accelerations, and often parameters relating to motion with respect to other vehicles or to a given reference point. Several (at least four) methods exist for measuring the location and velocity of aerospace vehicles:

- The simplest method of measuring flight altitude and velocity is to use aerial medium parameters such as static and dynamic pressure. This can be done cheaply and reliably by many kinds of onboard devices that have been installed on practically all airplanes since the beginning of aviation. Altitude as measured by barometric pressure and air speed are probably the most reliable instrumental data about flight mode and are available aboard any flying vehicle. Unfortunately, aerial medium parameters are rather unstable and sensors based on them are inaccurate (see Chapter 2).
- One basic method of aerospace vehicle positioning involves the exploitation of natural physical fields and landmarks. This involves the use of magnetic compasses to navigate aerospace vehicles using the Earth's magnetic field, though magnetic anomalies lessen the accuracy of such compasses (see Chapter 7). The most stable natural physical field is the disposition of stars observed in the sky or in space, and star sensors are widely used for both navigation and attitude control of aircraft and, especially, space vehicles. Great progress in charge-coupled devices (CCDs) has presented a major opportunity for creating precise, reliable, and inexpensive star and sun sensors. Existing star sensors and those under development are highly appropriate tools for broad-spectrum aerospace vehicles, especially spacecraft and deep space probes.

  It should also be stated that during recent decades such exotic and special natural physical fields as terrain features, gravitational anomalies, ground temperature, and the ratio of diffusely reflected to incident electromagnetic radiation (terrestrial albedo) have attracted attention as potential bases for navigation by map-matching. Any natural physical field can be used to position a vehicle if both a map of this field and sensors for the current values of the related physical parameters are available on board. The great advantage of map-matching navigation systems for military applications is that they are absolutely autonomous and totally secure from artificial jamming. Of course, visual landmarks also have an important role in vehicle positioning, and for aircraft this method has been termed "pilotage."

- The use of artificial physical fields for vehicle positioning became possible when powerful radio stations were developed and reliable, complex radio equipment appeared. The first long-range (e.g., LORAN and CHAYKA) and short-range (e.g., VOR/DME and TACAN) radio navigation systems were created in the 1940s and 1950s.

  Gradually, almost all aircraft took advantage of the availability of such radio navigation systems, but their accuracy remained low due to some fundamental limitations. Long-range navigation systems with a network of ground stations needed to use radio waves in the thousand-metre range (about 300 kHz) to service users beyond the horizon. Because the waves were so long, accurate measurement was impossible, but considerably improved accuracy came after the development of satellite navigation systems (SNS), which use radio waves in the ten-metre range (about 30,000 kHz) to shape the special artificial radio field anywhere on Earth or in terrestrial space where several navigation satellites are in the zone of direct sight. The SNS is one of the most outstanding engineering achievements in the field of navigation. It is based on the use of both advanced space and radio technologies, and its importance can be compared with, for example, the invention of the compass. Further improvements in radio navigation accuracy require the application of shorter radio wave ranges (especially millimetric radio waves) and this is already happening in advanced short-range navigation and landing radio systems and also in homing systems.

  The accuracy of any SNS can be approximately doubled if it is used in a differential regime that permits user coordinates to be corrected by a systematic correlation component. This may be achieved by the use of information coming from reference stations located nearby, that is, again by application of the short-range navigation principle. The accuracy and reliability of measurements can also be effectively increased by integrating the SNS with other onboard sensors, notably with the inertial system (see Chapter 6).
- The application of inertial navigation principles is the latest and most effective approach to the autonomous measurement of aerospace vehicle motion parameters. Accelerometers and gyroscopes are the main primary sensors in the realization of inertial navigation systems (INS). Three accelerometers with orthogonal sensing axes react to a craft's absolute linear acceleration along these axes and perform independent measurements of this acceleration and of its integral. The subsequent subtraction of a known gravitational component allows for the determination of relative acceleration or increments of linear velocity.

  Gyroscopes measure the attitude or angular velocity of a vehicle in the inertial reference frame. By tracking both the current angular velocity and the current linear acceleration measured relative to the moving frame, it is possible to determine the linear acceleration of the system in the inertial reference frame. Integration using the correct kinematic equations yields the inertial velocities of the vehicle, and integration again (using the original position as the initial condition) yields the inertial position.

  The growing interest in INSs shown in recent decades has resulted in the development of theory and practice essential to their successful implementation. The sensors themselves are constantly being improved and new types of such devices frequently appear on the market (see Chapters 5 and 6).

### 1.1.5 REFERENCE FRAMES ACCEPTED IN AVIATION AND ASTRONAUTICS

Several systems of axes that differ in several respects are employed to investigate aerospace vehicle motion. The origin of a set of coordinates may be located at the center of a celestial body (including the Earth, the Sun, or the Moon), at the surface of a celestial body, or at any point on an aerospace vehicle. Classification can also be performed by the position of the main flat, the orientation of the axes, and the readout of angles (Recommended Practice for Atmospheric and Space Flight Vehicle Coordinate Systems 1992).

The system of coordinates relative to which the position and attitude of an aerospace vehicle is referred has to be chosen with the vehicle flight program taken into account. It is advisable to consider which reference frame provides the simplest technical realization of this program, the simplest equations of vehicular motion, and how these equations can be linearized. Many systems of axes exist; for example, at the center of mass of the vehicle, at the center of mass of the Earth, or at any other point. The axes of the reference frame can be fixed in inertial space, or they can move in this space in a known manner. The following are the main orthogonal-axis systems used to investigate aerospace vehicle motion and in developing the appropriate equations:

- The *body-fixed frame* (*b-frame*) is connected to the vehicle and moves with it. The origin of this frame is usually at the center of gravity (CG) of the vehicle, and its axes are the positive $x_b$-axis, or longitudinal axis, which lies fore and aft; the positive $y_b$-axis, or pitch axis, which lies in the right wing on the horizontal plane; and the positive $z_b$-axis, or yaw axis, which is vertical. This axis system is denoted by $(x_b, y_b, z_b)$. The Euler angles $(\psi, \theta, \varphi)$ are commonly used to define a missile's attitude: $\varphi$ is roll about the $x$-axis, $\theta$ is pitch about the $y$-axis, and $\psi$ is yaw about the $z$-axis (Figure 1.1).

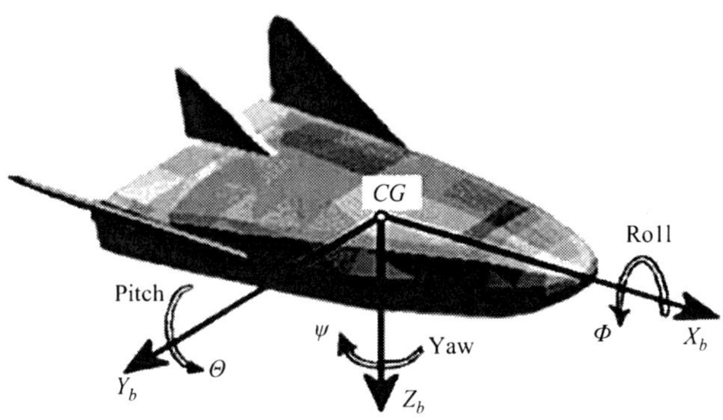


**Figure 1.1.** The body-fixed frame.

Because measuring equipment and transducers are mainly located aboard the aerospace vehicle, their readouts can be processed most easily in the $b$-frame. This simplifies the equations of motion, because the vehicle's products of inertia become zero. It is essential also in most aircraft and missiles that the $x$-$z$-plane be a plane of symmetry.

For some vehicles, the CG changes during flight and measurements may become inaccurate if this roaming point coincides with the *b*-frame center. However, the origin of the coordinates can be fixed at any stable point in the vehicle. For example, the leading edge of the root chord, the tip of the nose, or the aerodynamic center of the wings may serve as reference points.

- In the *wind frame* (*w-frame*), the *x*-axis points directly into the relative wind. The *z*-axis remains in the reference plane, but it rotates so that it remains perpendicular to the *x*-axis. The *y*-axis completes a right-handed system. The mutual position of *w*-axis and *b*-axis can be defined by the attack angle $\alpha$ and the sideslip angle $\beta$. The transformation from *b*-frame to *w*-frame consists of two rotations. First, the body axes are rotated about the *y*-axis through the angle of attack $\alpha$, then the axes are rotated about the *z*-axis through the angle of sideslip $\beta$, yielding the wind axes.

  The main reason for using the wind axis system is that it makes calculating aerodynamic forces more convenient. For instance, lift is by definition perpendicular to the relative wind, while drag is parallel. So, both lift and drag resolve into a force parallel to one of the *w*-frame axes.
- The *flight path frame* (*fp-frame*) differs from the *w*-frame in that its *x*-axis is aligned with the velocity vector $V$ of the vehicle, this being more convenient for navigational purposes. The relationship between the *b*-frame and the *fp*-frame can be defined by the flight path angle $\Theta$, the bank angle $\Psi$, and the aerodynamic angle of roll $\gamma$.

  The distinction between the *fp*-frame and the *w*-frame is essential for subsonic aircraft, whose speed may be commensurate with wind speed. For rockets and other high-speed vehicles, the *fp*-frame and *w*-frame are essentially indistinguishable. In this case, the attack and sideslip angles may be defined as the *b*-frame inclination with respect to the *fp*-frame.
- An *inertial frame* (*i-frame*) is fixed in space and is subject to Newton's laws of motion. Its origin is at the earth's center, its *z*-axis is perpendicular to the equatorial plane, its *x*-axis lies in the equatorial plane in a direction that can be specified arbitrarily but as a rule is directed to the point of the vernal equinox, and its *y*-axis completes the right-handed coordinate system.

  The parameters of space vehicle orbits are normally given in the *i*-frame: $i$ is the inclination of the orbit, $\omega$ is the argument of the perigee, $\theta_a$ is the true anomaly, and $u$ is the argument of the breadth. Other *i*-frames are used, for example, when one axis is directed toward the sun and the other toward any star, or even when two axes are referred to one star. Such coordinate systems are sometimes called *sighting*.
- The *Earth-fixed frame* (or *Earth-surface frame*) is used to define location. Its *z*-axis is coincident with the Earth's polar axis, while the other two axes are within the equatorial plane. These three axes are defined as down, north, and east, the origin being located on the earth's surface at a point with zero altitude. This coordinate system is also known as north-east-down (NED). Any vehicle trajectory in NED can be characterized by its heading angle $\Psi$ (about the *z*-axis) and flight path angle $\theta_t$ (about the *y*-axis), which show the relationship with the *fp*-frame (Figure 1.2).
- The *Earth-centered frame* (*e-frame*) has the same axes, but its origin is located at the Earth's center. Thus the point (0,0,0) denotes the center of the Earth. The *e*-frame is especially suitable for controlling the motion of satellites and for global positioning.

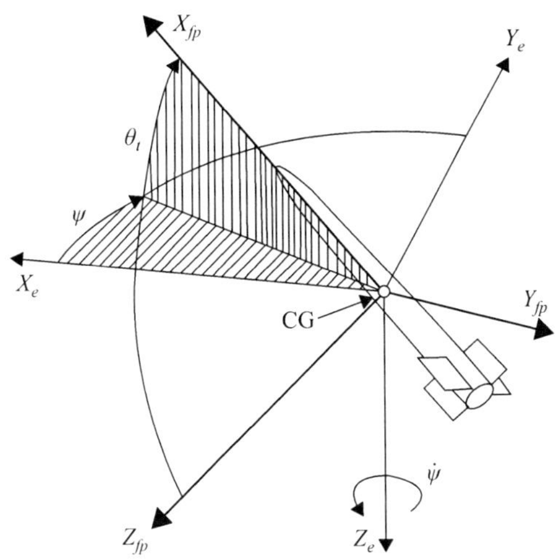


**Figure 1.2.** The earth-fixed frame.

- The *launch-centered inertial frame* is inertially fixed and is centered at the launch site at the instant when a rocket is launched. In this system, the $x$-axis is commonly taken to be in the horizontal plane and in the direction of launch, the positive $z$-axis being vertical, and the $y$-axis completing the right-handed coordinate system. This frame is suitable for defining a ballistic missile's position and angular orientation in space.
- *The orbital base frame* (*o-frame*) is suitable for satellites moving in near-Earth orbits. In this system, the $y$-axis is a local vertical, that is, a line uniting the center of mass of the vehicle with the center of mass of the Earth. The $x$-axis is transverse to the direction of motion of the satellite and so is perpendicular to the local verticals and to the surface of the orbit. The $z$-axis is a vector normal to the surface of the orbit such that the system becomes right-handed.

  It is advisable to provide overlapping of one or two axes of the $o$-frame with the $b$-axis beginning at the satellite CG. In this case, simplified equations of angular motion exist, and the influence of moving parts inside the satellite during changes in the positions of the axes is minimal. During small angles of deviation of the satellite from the position of the given orientation in the $o$-frame, aviation terminology is used, and the Euler angles may be called roll angle $\gamma$, yaw angle $\Psi$, and pitch angle $\theta$.
- *The geodetic frame* (*g-frame* or *geodetic data*) is used in navigation, geodesy, surveying by cartographers, and satellite navigation systems to translate positions indicated on their devices to their real position on the Earth. This is needed because the Earth is not a perfect sphere. The difference in coordinates between two particular data can vary from one place to another and can be up to hundreds of metres or even several kilometres. Different data use different estimates for the precise shape and size of the Earth. The most commonly used geodetic data corresponds to WGS-84 (the World Geodetic System) adopted by the International Civil Aviation Organization (ICAO; Office of GEOINT Sciences 1984) and F. Krasovsky's ellipsoid, which is popular in Russia.

All the above-named coordinate systems are rectangular and right-handed, where three Cartesian axes are perpendicular to each other. The relevant coordinates are of the form $(x, y, z)$. The $xy$-, $yz$-, and $xz$-planes divide the three-dimensional space into eight subdivisions known as octants. Only the first octant of three-dimensional space is labeled, mostly because it contains all the points whose $x$, $y$, and $z$ coordinates are positive.

Besides Cartesian frames, some types of spherical and polar coordinate systems may be considered. They are especially suitable for the investigation of radar sensors and systems.

Any coordinates may be easily converted from one frame to another on the basis of the formulae of recalculation.

## 1.2 CHARACTERISTICS AND CHALLENGES OF THE ATMOSPHERIC ENVIRONMENT

### *1.2.1 COMPONENTS OF THE EARTH'S ATMOSPHERE*

The atmosphere fills the expanse above the surfaces of land and water. Its density decreases gradually as the distance from the surface increases, and becomes negligible at a great altitude that may be considered the conditional boundary between the atmosphere and outer space.

Earth's atmosphere consists of air that has a density of 1.23 kg $m^{-3}$ and pressure of 760 mm of mercury (or 29.9 in. of mercury; other equivalents being $1.014 \times 10^5$ N $m^{-2}$, 101.4 kPa, and 1,033 gf $cm^{-2}$) at sea level at latitude 45°. The air pressure at a height of 80 km is $10^{-5}$ of that at sea level; and at 400 km (the outer boundary of the Earth's atmosphere) it decreases by $10^{-10}$. The chemical constituents of the air are nitrogen ($N_2$) 78%; oxygen ($O_2$) 21%; argon (Ar) 0.93%; carbon dioxide ($CO_2$) 0.03%; and other rare gases (Ne, He, Kr, Xe), ozone ($O_3$), and hydrogen ($H_2$) in very small quantities. The total mass of air is estimated to be $5.136 \times 10^{15}$ tons (Allen 1973).

The atmosphere is conventionally divided into several layers, each having different properties. Four main zones exist: the troposphere (up to 11 km where the temperature falls as altitude increases), the stratosphere (from 11 to 50 km where the temperature rises with altitude), the mesosphere (50–85 km where the temperature again falls), and the thermosphere (85–800 km where the temperature again rises). The tropopause, mesopause, and thermopause are rather thin layers that share a mixture of the characteristics of the main layers that surround them (Prolls 2004).

Ultraviolet radiation converts ordinary oxygen ($O_2$) into ozone ($O_3$), which is located in the stratosphere with maximal concentration at 25–28 km. This is why the stratosphere is sometimes called the ozonosphere, and it is in this layer that most ultraviolet solar radiation is absorbed.

The oceans provide great areas over which aircraft may fly, but where the characteristics of the atmosphere need to be measured by the numerous sensors installed on such aircraft. The main domain for aircraft flight is of course the troposphere and occasionally the stratosphere. The mesosphere can be considered the portal to space through which spacecraft must pass at launch and re-entry.

The ionosphere is markedly affected by solar activity (Warneck 1999). The great influence of the ionosphere on radio wave propagation means that considerable attention must be given to this upper layer in the atmosphere in the design of relevant sensors.

### 1.2.2 STATIONARY MODELS OF THE ATMOSPHERE

Several models of atmospheric density and pressure are used around the world. Such models relate atmospheric pressure to height and allow altitude to be measured with rather simple barometric devices. Airspeed can also be measured via the dynamic pressure resulting from the motion of the aircraft. The accuracy of such simple measuring instruments based on the barometric approach depends directly on the quality of the relevant model, the simplest being the so-called isothermal model described by the following formula:

$$p_H = p_0 \exp(-gH/R_a T_a) \tag{1.1}$$

where $p_H$ and $p_0$ are the air pressures at height $H$ and at sea level ($H = 0$); $g$ is the gravitational acceleration at altitude $H$; $R_a$ and $T_a$ are the ideal gas constant of air and its absolute temperature for the isothermal atmosphere.

Isothermal conditions are not uniform, and air temperature generally varies as height increases. As the vertical gradient of temperature is not constant, especially in the stratosphere, the conditional standard atmosphere has been adopted. According to this international agreement, the temperature at sea level is taken as constant and equal to 15° C; a constant vertical temperature gradient of −6.5° C for each 1 km of height is adopted for the range 0–11 km; and a constant temperature of −56.5° C is assumed for any height above 11 km. It is considered that $p_0$ = 760 mm of mercury, and that the relative humidity is 0%. Unfortunately, neither this nor any other stationary model can accurately predict atmospheric parameters at any point in a flight path.

### 1.2.3 ANISOTROPY AND VARIABILITY IN THE ATMOSPHERE

Atmospheric anisotropy is affected not only by the dependence of pressure and other parameters on height, but also by geographical coordinates that play a part in determining the average parameters of atmospheric and wind fluctuations. Average air temperature depends on the latitude, and at low (tropical) latitudes the average temperature may be about 40° C higher than at high (polar) latitudes. Differences in temperature cause differences in air pressure and winds. Heat exchange between the oceans and the atmosphere is irregularly distributed over the Earth's surface. All these effects bring anisotropy and instability to atmospheric processes.

Oceans and other waters release water vapor into the atmosphere and the average relative humidity may reach 90–95% in some places on the Earth's surface, depending on air temperature and pressure. Clouds of various types are the visible displays of vapor concentration in the atmosphere. The humidity index strongly influences the way the sun's rays heat land and water. All these and many other factors make the task of accurate weather prognostication very complex, and the relevant problems become almost insoluble. The real condition of the atmosphere is always a stochastic phenomenon to a greater or lesser degree, and mathematical models provide only approximations based on average conditions.

Gusts of wind are important characteristics of the atmosphere and numerical models exist that describe the strength of the wind, the dynamics of wind change, and the structure of the space in which they occur. The speed of wind gusts may exceed the average speed of the wind by 1.5 to 2.5 times and sometimes even more.

By analyzing weather statistics over many years and over different areas of the Earth's surface, it is possible to take into account the following limiting values:

Lowest temperature: −89°C (Antarctica, 1983)
Highest temperature: 58°C (Libya, 1922)
Lowest atmospheric pressure in the Northern Hemisphere (not taking into account tropical cyclones and tornados): 741 mm of mercury (Aleut cyclone)
Highest atmospheric pressure: 804 mm of mercury (Siberian anticyclone) (Gresswell 1967).

Wind speeds at sea level can exceed 60 m $s^{-1}$ in massive storms, and at high altitudes may be even greater. In particular, the jet stream, at altitudes of 7–18 km, can cover great areas of hundreds or thousands of square kilometres. The highest recorded speed of the jet stream is 100–150 m $s^{-1}$. The distribution of average wind speeds and directions, taking into account the height, area, and season, has been thoroughly investigated and described in numerous reference books (CIRA 1965).

The normal speed of sound propagation at sea level in dry air at 0°C is 331 m $s^{-1}$. At a temperature of 15°C its value at sea level is 340 m $s^{-1}$, at 10 km it becomes 294 m $s^{-1}$, at 50 km it returns to 338 m $s^{-1}$, and at 80 km it falls to 258 m $s^{-1}$. The speed of sound depends also on the humidity of the air.

### *1.2.4 ELECTRICAL CHARGES IN THE ATMOSPHERE*

Electrical charges appear in the atmosphere as the result of the relative motion of air masses (static electricity), solar radiation, and other causes. Two thousand thunderstorms disturb the Earth's atmosphere at any moment, which means there are around 50,000 storms on Earth every day. Lightning is a high-energy electrical discharge partly in the form of electromagnetic radiation over a wide frequency range. Such a burst of energy can disturb electronics that incorporate no special protective features. Local ionization of atmospheric air may be artificially produced to measure airflow velocity.

### *1.2.5 ELECTROMAGNETIC WAVE PROPAGATION IN THE ATMOSPHERE*

The main source of energy reaching the Earth is in the form of solar radiation and this includes the entire light spectrum. Energy density at the upper bound of the atmosphere is characterized by the solar constant and is equal to 2 cal $cm^{-2}$ $min^{-1}$ (8.4 j $cm^{-2}$ $min^{-1}$ or $1.4 \times 10^3$ W $m^{-2}$). The planetary albedo of the Earth, the amount of energy reflected back to space, is 30–40%.

Fortunately, atmospheric air is an excellent medium for radio wave transmission. However, in some frequency bands the absorption of electromagnetic energy increases, and this has to be taken into account in radar design. The loss of radio wave energy in the ionosphere is almost complete only at $\lambda > 3$ m but is negligible for ten-metre waves. However, the troposphere allows the passage of long radio waves ($\lambda \cong 3$ cm and $\lambda \cong 8$ mm) and essentially absorbs waves of $\lambda < 3$ cm, where molecules of water vapor and oxygen have resonances. In hard rain and fog the conductivity falls, especially at $\lambda < 5$ cm.

Radio noise invariably reaches an antenna along with any useful signal, and may be atmospheric or galactic. At $\lambda < 10$ cm atmospheric noise is more powerful than galactic noise.

Two varieties of atmospheric noise are especially intense at $\lambda = 1.35$ cm and $\lambda = 0.5$ cm, where the resonance radiation of vapor and oxygen molecules takes place.

The ionosphere reflects short radio waves, which is why this band can be used for long-distance radio communication on Earth. Unfortunately, the top of the atmosphere acts as a lens for radio waves and forces them to spread when they are deflected. This affects all signals sent from satellites and other space vehicles to aircraft.

### *1.2.6 GEOMAGNETISM*

The Earth's magnetic field can be thought of as a simple dipole located at the center of Earth and at 11° to the axis of rotation. This inclination accounts for the displacement of the magnetic pole from the geographical pole. However, some small-scale and large-scale anomalies are superimposed on this main geomagnetic field. The most significant large-scale anomalies are located in the areas above Siberia (increasing) and the Atlantic side of South America (decreasing). They strongly influence the motion of charged particles in the belts of radiation up to heights of thousands of kilometres above the Earth.

Detailed maps of the geomagnetic field have been obtained through experimentation, but this field is unstable for many reasons. Electric currents in the upper atmosphere (ionosphere and higher) cause the values of some components of the field to vary, and become especially large during auroras, when geomagnetic storms take place. Slow geological processes also cause the values of the main components of the field to vary.

The magnetic field of the Earth is always compressed on the side toward the Sun due to the stream of solar plasma (solar wind). The area where the geomagnetic field is deformed by this compression but generally preserves the direction of the lines of force is called the magnetosphere. Upon entering Earth's magnetosphere the ions (protons) of the solar wind curve toward the West, electrons toward the East.

These flows of separated, charged particles cause a strong current at the upper bound of the sunward side of the magnetosphere. In this area the solar wind decelerates, and its great kinetic energy converts into thermal energy, which causes the plasma temperature to shoot up to more than ten million degrees. This happens at a distance of tens of thousands of kilometres from the Earth's surface. At the opposite side (removed from the sun), the tail of the geomagnetic field extends millions of kilometres beyond the Earth.

The geomagnetic field, together with the ionosphere, performs the important function of smoothing the harmful influence of radiation from the Sun on humans and many kinds of man-made hardware, especially sensors. When the Sun flares, the flow of plasma and radiation increases manifold. Short wave radiation arrives at the Earth in 8–10 minutes and is mainly absorbed by the atmosphere, whereas the plasma (solar wind), which moves at a velocity of 400–1,000 km $sec^{-1}$, appears in 2–3 days, most of it being blocked by the magnetosphere. This natural defense helps aircraft more than it does spacecraft, which need artificial protection. The delay in the arrival of the gust of solar wind after a solar flare may therefore be predicted by the reception of a pulse of electromagnetic waves and so used to forecast geomagnetic storms.

The ordinary duration of a solar flare is only 1–2 hours, but a large amount of energy is released at this time, which causes geomagnetic storms to last much longer. Severe geomagnetic storms cause communication problems, greatly increase drag on spacecraft, cause electronic circuits to malfunction, and cause some people to become depressed.

Solar flares are not the only source of hard radiation in near-Earth space. The Earth has two radiation belts where high energy particles, mainly protons, are located. The internal radiation belt begins at an altitude of 500–600 km in the Western Hemisphere and from about 1,500 km in the Eastern Hemisphere, up to 5,000–10,000 km. Above the southern part of the Atlantic Ocean, the lower edge of this belt is as low as 300 km due to the magnetic anomaly. This internal radiation belt does not cover the whole Earth, but only about as far as latitudes 45° north and south (Beregovoy et al. 1989; Connerney 1993). The external radiation belt covers the equatorial zone at an altitude of more than $10^4$ km and down to 300 km at latitudes 55°–70°. The separation of radiation belts into internal and external belts is a convention; actually, these belts overlap.

### *1.2.7 THE PLANETARY ATMOSPHERE*

The atmosphere of a planet is its covering of gas, and is defined by its chemical constituents and by its mass, density, and temperature distributions. On the basis of most of these parameters, the planets in our solar system can be separated into two groups. The four comparatively small planets closest to the Sun form the Earth group. The next four great planets belong to the Jupiter group. The last planet, Pluto, has very specific features and cannot be associated with any group—whether it is a planet at all has become a matter for discussion.

The planets in the Earth group have, on average, thinner atmospheres than do the larger planets. Mercury, with its small mass (5% of the mass of the Earth) and the high temperature of its sunward side, has an extremely thin atmosphere, with a surface pressure of only $2 \times 10^{-10}$ Pa. The atmospheric pressure at the surface of Mars is about 0.6 kPa ($170^{-1}$ that of Earth), and on the surface of Venus the pressure is –9 Mpa (100 times greater than that of Earth). The main component of both atmospheres is carbon dioxide. A powerful hothouse effect acts to maintain the high temperature of Venus' lower atmosphere at around 740 K. A major feature of the weaker Mars atmosphere is the great fluctuation of temperature between day and night and between equator and winter pole, sometimes exceeding 100 K. The average high temperature may be 22°C and the low –70°C (Cadle 1966; Davis 2005; Pikelner 1976).

The giant planets, Jupiter, Saturn, Uranus, and Neptune, have massive atmospheres of gases with small molecular weights, the main components being hydrogen and helium for Jupiter and Saturn, and methane and ammonia for Uranus and Neptune. The atmospheric pressure at the surface may exceed 100 gPa.

## 1.3 CHARACTERISTICS AND CHALLENGES OF THE SPACE ENVIRONMENT

### *1.3.1 GENERAL CONSIDERATIONS*

All space beyond the Earth and its atmosphere is considered as cosmic or outer space. It can be divided into near-Earth space, circumsolar space, and deep space, this last category including interstellar space, our galaxy, and even other galaxies. Space boundaries are more a matter of rhetoric than of reality. The planetary atmosphere fades into interplanetary space, but this can be considered the continuation of the Sun's atmosphere, which fades into deep space.

### *1.3.2 NEAR-EARTH SPACE*

Currently, near-Earth space is the most explored area of the cosmos. Thousands of artificial satellites circle the Earth in orbits of different altitudes, ellipticities, and inclinations. Launching into a low-earth orbit (LEO), an altitude of 150–300 km, requires minimal energy, but such orbits cannot be maintained due to the presence of residual atmosphere in near space. Geostationary orbits (GEOs) at an altitude of about 36,000 km are also thickly settled, but higher orbits are not popular. Some satellites with essentially elliptical orbits come into this zone only in apogee. Actually, such altitudes correspond with the notional upper bound of near-Earth space.

Space is essentially a vacuum. The remnants of the atmosphere disappear quickly at altitudes greater than 100 km, as is shown in Table 1.1 (Beregovoy et al. 1989).

**Table 1.1.** Atmospheric densities

| Altitude, km | 100 | 200 | 300 | 400 | 600 | 800 | 30,000 |
|---|---|---|---|---|---|---|---|
| Density, kg × m$^{-3}$ | $5 \times 10^{-7}$ | $3 \times 10^{-10}$ | $2 \times 10^{-11}$ | $3 \times 10^{-12}$ | $10^{-13}$ | $10^{-14}$ | less than $10^{-21}$ |

The vacuum of space affects a spacecraft's equipment by changing the conditions of heat exchange, by the appearance of corona discharges, by volatilizing lubricants, and by sublimating materials and varying their mechanical qualities.

Among the numerous tangible objects in near-Earth space, approximately 5% are operating satellites, 12% are "dead" satellites, 18% are the last stages of carrier rockets and their fragments, and 65% are the wreckage of burst rockets, satellites exploded by design, and other debris (Reynolds, Fisher, and Rice 1983).

Space garbage is the wrongful consequence of human activity in the space age and it poses considerable danger to any vehicle moving in near-Earth space. The relative speed of collision with such numerous particles of former artificial artifacts and waste may be more than 10 km s$^{-1}$, which makes collision with even a 1 cm particle very dangerous. Such objects number more 300,000, whilst the number of objects with dimensions less than 1 cm is more than $10^6$. The number of objects with dimensions of more than 10 cm is about 8,000 and their total mass is more than 3,000 tons. These fragments are described in catalogs such as that published by NORAD. The garbage distribution is rather smooth at different altitudes from 400 km to 1,500 km, having a density in the order of magnitude of $F_g = 10^{-10}$ m$^{-2}$ h$^{-1}$. The probability of a space vehicle collision with a fragment of space garbage is given by the formula

$$p_c = 1 - \exp(-F_g A_v T_f) \cong F_g A_v T_f \qquad (1.2)$$

if $p_c \ll 1$. Here $A_v$ is the vehicle midship area in m$^2$, and $T_f$ is the flight duration in hours.

For example, a vehicle with a midship area of 1 m$^2$ during a flight lasting 10 years or $8.76 \times 10^4$ h has a probability of collision of approximately $10^{-5}$, this probability depending somewhat on the altitude and inclination of the vehicle orbit.

The partial purification of garbage from near-Earth space takes place mainly during periods of the sun's full activity every 11 years. The growth of space garbage may be stopped by tightening international regulations on space activity and by improving technical means of verification.

### 1.3.3 CIRCUMSOLAR (NEAR-SUN) SPACE

The solar system includes nine primary planets, six of which have their own natural satellites, and many meteoroids with circumsolar orbits. Also, some transient space bodies come into the area of solar gravity, are not absorbed, and move away. Meteoroids can be classified by mass, constituents, structure, and orbits. One of these classes encompasses the comets, which have rather small nuclei and extended tails of sparse matter.

Within the comparatively great expanse between the orbits of Mars and Jupiter lies the asteroid belt. There are a total of more than 50,000 asteroids, the three largest being Ceres (770 km in diameter), Pallada (490 km), and Vesta (384 km). The large asteroids, comets, and other meteoroids with consistent orbits may be interesting bodies for investigation, but small meteoroids and micrometeoroids present the main danger to space vehicles.

### 1.3.4 MATTER IN SPACE

Only part of the matter in the cosmos is concentrated in the stars and their planets, which occupy not more than $10^{-30}$ percent of the total volume of the observable universe. The remaining matter, at least several percent and probably more than half, is distributed sparsely throughout space. Typical values of this density are several atoms per $m^3$ in space between galaxies, about one atom per $cm^3$ in anabranches of our galaxy, and several atoms per $cm^3$ in interplanetary space.

At such low densities, matter behaves differently from solid substances on Earth. Because almost all atoms in space are ionized (especially near the bright stars), electrical and magnetic forces become essentially more important than short-range forces. Most atoms and ions move chaotically in local frames of reference, and at velocities ranging from several hundred kilometres per second to almost the velocity of light (Sellers 2000).

The average kinetic energy of atoms' chaotic motion has values from $10^{-2}$ to more than $10^2$ electron-volts. In addition, there are small numbers of particles like electrons, protons, and atomic nuclei of heavy elements that have energies up to $10^{20}$ electron-volts. During solar flares, the sun emits particles with energies of hundreds of millions of electron-volts. Some high-energy particles, of 1 GeV and more, have galactic or extragalactic origin, these forming the cosmic rays. The exact mechanism of particle acceleration in cosmic rays is still unknown, but it is certain that this acceleration occurs not in stars but in outer space. The existence of such acceleration processes shows that humans still have much to learn about the nature of space.

### 1.3.5 DISTANCES AND TIME SCALES IN DEEP SPACE

The average distance from the Earth to the Sun, one astronomical unit (AU) of distance, is $149.6 \times 10^6$ km.

Light travels 1 AU in 499 sec = 8.32 min.

The distance from the Earth to the Moon is $0.356–0.407 \times 10^6$ km, to Venus $40–260 \times 10^6$ km, and to Mars is $80–380 \times 10^6$ km.

The average distance from the Sun to Mercury is 0.39 AU, to Venus 0.72 AU, to the Earth 1 AU, to Mars 1.52 AU, and to Jupiter 5.20 AU.

The average distance from the Sun to the farthest planet, Pluto, is 39.5 AU.

A light-year is equal to $9.46 \times 10^{13}$ km or 63,239 AU.

The distance from the Sun to Alpha Centauri is 4.3 light-years; to Alpha Canis Major (Sirius) 9 light-years; and to Alpha Gemini (Castor) 45 light-years.

One parsec (pc) is 206,265 AU or $3.086 \times 10^{14}$ km or 3.26 light-years.

An important unit of length used in astronomy is parsec (pc). One parsec is 206,265 AU or $3.086 \times 10^{14}$ km or 3.26 light-years.

The diameter of our galaxy is 25 kiloparsecs (kpc) or 81,500 light-years.

The Sun is 10 kpc from the galactic center.

## REFERENCES

Allen, C. W. 1973. *Astrophysical Quantities*. London: Athlone.

Beregovoy, G. T., V. I. Yaropolov, I. I. Baranetskiy, V. A. Visokanov, and Ya. T. Shatrov. 1989. *Guide to the Safety of Space Flight*. Moscow: Mashinostroenie. (In Russian.)

Bradshaw, A., and J. M. Counsell. 1992. "Design of Autopilots for High Performance Missiles." Proceedings of the Institution of Mechanical Engineers 206 (12): 75–84.

Cadle, R. D. 1966. *Particles in the Atmosphere and Space*. New York: Reinhold.

Chatfield, A. B. (1997). *Fundamentals of High Accuracy Inertial Navigation*. Reston, VA: American Institute of Aeronautics and Astronautics. DOI: 10.2514/4.866463.

CIRA. 1965. *COSPAR International Reference Atmosphere*. Amsterdam: North-Holland.

Connerney, John E. P. 1993. "Magnetic Fields of the Outer Planets." *Journal of Geophysical Research* 98: 18659–79. DOI: 10.1029/93JE00980.

Davies, M., ed. 2003. *The Standard Handbook for Aeronautical and Astronautical Engineers*. McGraw-Hill.

Davis, A. M. 2005. *Meteorites, Comets, and Planets*. Amsterdam: Elsevier.

Etkin, B. 1982. *Dynamics of Flight: Stability and Control*, 2nd edition. New York: John Wiley & Sons.

Fleeman, E. L. 2001. *Tactical Missile Design*, AIAA Education Series. Reston, VA: American Institute of Aeronautics and Astronautics.

Gorbatenko, S. A., E. M. Makashov, Yu. F. Polushkin, and L. V. Sheftel. 1969. *Flight Mechanics*. Moscow: Mashinostroenie. (In Russian.)

Gresswell, R. Kay. 1967. *Physical Geography*. New York: Praeger.

Grewal, M. S., R. W. Lawrence, and A. P. Andrews. 2007. *Global Positioning Systems, Inertial Navigation, and Integration*. New York: Wiley Interscience. DOI: 10.1002/0470099720.

Kayton, M., and W. R. Fried. 1997. *Avionics Navigation Systems*, 2nd edition. New York: John Wiley & Sons. DOI: 10.1002/9780470172704.

Lawrence, A. 2001. *Modern Inertial Technology: Navigation, Guidance, and Control*, 2nd edition. Mechanical Engineering Series. New York: Springer.

Lin, Ching-Fang. 1991. *Modern Navigation, Guidance, and Control Processing*. Vol. 2. Englewood Cliffs, NY: Prentice Hall.

McLeon, D. 1990. *Automatic Flight Control Systems*. New York: Prentice Hall.

McRuer, D., Dunstan Graham, and Irving Ashkenas. 1973. *Aircraft Dynamics and Automatic Control*. Princeton: Princeton University Press.

Merhav, S. 1998. *Aerospace Sensor Systems and Applications*. New York: Springer.

Office of GEOINT Sciences. 1984. "World Geodetic System 1984 (WGS 84)." http://earth-info.nga.mil/GandG/wgs84/

Ohkami, Y., N. Tomita, S. Nakasuka, and S. Matsunora. 2002. *Introduction to Space Station*. Tokyo: University of Tokyo Press. (In Japanese).

Pikelner, S. B., ed. 1976. "Physics of the Cosmos." 43–79. Moscow: *Bolshaya Sovietskaya Encyclopedia.*

Recommended practice for atmospheric and space flight vehicle coordinate systems. 1992. *R-004-1992.* Washington, DC: ANSI/AIAA.

Reynolds, R. C., N. Y. Fisher, and E. E. Rice. 1983. "Man-Made Debris in Low Earth Orbits." *Spacecraft and Rockets* 20 (3): 279–85. DOI: 10.2514/3.25593.

Rogers, R. M. 2000. *Applied Mathematics in Integrated Navigation Systems.* Reston, VA: American Institute of Aeronautics and Astronautics.

Sellers, J. J., with W. A. Astore, R. B. Griffin, and W. J. Larson. 2000. *Understanding Space: An Introduction to Astronautics.* New York: McGraw-Hill.

Siouris, G. M. 2004. *Missile Guidance and Control Systems.* New York: Springer.

Stevens, B. L., and F. L. Lewis. 2003. *Aircraft Control and Simulation*, 2nd edition. New York: Wiley-Interscience.

Prolls, Von Gerd W. 2004. *Physics of the Earth's Environment.* Berlin: Springer.

Warneck, Peter. 1999. *Chemistry of the Natural Atmosphere*, 2nd edition. San Diego, CA: Academic Press.

Wertz, J. R., and W. J. Larson, eds. 2000. *Reducing Space Mission Cost.* Torrance, CA.

CHAPTER 2

# Air Pressure-Dependent Sensors

Ben Evans
*Swansea University, United Kingdom*

Joe Watson
*Swansea University, United Kingdom (Retired)*

## 2.1 BASIC AIRCRAFT INSTRUMENTATION

Although most modern aircraft currently feature electronic displays, often referred to as "glass cockpits," in place of traditional mechanical instruments, many of the actual sensors that provide the initial transduction processes have remained essentially unchanged since the early days of aviation. In particular, the measurement of altitude and airspeed may rely on sensors responding to air pressure, and the relevant sensors utilize both barometric and manometric technologies, as will be explained in this chapter. Many other sensing devices are utilized in modern aircraft, including accelerometers and inertial navigation systems depending on various forms of gyroscopes and acceleration sensors, and numerous different magnetic, radar, radio-navigation, and other sensors and systems. These will be treated in later chapters, and only sensors depending on air pressure will be considered here. The chapter begins with an overview of some of the fundamentals of aerodynamic theory necessary for a detailed study of such sensors.

## 2.2 FUNDAMENTAL PHYSICAL PROPERTIES OF AIRFLOW

Before considering the workings of air pressure-dependent sensors, some of the terminology and definitions associated with aerodynamics used throughout this chapter must be presented.

Sources of aerodynamic forces must also be considered because these play important roles in understanding the physics behind the workings of all pressure-dependent sensors.

### 2.2.1 FUNDAMENTAL AIRFLOW PHYSICAL PROPERTY DEFINITIONS

The "state" of a gas at any point in space can be characterized by the four fundamental properties of any gas: pressure, density, temperature, and velocity, and each has a formal definition, as follows.

#### 2.2.1.1 Pressure

The concept of air pressure is very familiar because everyone experiences the resultant force on the body. The force due to this pressure is actually exerted whether the body is at rest or in motion. As examples, the force due to wind is well known; when a hand is placed outside the window of a moving car, this also results in a force on that hand. However, the body also experiences the quasi-steady atmospheric pressure on it even when stationary. Hence, a more considered definition of air pressure must be provided prior to considering its measurement, as follows (Anderson 2000).

> *Pressure* is the *normal* force per unit area exerted on a surface due to the time rate of change of momentum of the gas molecules impacting on that surface.

In general, the air pressure at any point $X$ within a gas (using coordinates whose origin is where that gas comes into contact with the surface of a body) is:

$$p = \lim \left( \frac{dF}{dA} \right) \qquad dA \to 0 \tag{2.1}$$

where $dA$ is an incremental area around $X$ and $dF$ is the normal force, due to molecular impact, on one side of $dA$ as illustrated in Figure 2.1.

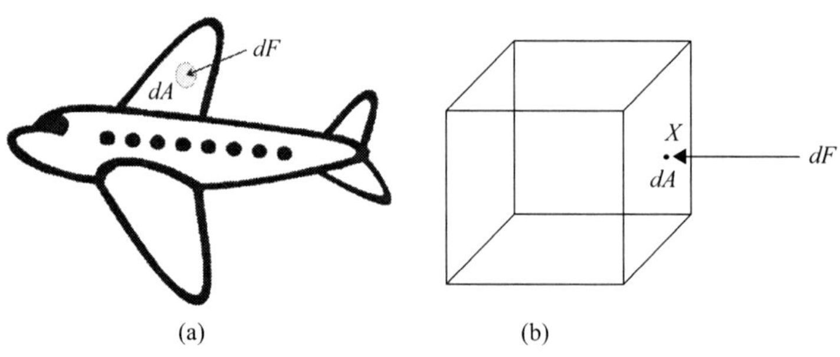


**Figure 2.1.** Normal force $dF$ action on an incremental area $dA$ of (a) a physical wing surface (b) a 'virtual' volume element.

Pressure is one of the most fundamental and important variables in aerodynamics and although there are a variety of additional pressure descriptions that are useful in practice, they all stem from this fundamental definition. Common units of pressure include pascals (Pa), which are newtons per square metre ($Nm^{-2}$); pound-force per square inch (psi); bar (or more often, millibar); and atmospheres (atm). A list of conversion factors between these units is provided in Table 2.1.

**Table 2.1.** Conversion factors between common units of pressure

| | Pascal (Pa) (newtons per square metre) | Millibar (mB) | Atmosphere (atm) | Pound-force per square inch (psi) |
|---|---|---|---|---|
| 1 Pa | 1 N $m^{-2}$ | 0.01 | $9.8692 \times 10^{-6}$ | $145.04 \times 10^{-6}$ |
| 1 mB | 100 | — | 0.00098692 | 0.014504 |
| 1 atm | 101,325 | 1013.25 | — | 14.696 |
| 1 psi | 6,896.55 | 68.9476 | 0.06804 | — |

### 2.2.1.2 Air Density

Throughout this book, air density will be designated by the symbol $\rho$ (rho) and, as with pressure, it is subject to a simple definition (Anderson 2000).

The *density* of any substance (including a gas) is the mass of that substance per unit volume.

As was the case with the definition of pressure at a point, $X$, the density $\rho$ at point $X$ can be defined as:

$$\rho = \lim\left(\frac{dm}{dv}\right) \qquad dv \to 0 \tag{2.2}$$

where $dv$ is an elemental volume around $X$, and $dm$ is the mass of gas inside $dv$ as illustrated in Figure 2.2.

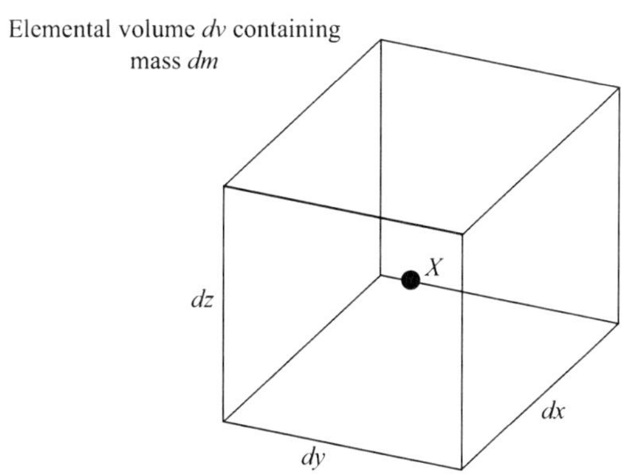


**Figure 2.2.** An elemental volume *dv* containing mass *dm*.

Again, there are various units that can be used in the measurement of air density, but the most common unit of kg $m^{-3}$ will be used throughout this chapter.

### 2.2.1.3 Temperature

To fully appreciate the significance of the temperature of a gas, it should be recalled that any gas flow consists of a collection of particles (molecules and/or atoms) that are in constant motion, and are moving through space and frequently colliding with one another as depicted in Figure 2.3. In this figure, each molecule has a position vector $\boldsymbol{r}$ and a velocity vector $\boldsymbol{c}$. This is so even when the bulk speed of the gas is zero—see Section 2.2.1.4.

It is possible to determine the average kinetic energy of a single particle (at least in principle) by considering its motion over some finite period of time during which it undergoes a series of collisions with other particles and, possibly, the container in which the gas rests or the body around which it flows. The temperature $T$ of the gas is directly proportional to the average particle (or molecular) kinetic energy. These concepts result in a formal definition of temperature (Anderson 2000):

Temperature is a measure of the average kinetic energy of the particles in a gas.

If $KE$ is the mean molecular kinetic energy, then the temperature is given by $KE = \frac{3}{2}kT$ where $k$ is the Boltzmann constant ($k = 1.38 \times 10^{-23}$ J/K ). In this definition, $J$ is the unit of energy, the joule, and $K$ is the unit of temperature, the kelvin. In addition to this unit of temperature, the kelvin $K$, other commonly used units are the degree Celsius, °C, the degree Rankine, °R, and the degree Fahrenheit, °F.

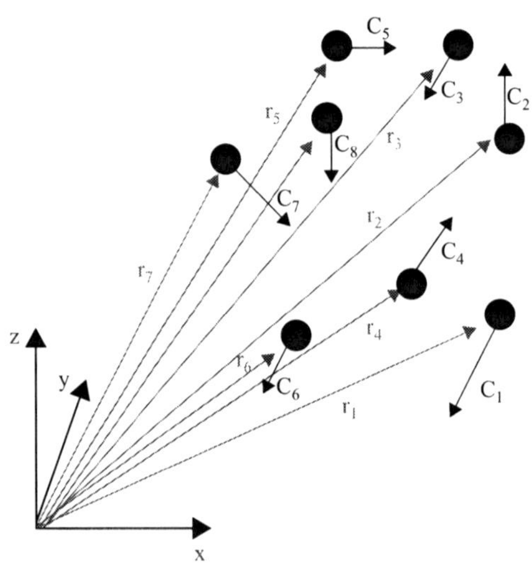


**Figure 2.3.** An assembly of molecules moving with continual random motion.

Note that $\Delta 1 K \equiv \Delta 1\ C$ and $\Delta 1\ R \equiv \Delta 1\ F$. Other useful relationships are:

$$C = \frac{5}{9}(F - 32) \quad \text{and} \quad F = \frac{9}{5}C + 32 \tag{2.3}$$

for conversion between degrees Fahrenheit and degrees Celsius. A chart showing the relationships of these units is given in Figure 2.4.

The property of temperature is one of the four fundamental properties of air, and it is at supersonic and hypersonic speeds (considered later) that it becomes most important when considering the operation of air pressure-dependent sensors.

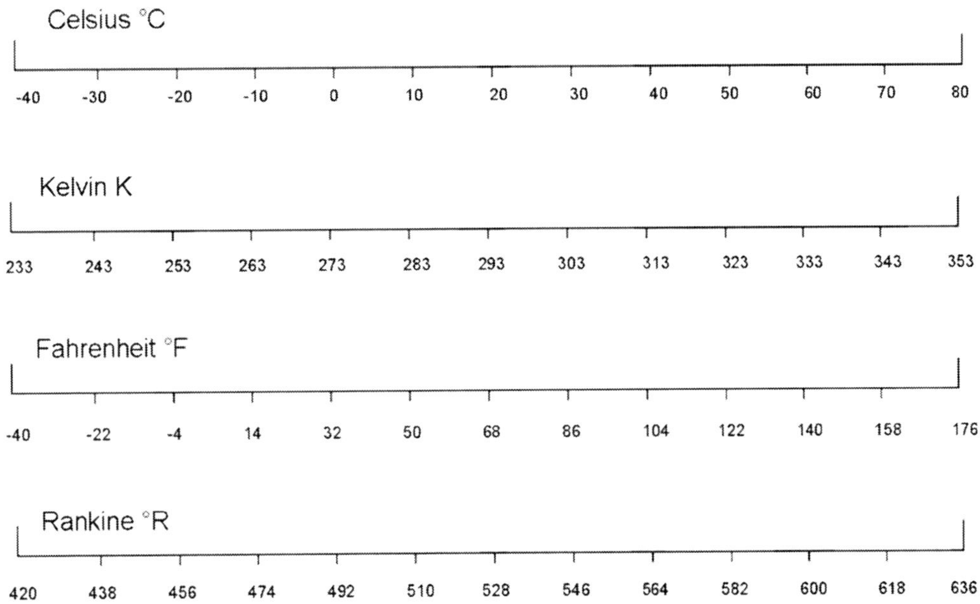


**Figure 2.4.** Relationships between four temperature measurement conventions.

### *2.2.1.4 Flow Velocity*

As with the properties considered so far (p, $\rho$, and T), the velocity of air within a flow field varies from point to point. The air velocity at any point $X$ within the flow field can be understood by considering an infinitesimally small element of the gas focused on that point and determining how this element moves with time. Both, its speed and direction of travel will change as it moves from point to point within the flow field. This concept leads to a definition of flow velocity (Anderson 2000).

> The velocity at any fixed point $X$ in a flowing gas is the velocity of an infinitesimally small fluid element as it sweeps through $X$.

Because a gas can be considered as a collection of particles, what is actually measured is the mean velocity of all the particles contained within this infinitesimally small fluid element. The particles themselves need not all be travelling with the same velocity as the fluid that they constitute. This is demonstrated in Figure 2.5.

Note that the velocity of the undisturbed air at a distance from an object being studied is often referred to as the *freestream velocity*, or *freestream speed*. When the term *freestream speed* is used, it may be taken to be equivalent to the airspeed of an aircraft.

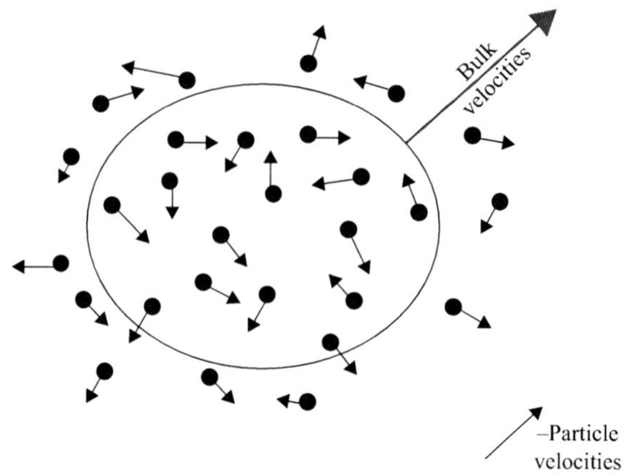


**Figure 2.5.** Particle velocities and bulk gas velocity.

This leads naturally to the concept of streamlines. Provided that the flow field is steady, the fixed paths of fluid elements through the flow field can be traced and it is these paths that are termed streamlines. The visualization of streamlines is very helpful, and sketching the streamlines of particular flow fields relevant to pressure-dependent sensors assists in the understanding of their operation. For interpreting streamline plots, a general rule is that streamlines that are packed closely together indicate high velocities and widely spaced streamlines indicate low velocities. Figure 2.6 indicates a typical streamline pattern over a simple 2D aerofoil shape (a) and the physical manifestation of streamlines using smoke traces in a wind tunnel (b).

### 2.2.2 *THE EQUATION OF STATE FOR A PERFECT GAS*

Returning to the image of a gas as a collection of particles in random motion, as in Section 2.2.1.3, consideration must be given to the nature of the interaction between these particles. In fact, each particle has an *intermolecular force field* that originates in the complex interactions of the electromagnetic properties of the electrons and nuclei (Sone 2007). The intermolecular force field of any given particle extends a long distance into the space around it when compared with the size of the particle itself. Close to the particle, this intermolecular force field is repulsive and very strong, but at greater distances becomes an attractive force that diminishes and becomes very weak at large distances from the molecule. This implies that particles at large separations from each other experience little intermolecular force, which leads to the definition of a perfect gas (Anderson 2000):

A perfect gas is one in which intermolecular forces are negligible.

The previous discussion of intermolecular force fields implies that gases in which the particles are widely spaced (low density) approach the definition of a perfect gas. Typically, the molecules in stationary air at room temperature and pressure are separated by approximately 10 molecular diametres ($\sim 10^{-9}$ m). This is deemed acceptable for the air to be treated as a perfect gas. It also transpires that the air in the vast majority of aerodynamic flows over aircraft, under subsonic and supersonic conditions, can also be treated as a perfect gas.

The advantage of being able to treat air as a perfect gas is that there exists a very simple relationship between $p$, $\rho$, and $T$ for perfect gases known as the *equation of state*, given by

(a)

(b)

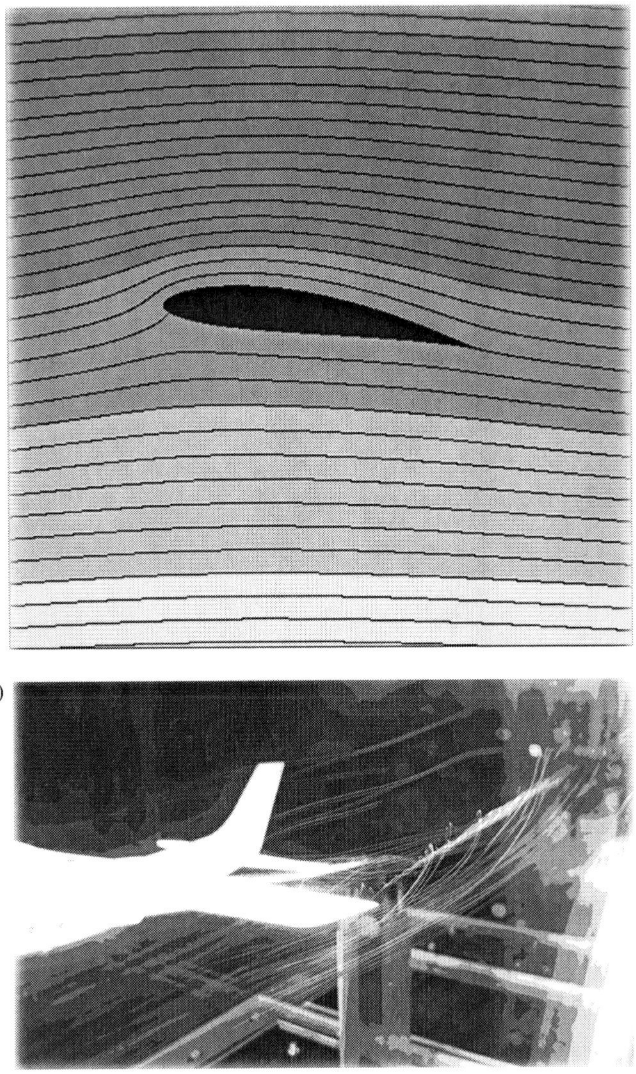

**Figure 2.6.** Streamlines (a) and smoke traces (b).

$$p = \rho RT \quad (2.4)$$

where $R$ is the specific gas constant for air at standard temperature and pressure, which is 287 J kg $K^{-1}$.

### *2.2.3 EXTENSION OF DEFINITIONS: TOTAL, DYNAMIC, STATIC, AND STAGNATION*

Although the four physical properties so far encountered are the most fundamental ones, there are further, more subtle definitions and relationships that must added to the inventory.

The pressure at "points in the flow field" should strictly be referred to as *static pressure*. This is the pressure that would be experienced by a measuring device moving with the fluid element under consideration. An alternative approach to understanding the definition of static pressure would be to consider the molecular velocities of the particles comprising the flow as having a mean, or "bulk," component equal to the velocity of the flow and a random superimposed fluctuating component. It is the momentum transfer to or across the surface in question, resulting from the random component of the molecular velocity that is responsible for the static pressure. This implies that if a body is stationary relative to the fluid (i.e., the bulk velocity is zero) then the pressure experienced is simply the static pressure.

A second type of pressure commonly used in aerodynamics is *total pressure*. For *incompressible flow*, in which the density is unchanging in space (the conditions required for compressibility will be considered in the following section), it is simply necessary to consider the total pressure at any point to be the sum of components resulting from the random velocity fluctuations of particles (static pressure $p$) and also from the mean velocity $V$ of those particles, the *dynamic pressure*. The total pressure is therefore:

$$\underbrace{p_0}_{\text{Total pressure}} = \underbrace{p}_{\text{Static pressure}} + \underbrace{\frac{1}{2}\rho V^2}_{\text{Dynamic pressure}} \tag{2.5}$$

The relationship of Equation (2.5) can be easily derived by considering the conservation of energy along a streamline in a flow, and this is the *Bernoulli equation* for incompressible flow.

In general, the total pressure at a point can be thought of as being the value of the static pressure were the fluid element surrounding that point to be brought isentropically[1] to rest. This definition allows the extension of the total pressure concept to include compressible flow, in which the air density does vary in space.

The *stagnation pressure* (or *ram pressure*) is the static pressure at *stagnation points*, which are points of zero velocity. The stagnation pressure is only equivalent to the total pressure if the fluid element reaching the stagnation point has been decelerated from the freestream speed isentropically.

### *2.2.4 THE SPEED OF SOUND AND MACH NUMBER*

#### *2.2.4.1 The Speed of Sound*

Sound waves allow disturbances in the properties of air to propagate through space. A sound wave is simply a discontinuous jump between two gas states, as shown in Figure 2.7.

It can be shown [1*] by a straightforward application of the principles of the conservation of mass and momentum across this moving sound wave that the wave speed $a$ must be given by:

$$a = \sqrt{\left(\frac{dp}{d\rho}\right)} \tag{2.6}$$

---

[1] An isentropic process is one which is both reversible and involves zero heat transfer. Generally, a parcel of air that does not experience a shock wave or significant viscous effects can be treated using isentropic assumptions (Anderson 2000). Further detailed study of isentropic aerodynamics processes is beyond the scope of this book.

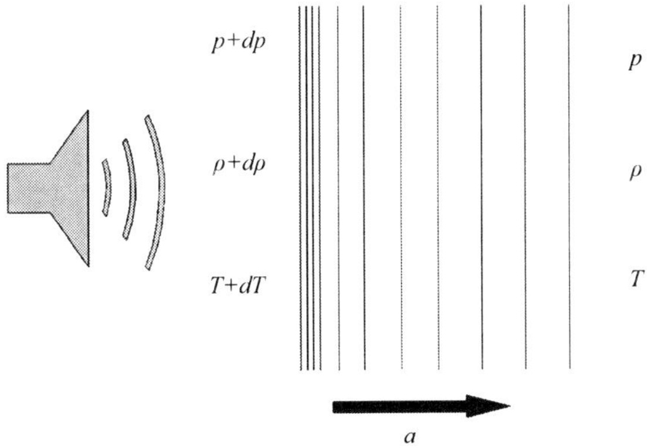


**Figure 2.7.** Model of a sound wave moving through a stationary gas.

This immediately indicates that the speed of sound (the speed of a moving sound wave) is related to the compressibility of the air since the differential term $d\rho$ appears in this relationship.

Further analysis using isentropic gas relationships (Anderson 2000) and the gas equation of state (2.4) leads to a simpler relationship for the speed of sound in a perfect gas:

$$a = \sqrt{\gamma RT} \tag{2.7}$$

where $\gamma$ is given by $\gamma = \frac{c_p}{c_v}$, the ratio of the constant pressure and constant volume-specific heat capacities; $R$ is the specific gas constant; and $T$ is the temperature measured in kelvin. For the applications relevant to this book, $\gamma$ may be treated as a constant and equal to 1.4 unless otherwise specified.

For air treated as a perfect gas (with $R = 287$ J Kg K$^{-1}$) at 288 K (15°C), the speed of sound computes as 340 m s$^{-1}$ (761 mph). Note that as with $p$, $\rho$, and $T$, the speed of sound $a$ is another fundamental physical point property in an airflow.

### *2.2.4.2 Mach Number and Compressibility*

The definition of the speed of sound as discussed earlier leads naturally to a related point property of a flow, the *Mach number*. The Mach number at a given point in a flow is a nondimensional property simply given by the ratio of the magnitude of the local bulk velocity and the local speed of sound:

$$M = \frac{V}{a} \tag{2.8}$$

The nondimensional Mach number at infinity (i.e., the freestream flow field Mach number), $M_\infty$ of an aerodynamic flow is a powerful measure of the regime of flow under consideration. This classification of flow regime gives an important indication of how the flow should be analyzed.

Figure 2.8 shows a typical breakdown of these regimes into subsonic, transonic, supersonic, and hypersonic. Although the boundaries between these regimes are depicted as being absolute in Figure 2.8, in reality there are significant blurs between them. For example, a flow will gradually transition from having supersonic to hypersonic properties as the freestream flow speed is increased.

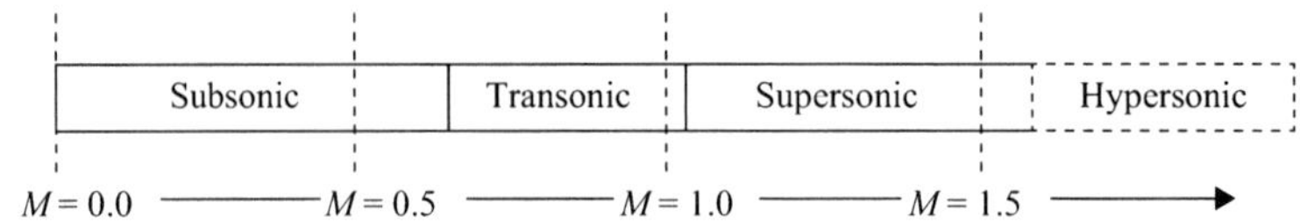


**Figure 2.8.** Aerodynamic speed regimes.

Finally, it has already been noted that whether or not a flow can be treated as incompressible is important in terms of how that flow must be analyzed. A useful rule of thumb in this context is that flows with $M\infty < 0.3$ may be treated as incompressible. Compressibility effects must be considered for all flows with $M\infty > 0.3$. The implication of this is that for many of the instruments considered in future sections of this chapter, it will be necessary to discriminate between incompressible and compressible flows in explanations of their operation.

### 2.2.5 THE SOURCE OF AERODYNAMIC FORCES

A knowledge of $p$, $\rho$, $T$, and $V$ at each point in a flow fully defines the flow field:

$$\left.\begin{aligned} p &= p(x,y,z) \\ \rho &= \rho(x,y,z) \\ T &= T(x,y,z) \\ V &= V(x,y,z) \end{aligned}\right\} \text{flow field} \qquad (2.9)$$

Probably the most important consequence to a body of a flow field acting upon it is the generation of a force. It is this aerodynamic force that drives the action of all the sensors discussed in this chapter.

Simplistically, the sources of all aerodynamic forces acting upon a surface may be separated into:

1. Pressure distribution on the surface
2. Shear stress (friction) on the surface.

In Section 2.2.1.1, it was shown that pressure acts normal to a surface as depicted in Figure 2.9 and as it is a point property it varies across the surface. In general, integration of this pressure across all the surfaces bounding a body results in a net force.

The second source, shear stress, $\tau_w$, is the force acting per unit area tangentially on a surface due to friction, as depicted in Figure 2.10. Again, this stress varies across the surface and integration over this surface leads to a net force. In the case of shear stress, most of the net force is typically in the direction of the freestream flow. For example, in the case of an aircraft, shear stress will contribute significantly to drag but have a negligible contribution to lift.

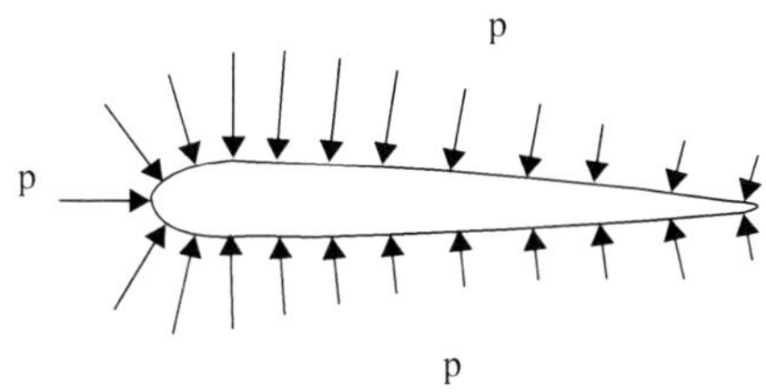


**Figure 2.9.** Pressure acting on a body in a flow field.

Regardless of the complexity of the body in question or the flow field around it, it is these two sources of force generation that form the basis of the flow's ability to interact with the body.

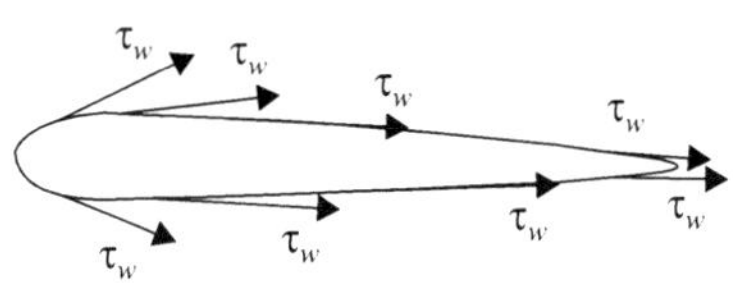

**Figure 2.10.** Shear stress acting on a body in a flow field.

## 2.3 ALTITUDE CONVENTIONS

There is no single absolute definition of altitude (sometimes referred to simply as "alt" in aeronautical publications) since an aircraft's vertical position may be measured relative to a range of reference points. Referring to Figure 2.11, some common altitude definitions are as follows (United Kingdom Aeronautical Information Publication AIRAC 09/2010).

- *Altitude MSL* is the height $H$ above **M**ean **S**ea **L**evel and is also known as the *True Altitude* in Western terminology.[2]
- *Altitude AGL* is the height $H_g$ **A**bove **G**round **L**evel and is the vertical distance between an aircraft and the ground (or water) surface over which it is flying. It is also known as the *Absolute Altitude* in Western terminology[2] and is the vertical distance of the aircraft above the terrain immediately below. *Height* is also used interchangeably with absolute altitude.
- *Relative Altitude* is the vertical distance relative to a reference *elevation*, usually the takeoff point; that is, the altitude $H_0$ relative to the height of the takeoff runway above MSL.
- *Indicated Altitude* is the altitude that actually appears on the indicating instrument. This includes any error present in the sensor.

[2] It must be noted here that there is a discrepancy between *True Altitude* and *Absolute Altitude* in Western and Eastern (notably Russian) practice, and the definitions are in fact interchanged. For example, in Russian the *Absolute Altitude* (абсолютная высота) means the *Altitude MSL*; and the *True Altitude* (истинная высота) means *Altitude AGL*. This is why the present volume uses the self-explanatory terms *Altitude MSL* and *Altitude AGL* throughout.

- *Pressure Altitude* is the indicated altitude when an altimeter is set to an agreed baseline pressure setting, usually the MSL pressure under International Standard Atmosphere conditions as introduced in Chapter 1. That is, the altimeter is set to read zero at 1013.25 millibars or 29.92 inches of mercury (Hg). Pressure altitude is used primarily in high-altitude flight where all aircraft use the same altimeter setting for vertical separation reasons.
- *Density Altitude* is the vertical distance above mean sea level (MSL) at which the existing atmospheric density would appear under the International Standard Atmosphere conditions. It can be thought of as the pressure altitude corrected for nonstandard temperature. It is of particular importance in the computation of predicted aircraft and engine performance.

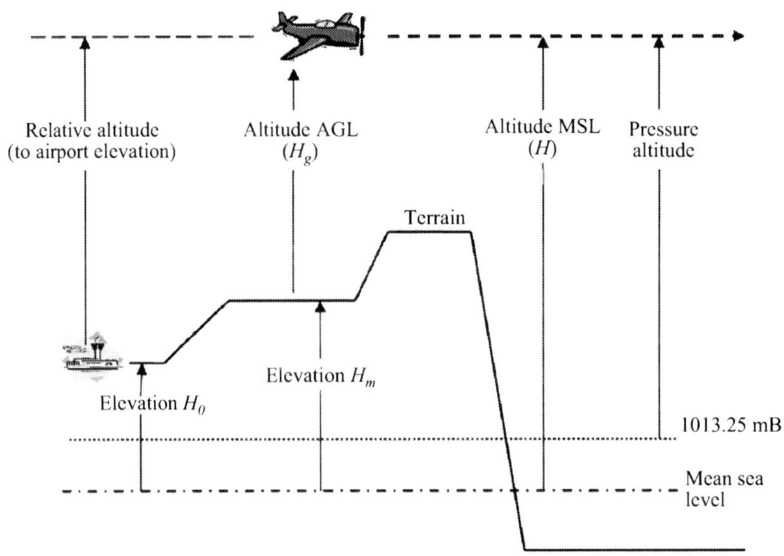


**Figure 2.11.** Some altitude definitions pictorially.

A pilot will use the most appropriate altitude definition as his primary source of reference depending on the phase of the flight. For example, under general cruise conditions, well clear of the ground, true altitude or pressure altitude might be the most useful references. Over high terrain, absolute altitude (or height) is clearly more important. On the approach to landing, the relative altitude (measured from the runway threshold) will become the most important reference.

Any device used to measure the altitude of an aircraft is called an *altimeter* of which there are several types (Pomukaev, Seleznev, and Dmetrechenko 1983), the most common using the barometric method that depends on the atmospheric pressure $p$, which is in turn dependent on the altitude $H$ (Pallett 1985). There are, however, several other methods for measuring flight altitude (Kershner 1984).

## 2.4 BAROMETRIC ALTIMETERS

Fundamentally, a barometric altimeter responds to changes in air pressure, not to altitude *per se*. However, the relationship between change in air pressure and change in altitude is well known and this will be considered later. Unfortunately, air pressure also changes over time, spatially, and with ambient temperature, all according to weather conditions, which means that the barometric altimeter must be capable of being set to the prevailing conditions at takeoff and thereafter during a flight. For a typical flight, and referring to Figure 2.11, the altitude MSL at the takeoff point (or elevation, $H_0$ in the diagram) must first be known so that the altimeter can receive an initial setting by the pilot (United Kingdom Aeronautical Information Publication AIRAC 09/2010). There are two ways in which this can be achieved. Firstly, the elevation of the takeoff airport or airstrip is marked on charts and the altimeter can be manually set to display this reading. Secondly, the actual air pressure can be obtained via radio from various sources including automated ones such as ATIS (Automatic Terminal Information Service). This reading can also be entered into the altimeter manually via a port in the dial, usually called the *Kollsman Window*, which is calibrated in terms of either millibars or inches of mercury. This is shown in Figure 2.12. Normal practice is to adjust it according to further radio reports as the flight progresses so that the *Indicated Altitude* can be corrected at appropriate intervals.

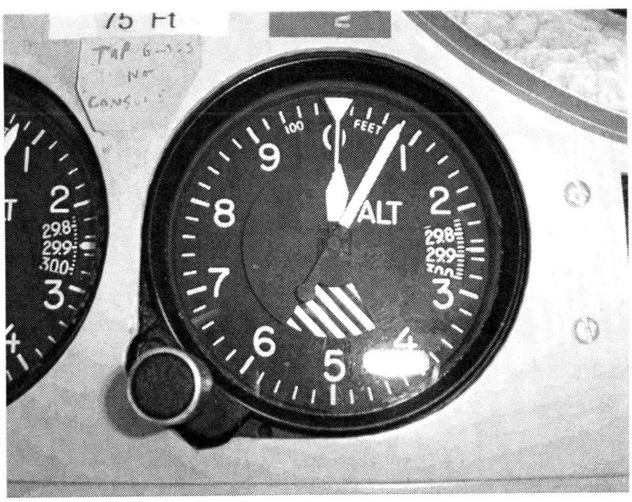


**Figure 2.12.** A typical altimeter with a Kollsmann window. (http://en.wikipedia.org/wiki/File:Aircraft_altimeter.JPG)

Recapping, the atmospheric density $\rho$ and temperature $T$ change with altitude, and it is largely these that determine the indicated altimeter readings. To compensate for changes in them automatically would be very difficult, so for calibration reasons a *Standard Atmosphere* has been defined by the ICAO (International Civil Aviation Organization) as the condition at MSL when the ambient temperature $T_0$ is 15°C and the average air pressure $p_0$ is 1013.25 millibars (mB) or 29.92 inches of mercury. In SI units, $T_0 = 288.15$ kelvin (K); and $p_0$=101,325 pascals (Pa) (1 millibar = 100 Pa). Furthermore, the rate at which the air temperature decreases with increasing altitude, or *lapse rate*, *L*, has also been defined as 1.98°C per 1000 feet (or 300 metres). This is also known as the *thermal gradient*, *L*. The altitude referred to in these standard conditions is called *Density Altitude* (see Section 2.3) and in reality, it is rarely met because of constantly changing weather.

## 2.4.1 *THEORETICAL CONSIDERATIONS*

The vertical extent of the atmosphere in which most aircraft fly can be divided into the troposphere and stratosphere, and these are separated by the tropopause. The troposphere extends vertically to an altitude of approximately 11 km and within it the temperature decreases. Above 11 km, a temperature inversion exists and within the stratosphere, the temperature begins to increase with altitude. This is due to the presence of ozone within the stratosphere, which absorbs ultraviolet radiation from the sun, re-emitting it as heat.

The temperature distribution in the standard atmosphere is shown in Figure 2.13.

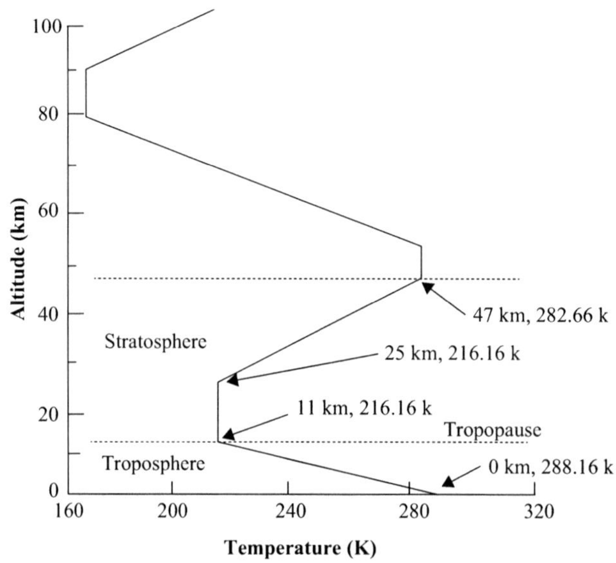


**Figure 2.13.** Temperature distribution in the standard atmosphere.

### *2.4.1.1 The Troposphere*

Basic hydrostatic principles give the rate of change of air pressure with respect to altitude:

$$\frac{dp}{dH} = -\rho g \tag{2.10}$$

where $\rho$ is the air density and $g$ is the gravitational acceleration taken as constant at the MSL value.

Air pressure is dependent upon both the temperature and density of the air according to the ideal gas law, Equation (2.4):

$$p = \rho RT$$

Combining Equations (2.10) and (2.4) gives:

$$\frac{-dp}{p} = \frac{g}{RT}dH \tag{2.11}$$

At this point it must be stated that the atmospheric temperature decreases with altitude only up to about 11 km. This is the height of the troposphere, within which almost all weather phenomena take place. So, for any altitude $H$ below 11 km, the temperature will be:

$$T = T_0 - LH \tag{2.12}$$

where $T_0 = 288.15°K$, the standard atmosphere value, and $L$ is the average thermal gradient or lapse rate, which has a tabulated value for every altitude.

Combining these equations and integrating eventually gives the standard barometric formula for the dependency of atmospheric pressure $p$ on altitude $H$, which is:

$$p = p_0\left[1 - \frac{LH}{T_0}\right]^{\left(\frac{g}{LR}\right)} \tag{2.13}$$

Here, the values of $g$ and $R$ would be termed $g_0$ and $R_0$ at MSL.

Extracting the altitude $H$ from Equation (2.13) gives the following hypsometric formula for altitude in the troposphere, $H_T$:

$$H_T = \frac{T_0}{L}\left[1 - \left(\frac{p}{p_0}\right)^{\frac{LR}{g}}\right] \tag{2.14}$$

Actually, this expression can also be used for nonstandard conditions (which are the norm) using appropriate measured values for the "constant" terms.

#### *2.4.1.2 The Stratosphere*

Within the stratosphere and below 25 km altitude, the temperature remains roughly constant and equal to the temperature at the tropopause, $T_{11}$ (see Figure 2.13). So, within the stratosphere, integration of Equation (2.11) results in the following expression for the dependency of atmospheric pressure, $p$, on altitude, $H$:

$$p = p_{11}\exp\left(\frac{-g(H - H_{11})}{RT_{11}}\right) \tag{2.15}$$

where the subscript "11" indicates values at the tropopause (altitude 11 km). Again, extracting the altitude $H$, from this relationship gives the formula for altitude in the stratosphere, $H_S$:

$$H_S = H_{11} + \frac{RT_{11}}{g}\ln\left(\frac{p_{11}}{p}\right) \tag{2.16}$$

Note that the above analysis assumes that both $g$ and $R$ remain constant within the troposphere and stratosphere. More accurate relationships for altitude and pressure may be derived by following the same procedures as shown here whilst including the dependencies of $g$ and $R$ on altitude. This is beyond the scope of this chapter.

### 2.4.2 BAROMETRIC ALTIMETER PRINCIPLES AND CONSTRUCTION

The principle of the barometric altimeter is depicted diagrammatically in Figure 2.14, where an aneroid chamber or *capsule* (1) is the basic sensor element. This capsule is partially evacuated, but is prevented from collapsing by a U-shaped spring (2). It is located inside a hermetically sealed box (3) which is connected via a pipe (4) to a *static probe* (5). This probe is installed outside the aircraft and is offset forward using an extension that may be on the front part of the fuselage or the leading edge of a wing. When the static pressure decreases as the aircraft climbs, the spring is able to open capsule (1) further. This displacement is transferred via a pivoted rod driving a cord (6) to the altimeter indicator (7). (It may also actuate some other readout or transmission device.) When the static pressure increases during descent, the opposite occurs. (Note that this diagram is intended to depict the aneroid barometric principle and not the construction of an actual altimeter.)

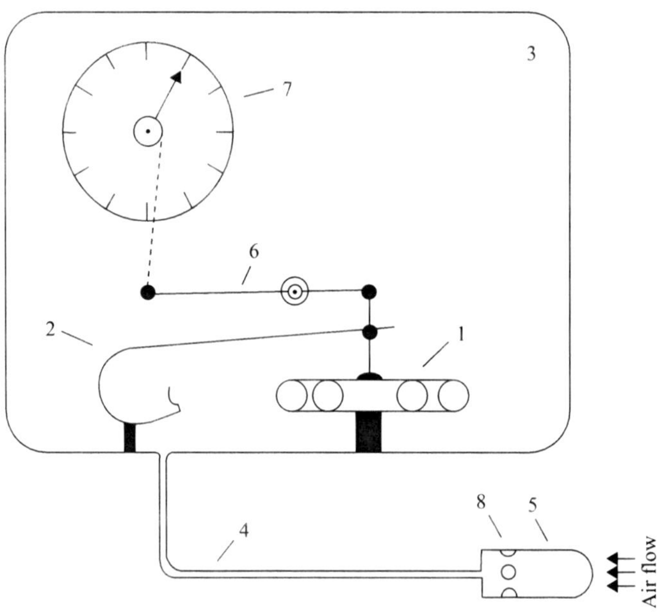


**Figure 2.14.** Principle of the barometric altimeter.

In Figure 2.14, the static probe (5) is a cylinder with a hemispherical head (for subsonic speeds). Apertures (8), at a distance $l$ from the tip, are used to connect the probe interior to the atmosphere. Distance $l$ is of critical importance for accurate operation of the altimeter system, and the pressure on the cylinder surface at the locations of the apertures should be as close as possible to the static pressure of the undisturbed atmosphere.

The distribution of dynamic pressure measured at the surface of a static probe of diameter *d* as a function of *l* is shown in Figure 2.15. This indicates that the distance of the probe apertures from the probe head should be about 4*d* to 5*d*, where the dynamic component of the total pressure approaches zero. The result is that the total measured pressure becomes close to the true static pressure in this region for subsonic flight. A probe for supersonic speeds has a steepled head, and in which case the distance *l* is typically set from 9*d* to 10*d*.

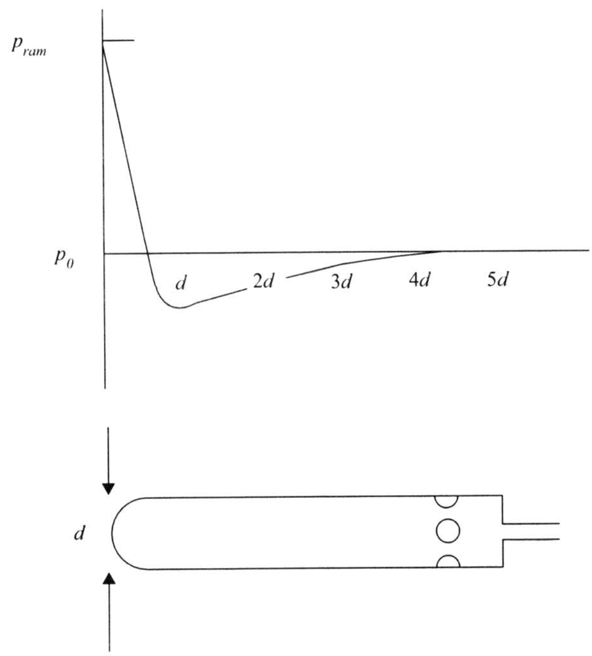


**Figure 2.15.** Distribution of total air pressure along a static pressure probe. (Airflow left to right along probe)

Alternatively, the static pressure may be derived from a port positioned somewhere on the fuselage of the aircraft. This is seen mainly on light aircraft, but tends to be less reliable than a static probe and is often used on larger aircraft only as an emergency alternative in the event of the static probe becoming blocked.

A cut-away diagram showing the actual construction of a basic altimeter is shown in Figure 2.16. Here, the small movement of the capsule is converted to a rotary displacement via a quadrant and appropriate gearing. The spring is seen to be U-shaped and connected to the capsule by a pair of wires.

A "double-hand" altimeter is used to increase accuracy, and the kinematics for this are shown in Figure 2.17. One full rotation of the long hand is equal to 1,000 ft of altitude, and the short hand indicates the number of thousands of feet, performing one rotation for the highest altitude for which the altimeter is designed.

The sensor for the altimeter shown is the double capsule (6). The counterbalance (5) balances all the moving parts and is pivotally connected with axle (15) by means of a separate rod. Bimetallic strips (2) and (3) are used to compensate for instrumental errors in the crank gears.

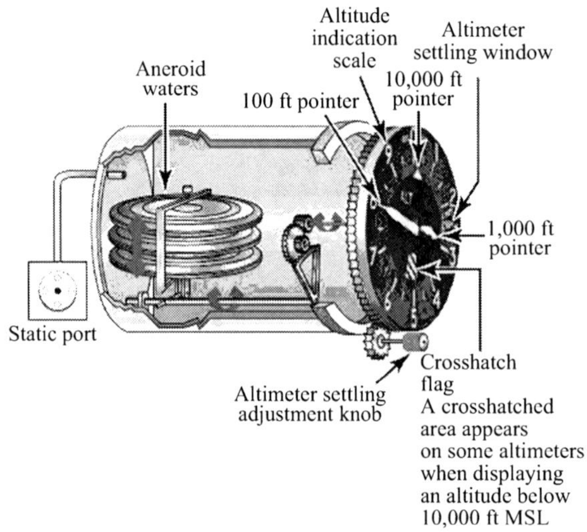


**Figure 2.16.** Internal mechanics of a barometric altimeter.
(*Source:* aviationknowledge.wikidot.com)

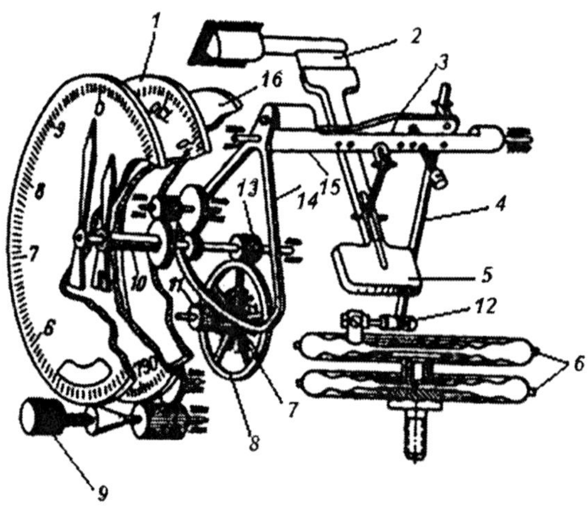


**Figure 2.17.** Construction of the double-hand barometric altimeter.

A geared quadrant (14) transfers rotation from axle (15) to the high-speed axle (13) via gearing (8). This rotation is then transferred from the high-speed axle to the low-speed axle using reduction gearing. A feature of the double-hand altimeter is an additional mechanism that allows manual setting to be performed. This is accomplished via the knob (9) that rotates the base structure (16) in conjunction with the capsule, crank gear, and two first stages of gearing. The other two stages of gearing remain motionless. Hence, the barometric scale (1) moves together with the base (16). As has been mentioned, the initial pressure may be preset at takeoff and modified during flight (particularly prior to landing) using a barometric scale, which is observed through the Kollsman Window.

### 2.4.3 BAROMETRIC ALTIMETER ERRORS

The errors inherent in all of the sensors detailed in this chapter may be grouped into methodical errors and instrumental errors (Hamaki 2003).

#### 2.4.3.1 Methodical Errors

##### 2.4.3.1.1 ERRORS CAUSED BY LANDSCAPE CHANGES

In Figure 2.11, $H_M$ is the height of the ground above mean sea level immediately below the aircraft during *en route* flight; and $H_0$ is the height of the ground above mean sea level at a reference location (such as the takeoff point). Hence, the difference between the altitude above the takeoff point to a point *en route* will be:

$$\Delta H = H_M - H_0 \tag{2.17}$$

This error must be determined using appropriate aeronautical maps or charts. The barometric altimeter cannot detect this error at all, of course, so proper flight planning is mandatory, especially where major changes in ground elevation occur such as in mountainous regions. During flight controlled by autopilot, such corrections may be calculated via an airborne computer. (A radar altimeter can measure absolute altitude, however, as will be seen in Chapter 3.)

##### 2.4.3.1.2 ERRORS DEPENDENT ON DEVIATIONS FROM THE STANDARD ATMOSPHERE

As has been mentioned, the standard atmosphere is rarely encountered in practice, and atmospheric pressures different from this are quite usual. Hence, at a reference point such as the takeoff point, the resulting indicated altitude on an altimeter calibrated for standard conditions will be in error according to the following expression:

$$\Delta H = RT_{av} \ln \frac{\Delta p_0}{p_0} \tag{2.18}$$

where $\Delta p_0$ is the difference between the actual initial atmospheric pressure at takeoff and the relevant standard pressure $p_0$. Also, $T_{av}$ is again the average temperature between the standard temperature and the real temperature at the takeoff point, $(T_0 + T)/2$.

If the conditions at the landing point are different from those at the takeoff point, the values used in Equations (2.14) and (2.16) can easily be changed to those relevant to the two points rather than using the standard-condition values. However, all this is actually accommodated by resetting the altimeter prior to landing using information received by radio from a suitable ground station, and by this means, these errors are completely eliminated.

#### 2.4.3.2 Instrumental Errors

These consist of errors caused by hysteresis and imbalance in moving parts, and frictional and structural changes due to temperature changes.

#### 2.4.3.2.1 ERRORS DUE TO FRICTION

To estimate errors caused by frictional forces in the driving gear, the pressure $\Delta p_{fr}$ required to overcome these frictional forces must be known. The altitude error due to friction, $\Delta H_{fr}$ can then be determined as follows:

$$\Delta H_{fr} = \frac{\Delta p_{fr}}{\xi_i} \tag{2.19}$$

where the air pressure gradient $\xi = \frac{dp}{dH}$ can be found by differentiating Equations (2.13) and (2.15), which lead to the following functions for air pressure gradient in the troposphere (H ≤ 11 km) and stratosphere (H ≥ 11 km) respectively,

$$\xi = -\frac{p_0 g}{RT_0}\left[1-\frac{LH}{T_0}\right]^{\frac{g}{LR}-1} \quad \text{for } H \le 11 \text{ km} \tag{2.20}$$

$$\xi = -\frac{p_{11} g}{RT_{11}}\exp\left(\frac{-g(H-H_{11})}{RT_{11}}\right) \quad \text{for } H \ge 11 \text{ km} \tag{2.21}$$

Thus, as the pressure gradient decreases with altitude, the error increases.

#### 2.4.3.2.2 ERRORS DUE TO TEMPERATURE CHANGES

The instrumental error due to temperature changes can be estimated by the formula

$$\Delta H = \beta R(T_0 - LH)T \tag{2.22}$$

where $\beta$ is the temperature coefficient of elasticity for the bellows material, and $T$ is the temperature.

Real altimeters appear to have a linear scale, whereas it will be obvious from the above that this is not an inherent attribute. It is for this reason that the mechanism linkages are designed to linearize the scale insofar as is possible, and this is assisted by the choice of capsule material.

## 2.5 AIRSPEED CONVENTIONS

The concept of flow velocity was introduced in Section 2.2.1.4 and that of Mach number appeared in Section 2.2.4.2. Both are measures of the rate at which air is flowing, this being clearly of critical importance to an aircraft in flight. The velocity of the airflow relative to the aircraft ultimately dictates the aircraft's aerodynamic behavior, whereas it is the rate at which the aircraft is moving relative to the ground that dictates the time taken for its flight between two locations.

The speeds of an aircraft relative to both the air and the ground (Mangalam 2003) are known respectively as the *Airspeed* (AS), $V$ and the *Ground Speed* (GS), $V_g$.

There are also the following airspeed subdivisions:

(a) The *True Airspeed* (TAS) is the airspeed of the aircraft through the air mass, corrected for various errors, $V$.

(b) The *Indicated Airspeed* (IAS) is the airspeed shown by the *airspeed indicator*, $V_i$. A formal definition is that the IAS is the airspeed of an aircraft as shown on its airspeed indicator calibrated to the equivalent airspeed in the standard atmosphere at MSL where the air density $\rho = 1.225$ kg m$^{-3}$ and is uncorrected for system errors. Here, one of the main system errors is the *position error* that arises from the difficulty of placing a static air pressure port at a point that is truly at static air pressure under all flight conditions. When all such errors are accounted for, the *Calibrated Airspeed* (CAS) or *Rectified Airspeed* (RAS) results.

(c) Any airspeed can be converted into a Mach number simply by dividing by the local speed of sound $a$:

$$M = V/a \tag{2.8}$$

Because an aircraft moves within the surrounding air mass, its track along the ground must accommodate the movement of this air mass. That is, the effect of the wind must be taken into account. Hence, the ground speed GS must be equal to the vector sum of the horizontal components of the airspeed and the windspeed $V_w$. That is,

$$\overline{V}_g = \overline{V}_w + \overline{V} \tag{2.23}$$

The vertical speed, or rate of climb or descent, is obviously of importance too, and is measured separately. Formally, $V_v = dH/dt$. Devices for measuring these various speeds are called *airspeed indicators, Mach number indicators (or Mach-meters)*, and *vertical speed indicators (VSI)* or *variometers*.

Manometric, aerodynamic, thermodynamic, thermal, turbine, ultrasonic, and inertial methods are all used to measure AS, IAS, and the Mach number $M$. Doppler, correlated, inertial, and radiation methods are used to measure the GS. For the purposes of this chapter, only the manometric method will be described in detail. The term *manometric* implies that the method is dependent on the comparison of two separate pressure sources. In the case of airspeed indicators the difference between these pressure sources is related to the airspeed.

## 2.6 THE MANOMETRIC AIRSPEED INDICATOR

Accurate knowledge of an aircraft's airspeed is of fundamental importance to a pilot not simply because it gives guidance in navigation but also because the rate at which the aircraft is moving through the air will dictate its aerodynamic responses. For example, the aircraft controls will increase in their responsiveness as airspeed increases, and flight at too low an airspeed might cause the wings to stall resulting in their lifting ability being dramatically diminished.

### *2.6.1 MANOMETRIC AIRSPEED INDICATOR PRINCIPLES AND CONSTRUCTION*

In the explanation of aneroid altimeter operation given in Section 2.4, the sealed aneroid capsule is partially evacuated and held from collapsing by a spring. Thus, as the static atmospheric pressure changes, the thickness of the capsule also changes. In other words, the atmospheric pressure is balanced against the pressure applied by the spring. If now the capsule is not sealed, but its internal volume is connected to another pressure source, the device will respond to the difference between these two pressures. This is the *manometric principle*, originally used to describe the difference in heights of liquid in a U-tube connected to two air pressure sources.

To realize a practical airspeed indicator, it is necessary to supply the capsule from a source at a pressure defined by the airspeed of the aircraft. To do this, a probe similar to that of Figure 2.14 can again be employed, but with a hole drilled through the "nose" of the probe as shown in Figure 2.18. Here, the incident air stream is brought to rest, and the air pressure at this point is the *stagnation* (or *ram*) pressure. This pressure is dependent on the velocity of the incident air—that is, the airspeed. Such a probe is called a *pitot tube.*

The two pressure sources—stagnation and static—can actually be separate, as in most light aircraft, where a pitot tube is often mounted below a wing and a static source is provided on the fuselage side. However, a more sophisticated arrangement is to combine the two in a single probe called a *pitot-static tube* as shown in Figure 2.18. Here, the hole at the stagnation point on the probe nose is connected to the capsule via another pipe, as shown. By this means, the stagnation air pressure is transmitted to the interior of the capsule, whereas the static pressure is still derived from holes in the probe side at an appropriate distance $l$ from the nose as described earlier. The difference between the two air pressures now determines the capsule thickness, and hence the pointer position, so no spring is needed. The instrument has now become a manometric *airspeed indicator.*

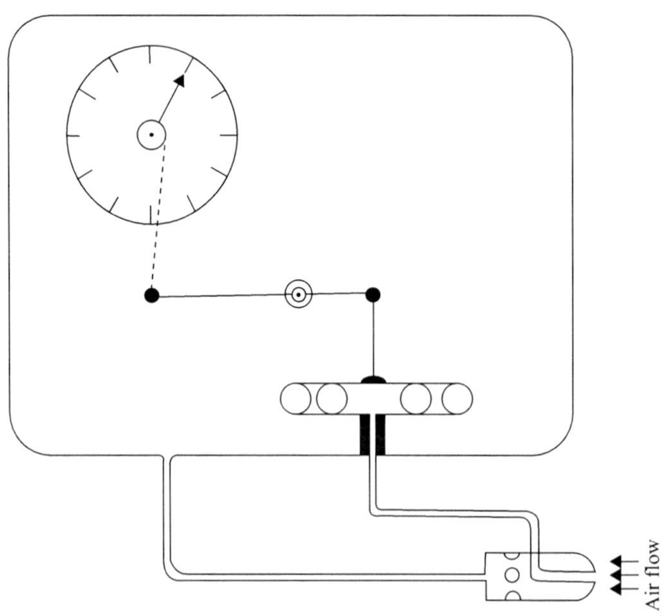


**Figure 2.18.** Principle of the manometric airspeed indicator.

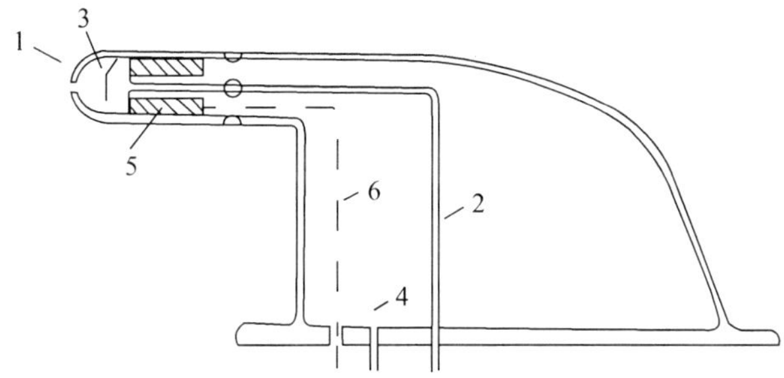


**Figure 2.19.** The basic layout of a pitot-static probe.

The basic layout of a practical pitot-static tube is shown in Figure 2.19. Here, the air flows into the pitot head (1) and is connected to the airspeed indicator via a pipe (2). Baffles (3) serve to prevent the ingress of rainwater and a drain (not shown) is positioned according to the eventual mounting position of the assembly, which may be under a wing, or at the aircraft nose, and so forth as exemplified in Figure 2.20. The static air pressure is conveyed via a second pipe (4). (Here, the casing is at static pressure whereas in some designs it is at pitot pressure and a pipe connects the static holes to the various instruments.) To prevent freezing, a heater coil is also provided (5) along with a cable (6) to the electrical system. Comprehensive practical details have been described by Pallett (Pallett 1985).

Inevitably, the simple explanations given above by no means represent the complete situation. For example, the position of the zero point of the dynamic pressure along the tube is actually a function of the airspeed. Hence, the pressure received by the static pressure apertures is different from the true static pressure and this has to be taken into account. Also, to establish the true airspeed, it is necessary not only to measure the stagnation and static air pressures, but also

**Figure 2.20.** Positions of pitot-static tubes on various aircraft.
(*Source:* **NASA archives**)

the temperature at the flight altitude, the latter because the air density, and hence the air pressure at altitude, are functions of temperature. However, for practical purposes, it is not possible to measure the temperature $T$ at the altitude of flight with sufficient accuracy, so appropriate compensation is needed.

At supersonic speeds yet more problems arise. For example, a compression shock wave appears in front of the probe and leads to further static pressure distortion. To measure the airspeed and Mach number under these conditions it is therefore necessary to measure the total pressure after the compression shock as well as the static pressure, all in undisturbed flow.

## 2.6.2 THEORETICAL CONSIDERATIONS

### 2.6.2.1 Subsonic Incompressible Operation

The theory of operation for a pitot-static tube for airspeed measurement is simplest in the case of subsonic incompressible airflow, that is, $M_\infty < 0.3$. Under these conditions, the total pressure is equivalent to the stagnation pressure and is given simply by the sum of the static and dynamic pressures. Recalling Equation (2.5):

$$p_0 = p + \frac{1}{2}\rho V^2$$

Rearranging Equation (2.5) gives a simple relationship for $V$, which in this case is the true airspeed,

$$V = \sqrt{\frac{2(p_0 - p)}{\rho}} \tag{2.24}$$

This implies that the true airspeed is a function of the difference between stagnation and static pressures. Using the ideal gas relationship to substitute for the air density in Equation (2.24) gives:

$$V = \sqrt{\frac{2RT(p - p_0)}{p}} \tag{2.25}$$

### 2.6.2.2 Subsonic Compressible Operation

The simple relationship of Equation (2.5) that stems from Bernoulli's equation only holds under incompressible flow conditions. To find the relationship between the stagnation and static pressure and airspeed under compressible subsonic conditions (i.e., $0.3 < M_\infty < 1.0$), it is necessary to begin with the isentropic gas relationships below, the derivation of which is provided in [1*]:

$$\frac{p_0}{p} = \left(1 + \frac{\gamma - 1}{2} M^2\right)^{\gamma/(\gamma - 1)} \tag{2.26}$$

where $\gamma$ is the ratio of specific heat capacities (1.4 for normal air). Rearranging Equation (2.26) gives a relationship for the Mach number:

$$M = \sqrt{\frac{2}{\gamma - 1}\left[\left(\frac{p_0}{p}\right)^{(\gamma-1)/\gamma} - 1\right]} \tag{2.27}$$

This value could be displayed in the cockpit for the pilot on a *Mach meter.*

Alternatively, recalling the definition of Mach number introduced in Equation (2.8), and after some algebraic manipulation, the following relationship for true airspeed appears:

$$V = \sqrt{\frac{2a^2}{\gamma - 1}\left[\left(\frac{p_0 - p}{p} + 1\right)^{(\gamma-1)/\gamma} - 1\right]} \tag{2.28}$$

where $a$ is the local speed of sound, requiring knowledge of the local air temperature to give $a = \sqrt{\gamma RT}$.

#### *2.6.2.3 Supersonic Operation*

The operation of a pitot-static tube requires a qualitatively different approach under supersonic conditions. This is due to the formation of a shock wave upstream of the pitot tube when $M\infty > 1$ as shown in Figure 2.21. A shock wave is a very thin region of ~$10^{-4}$ cm in thickness across which some very severe changes in the flow properties take place. A detailed explanation of shock wave formation is beyond the scope of this book but will be found in references (*System for Aircraft Performance, Safety, & Control*. IEEE Aerospace Conference, Big Sky, MT.; Thom and Godwin 2003).

As a fluid element flows through a shock wave, as shown in Figure 2.20 for a perfect gas:

1. The Mach number decreases.
2. The static pressure increases.
3. The static temperature increases.
4. The flow velocity decreases.
5. The total pressure $p_0$ decreases.
6. The total temperature $T_0$ stays the same if the perfect gas assumption is made.

Because of the significant frictional and thermal conduction effects present in shock waves, it is not possible to apply the isentropic relationship of Equation (2.26) to relate total and static pressures to airspeed. The total pressure measured at the nose of the pitot tube will **not** be the same as the total pressure within the freestream flow due to the fact that the fluid elements reaching the pitot tube will have passed through the shock wave. Consequently, a separate shock wave theory must be applied to relate the stagnation and static pressures measured by the pitot tube to the freestream Mach number under supersonic conditions. This theory is beyond the scope of

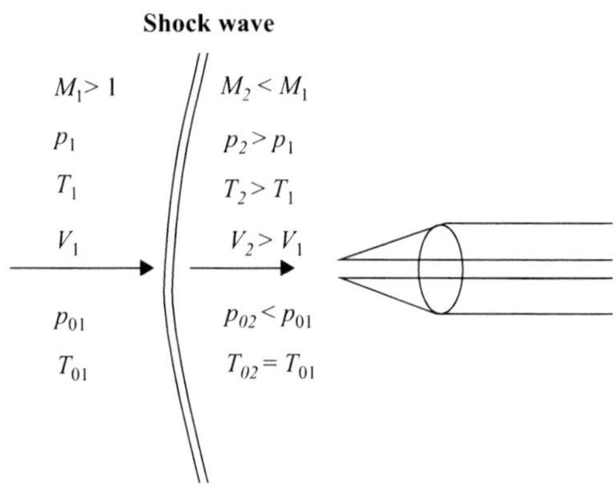


**Figure 2.21.** Pitot tube in supersonic flow.

this book, but the resulting formula, known as the *Rayleigh pitot tube formula* (Balachandran 2006) is provided here for completeness:

$$\frac{p_0}{p} = \left[ \frac{(\gamma + 1)^2 M^2}{4\gamma M^2 - 2(\gamma - 1)} \right]^{\gamma/(\gamma - 1)} \frac{1 - \gamma + 2\gamma M^2}{\gamma + 1} \tag{2.29}$$

This relationship can be used to calibrate a Mach metre or airspeed indicator for supersonic flight.

## *2.6.3 MANOMETRIC AIRSPEED INDICATOR ERRORS*

### *2.6.3.1 Methodical Errors*

The error inherent in Equation (2.5) arises by assuming the validity of Bernoulli's formula and that the air behaves as an "ideal gas." A derivation of Bernoulli's formula based on the principle of conservation of energy requires the assumptions that the air "does no work," that there is no heat transfer to or from the air (adiabatic flow), and that the air is incompressible. These assumptions are never completely valid because in the case of flow around an aircraft for example, heat may well be transmitted from the aircraft engine into the airflow; the air must "do work" to overcome friction in the aerodynamic boundary layer close to the aircraft skin; and small density changes do occur even below $M\infty = 0.3$. However, given appropriate positioning of the pitot-static sensor, the first two assumptions will have negligible effects on the airspeed error, and density variations will contribute less than 2% below $M\infty = 0.3$.

The error inherent in the Equation (2.28) occurs because the isentropic gas relationship has been assumed to be valid, and also that correct values for the ratio of specific heat capacities $\gamma$ and air temperature have been inserted. Again, given appropriate positioning of the pitot-static sensor (discussed below) such errors can be assumed to be less than 2%.

The Rayleigh Pitot tube formula introduced in Section 2.6.2.3 forms the basis for measuring airspeed under supersonic flight conditions and relies heavily on normal shock wave theory (Anderson 2000) based on a simplified model of shock wave behavior and isentropic flow theory. The combined errors due to these assumptions will translate into airspeed indicator error and will be the largest of the methodical errors associated with airspeed measurement.

*Compressibility errors* occur because an altimeter is calibrated to account for the compressibility of the air within the pitot tube under sea level conditions. The pitot-tube air compressibility at altitude will differ from that at sea level, so that an error is introduced. Compressibility errors will only be of significance above 10,000 ft altitude and at airspeeds greater that 200 knots (370 km $h^{-1}$). Often, a correction table is used to modify the indicated airspeed to take these compressibility effects into account.

### *2.6.3.2 Instrumental Errors*

It may be assumed that instrumental errors will form the bulk of the total error inherent in airspeed measurement.

#### 2.6.3.2.1 HYSTERESIS

Hysteresis is an error that is caused by the mechanical properties of the aneroid capsules located within the instruments. These capsules, used to determine pressure differences, have physical properties that resist change by acting to retain a given shape even though the external forces may have changed.

#### 2.6.3.2.2 BLOCKED PITOT TUBE

Pitot tubes are prone to becoming clogged by ice, water, insects, or other obstructions. For this reason aviation regulatory agencies (such as the British CAA and the American FAA) recommend that the pitot tube be checked for obstructions prior to any flight. To prevent icing, many pitot tubes are equipped with a heating element.

A blocked pitot tube will cause the airspeed indicator to register an increase in airspeed when the aircraft climbs, even though the actual airspeed is constant. This is due to the pressure in the pitot system remaining constant whilst the static pressure is decreasing. The reverse is true in descent.

#### 2.6.3.2.3 BLOCKED STATIC PORT

The most common cause of a blocked static port is airframe icing. The error that occurs will be the reverse of that which occurs with a blocked pitot tube, that is, the airspeed indicator will under-read during a climb and over-read during descent. On most aircraft with unpressurized cabins, an alternative static source is made available and can be brought into operation by the pilot.

#### 2.6.3.2.4 POSITION ERRORS

Position errors are introduced as a result of either an aircraft's measured static pressure being different from the air pressure remote from that aircraft, or incorrect measurement of the freestream total pressure due to the positioning of the pitot-static sensor and/or the static port. The static pressure error arises because air flowing past the static port might be different from the aircraft's true airspeed because of the local flow field induced by the aircraft itself. Both the total pressure error and static pressure error will be affected by factors such as airspeed, angle of attack, aircraft weight, acceleration, aircraft configuration, propeller wash, and rotor downwash in the case of helicopters. Position errors may be divided into fixed errors and variable errors. Fixed errors are defined as errors specific to a particular make of aircraft, whereas variable errors are caused by external factors such as deformed panels obstructing the airflow or particular situations which may overstress the aircraft. More details on the appropriate positioning of pitot-static tubes and potential solutions to position errors in airspeed appear in reference (NACA Technical Note 616).

## 2.7 THE VERTICAL SPEED INDICATOR (VSI)

The VSI, or *variometer*, is a device for measuring the vertical rate of climb or descent of an aircraft, $V_v$, and is calibrated in feet (or metres) per minute. An accurate measurement of climb and descent rate is of obvious importance to a pilot when transitioning between altitudes, especially when required to comply with air traffic control restrictions. It is also of critical importance to a pilot in planning and executing the final approach for landing.

Note that the term *vertical speed indicator* or *VSI* is most often used for the instrument when it is installed in a powered aircraft. The term *variometer* is most often used when the instrument is installed in a glider or sailplane.

### 2.7.1 VSI PRINCIPLES AND CONSTRUCTION

The principle of the VSI is shown in Figure 2.22, which is similar to Figure 2.18 except that its operation is based on introducing a lag in the pressure change inside the closed container volume

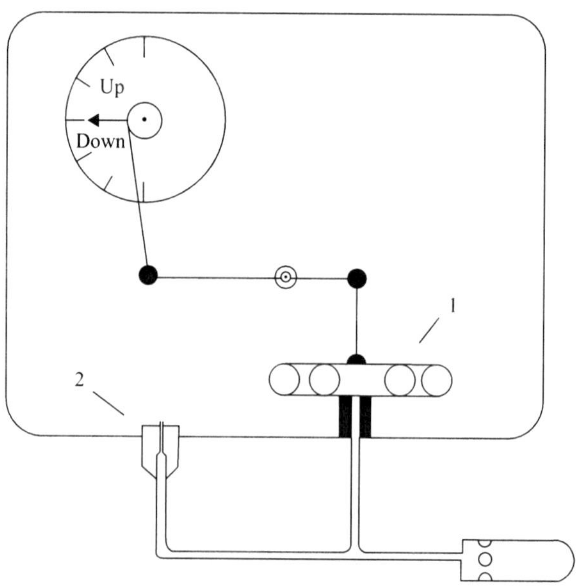


**Figure 2.22.** Principle of the VSI or variometer.

relative to a more rapid change in the static pressure as applied to the capsule interior. This is done by connecting the source of static pressure from a probe (or port) directly to the aneroid capsule (1), but also to the closed containment chamber via a small-diameter capillary tube (2). The capsule is again mechanically connected to an appropriate scale as for the altimeter and airspeed indicator.

During horizontal flight the pressures inside the capsule and the containment chamber are equal, and the indication is zero. As the aircraft climbs, the pressure in the capsule decreases accordingly, and so does that in the containment chamber, but more slowly because of air friction in the small-diameter capillary tube (which is of course a calibrated precision unit). Hence the capsule thickness decreases and the scale pointer rises to indicate the rate-of-climb. The rate-of-descent is similarly indicated by the reverse process. The pressure difference is therefore constant for a constant rate of climb or descent but changes as the rate of climb or descent varies. A typical VSI display is shown in Figure 2.23.

**Figure 2.23.** The VSI display.
(*Source:* **Wikipedia courtesy of Benet Allen**)

### 2.7.2 *THEORETICAL CONSIDERATIONS*

The VSI output, $V_v$, is a function of the time derivative of the barometric signal comparing the static pressure inside the containment chamber $p_c$ with that in the aneroid capsule, $p_a$

$$V_v = f\left(\frac{d}{dt}(p_c - p_a)\right) \tag{2.30}$$

#### 2.7.2.1 *Lag Rate (Time Constant)*

As the altitude changes, so will the external static pressure, and there will always be a lag between the aircraft changing its rate of climb or descent and the stabilizing of the pressures within the containment chamber and the aneroid capsule. Typically, this lag is some

6 to 9 seconds. Rough control and/or turbulence can extend this lag period and cause erratic and unstable rate indications. Some aircraft are equipped with an instantaneous VSI which incorporates accelerometers to compensate for the lag in the traditional VSI described here.

#### 2.7.2.2 *Sensitivity to Mach Number*

As has been seen in Sections 2.4.2 and 2.6.1, the positioning of the static apertures on the static tube is critical for an accurate measurement of the true freestream static pressure. The appropriate positioning of these apertures on the static tube is Mach number-dependent. If the aneroid capsule is connected to a static port on the aircraft fuselage, particular care must be taken in positioning the static port, particularly for operation in transonic flight conditions when transient shock waves might pass that static port resulting in large fluctuations from the correct freestream static pressure.

#### 2.7.2.3 *Sensitivity to Altitude*

Equations (2.20) and (2.21) from Section 2.4.3.2 show that the pressure gradient varies significantly with altitude in the troposphere and stratosphere, respectively. These equations are repeated below:

$$\xi = -\frac{p_0 g}{RT_0}\left[1 - \frac{LH}{T_0}\right]^{\frac{g}{LR}-1} \quad \text{for H} \leq 11 \text{ km} \tag{2.20}$$

$$\xi = -\frac{p_{11} g}{RT_{11}} \exp\left(\frac{-g(H - H_{11})}{RT_{11}}\right) \quad \text{for H} \geq 11 \text{ km} \tag{2.21}$$

A VSI operating over a wide range of altitudes must therefore be calibrated appropriately to take this variation into account.

### 2.7.3 *VSI ERRORS*

Dynamic errors may be reduced by decreasing the time constant $\tau$ and hence the lag in the system. This could be achieved by decreasing the capillary length or increasing its diameter. However, as $\tau$ decreases, so does the device sensitivity.

At low altitudes this time constant is likely to be several seconds. At high altitudes it increases because the air density decreases. During the measurement of a varying vertical speed, dynamic errors in the VSI reach large values.

It is important to provide heat insulation for VSIs in order to reduce errors caused by variations in temperature because the air viscosity in the capillary, which in turn affects the time constant of the system, is actually proportional to the air temperature. The relationship between viscosity and temperature is given by Sutherland's law:

$$\mu = \mu_{ref}\left(\frac{T}{T_{ref}}\right)^{\frac{3}{2}}\frac{T_{ref}+S}{T+S} \tag{2.31}$$

where $\mu$ is the dynamic viscosity of the air, the subscript "ref" indicates reference values ($\mu_{ref} = 1.716\times10^{-5}, T_{ref} = 273.15\,K$ for air), and $S$ is the Sutherland temperature 110.4 $K$. It is possible to use this relationship as a guide for calibrating a VSI in situations where the temperature in the system is likely to experience significant variation.

A further error is introduced due to departures from the standard atmosphere when Equations (2.20) and (2.21) are used to calibrate the system because the derivation of these equations is based on the assumption of a standard atmosphere. However, this error contribution is likely to be negligible in comparison with the dynamic errors described above.

## 2.8 ANGLES OF ATTACK AND SLIP

The direction of the airspeed vector in a coordinate system rigidly connected to the airplane axes is defined by the *angle of attack* $\alpha$ and the *slip angle* $\beta$ (Callegari et al. 2004). Both are useful indications of the aircraft's aerodynamic trim condition, particularly in situations such as crosswind landings.

Formally, the angle of attack is the angle that the wing chord makes with the vector representing the relative motion between the aircraft and the atmosphere, as illustrated in Figure 2.24. A knowledge of this angle gives the pilot an indication of whether or not the *critical angle of attack* is being approached, at which the wing(s) will stall, resulting in a potentially dangerous loss of lift, possibly leading to a full aircraft stall. In Figure 2.24 the airflow direction appears horizontal but this is for convenience and is not necessarily so. This leads to the concept of *relative airflow*, which is the flow direction relative to the wing chord.

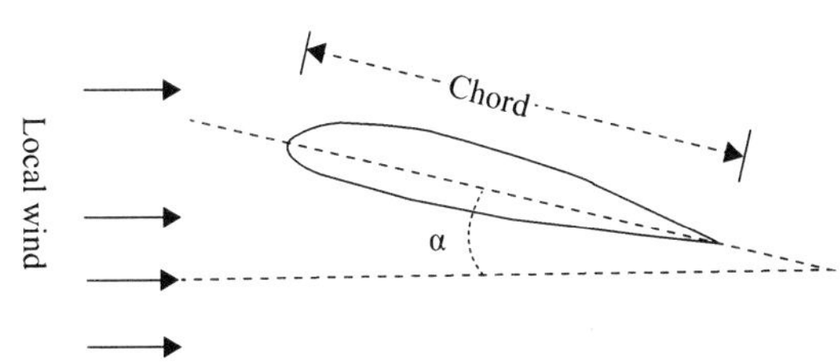


**Figure 2.24.** Aerofoil section showing angle of attack α.

*Slip* is similar in that it is a measure of the angle that the aircraft makes in the horizontal plane relative to the airflow (and is normally small). In principle, the slip angle can be thought of as similar to the angle of attack but rotated by 90°, for which reason the methods of measuring the angle of attack are applicable also to the measurement of slip angle.

There are two common methods of measuring the angle of attack, as follows (Sims 1996).

### 2.8.1 THE PIVOTED VANE

This method uses a small mass-balanced wind vane, which is a streamlined symmetrical aerofoil-section winglet used as a sensing device and having one axis of rotation about its length. Two pairs of such wind vanes can be installed on a probe as shown in Figure 2.25.

The angle-of-attack vanes are connected to form a single swept winglet that is free to move and in doing so operates a digital angle-data transmitter within the probe body. This winglet, being able to rotate about its long axis, takes up the direction of the relative airflow and the angular difference between this and the wing chord (or indeed any similar reference line) is measured via the angle-data transmitter. This is a measure of the angle of attack, $\alpha$.

As has been mentioned, the angle of attack is very important in determining the amount of lift generated, and if the critical angle of attack is actually reached, the aircraft will stall irrespective of its attitude or power setting.

The slip angle $\beta$ refers to the angle between the aircraft centerline and the relative airflow and is usually taken as positive when the relative wind is coming from the right of the aircraft. This angle may be unintentionally produced if insufficient rudder excursion is applied during a turn, or it may be intentional as when performing a crosswind landing. The slip angle vanes are similar to the angle of attack vanes and operate in the same manner, but are mounted as shown in Figure 2.25. Again, an angle-data transmitter is located within the assembly.

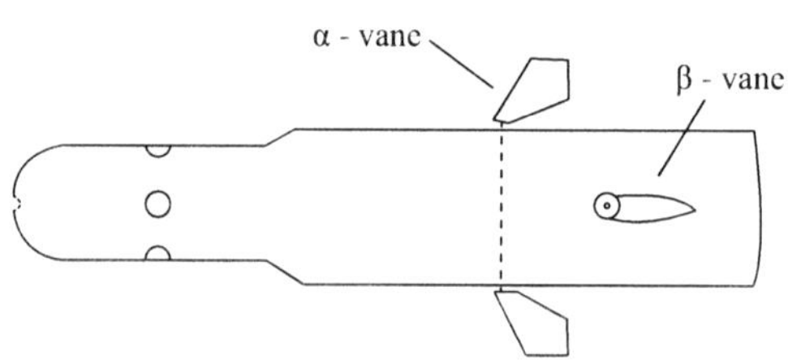


**Figure 2.25.** Basic pitot-static probe layout with attack and slip angle vanes (plan view).

The angle measured by a vane mounted on a probe, such as shown in Figure 2.25, will be influenced by distortion of the flow when the boom is inclined to that flow (the upwash effect) or by asymmetry of the vane due to imperfections in manufacture. It may also be influenced by bending of the boom support, and during maneuvering flight additional errors may be introduced because of further bending due to normal aircraft pitching accelerations (NACA Technical Note 4351).

Pivoted vane devices have been successfully employed across the Mach range from low supersonic to high subsonic, but their sensitivities to the influences mentioned above are likely to be Mach number-dependent, and also specific to particular aircraft installation methods.

### 2.8.2 THE DIFFERENTIAL PRESSURE TUBE

Both $\alpha$ and $\beta$ can also be determined by measuring the pressure difference between two static ports situated on the top and bottom surfaces of a wing, or the nose of an aircraft, or alternatively using a differential pressure tube probe as shown in Figure 2.26. That is, $\Delta p = p_2 - p_1 = f(\alpha \text{ or } \beta)$.

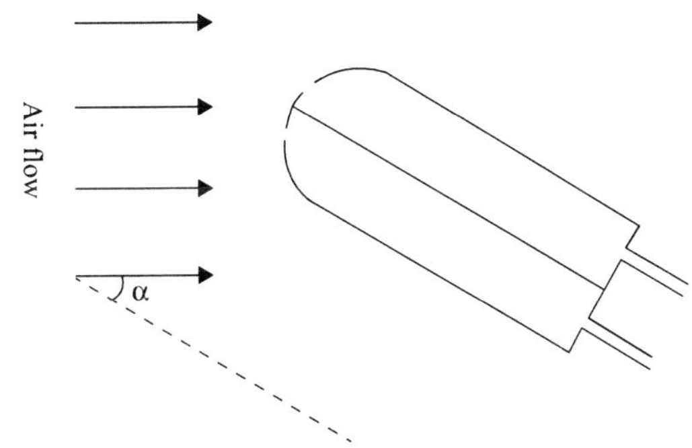


**Figure 2.26.** Basic differential port angle of attack probe layout.

The relative angles at which the two separated tubes are projected to the airflow will affect both the sensitivity and the range of operation of the sensor. Generally, the higher the operational Mach number of this sensor, the smaller the required relative angle between the two tubes for sufficient accuracy.

Differential pressure tubes for measuring both $\alpha$ and $\beta$ may be fitted to a single hemispherical device on the end of a boom (NASA Technical Note D-7461) with the pressure sources at various positions around the hemisphere, but using the same principles as described earlier. This method has been applied successfully up to Mach 1.8.

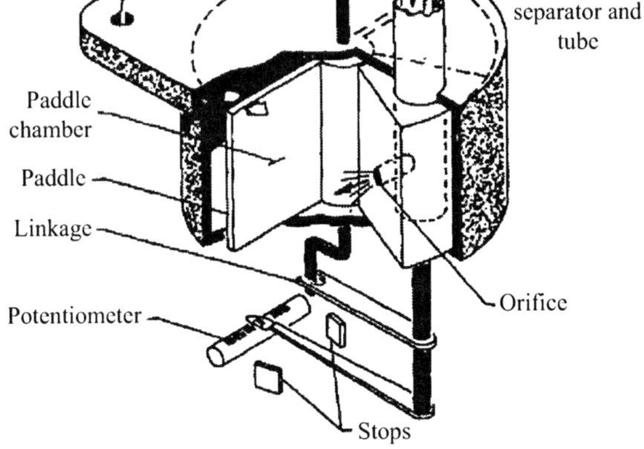


**Figure 2.27.** An example of a null-seeking differential pressure tube.

(*Source:* **NASA Technical Note 616 courtesy of Space Age Control, Inc)**

### *2.8.3 THE NULL-SEEKING PRESSURE TUBE*

As shown in Figure 2.27, the null-seeking pressure sensor is a modification of the differential pressure probe for measuring angles of attack or slip. It consists of a rotatable tube with two orifices disposed at equal angles to the axis of the tube, a pressure-sensitive device for detecting the pressure difference between these two orifices when the tube is inclined to the flow, a mechanism for rotating the tube to the null-pressure position, and instrumentation for measuring the angular position of the tube. The advantages of this approach over the differential pressure tube method are that the measurements are independent of total pressure and Mach number. Also, the differential-pressure sensing elements can be comparatively sensitive because the operating differential pressure signal will always approach zero provided the response of the system is sufficiently rapid.

## REFERENCES

Anderson Jr., J. D. 2000. *Introduction to Flight* (4th edition). WCB/McGraw-Hill.

Balachandran, P. 2006. *Fundamentals of Compressible Fluid Dynamics*. Upper Saddle River, NJ: Prentice-Hall.

Callegari, S., Talamelli, A., Zagnoni, M., Golfarelli, A., Rossi, V., Tartagni, M., and Sangiorgi, E. (2004). "Aircraft angle of attack and air speed detection by redundant strip pressure sensors. **Sensors"**, *IEEE Proceedings* 3: 1526–9.

Hamaki, I. 2003. *Position Error for the Airspeed Sensor of the Multi-Purpose Aviation Laboratory MuPAL-.ALPHA*. Technical Memorandum of National Aerospace Laboratory. **No.** 776. (in Japanese.)

Kershner, D. D. 1984. *Miniature Airflow Sensor. Langley Research Center*. Report Number: LAR-13065.

Mangalam, S. M. 2003. "Phenomena-based Real-time Aerodynamic Measurement (PRAM)."

NACA Technical Note 616, *The Measurement of Airspeed in Airplanes.*

NACA Technical Note 4351, *Summary of Methods of Measuring Angle of Attack on Aircraft.*

NASA Technical Note D-7461, *Flight Calibration Tests of a Nose-Boom-Mounted Fixed Hemispherical Flow-Direction Sensor.*

Novoty, A., and Straskraba, I. 2004. *Introduction to the Mathematical Theory of Compressible Flow.*

Pallett, E. H. J. Aircraft Instruments and Integrated Systems. *Avionics Communications* 1992.

Pomukaev, E. E., Seleznev, V. P., and Dmetrechenko, L. A. 1983. *Navigation Devices and Systems.* Moscow. (In Russian.)

Sims, P. J. 1996. "In-flight measurement of angle of attack", University of Wales, Swansea, United Kingdom, Ph.D thesis.

Sone, Y. 2007. *Molecular Gas Dynamics*. Boston, MA: Birkhauser. DOI: 10.1007/978-0-8176-4573-1.

System for Aircraft Performance, Safety, and Control. IEEE Aerospace Conference, Big Sky, MT.

Thom, T., and Godwin, P. 2003. *The Air Pilot's Manual* (5th edition, Volume 3). Air Pilot Publisher Ltd.

United Kingdom Aeronautical Information Publication AIRAC 09/2010.

## APPENDIX

Simplified Standard Atmosphere table (approximated)

| Altitude (metres) | Pressure (mB) | Temperature (°C) |
|---|---|---|
| −250 | 1,044 | 17 |
| 0 | 1,013 | 15 |
| 250 | 984 | 13 |
| 500 | 955 | 12 |
| 750 | 926 | 10 |
| 1,000 | 899 | 8 |
| 1,500 | 846 | 5 |
| 2,000 | 795 | 2 |
| 2,500 | 747 | −1 |
| 3,000 | 701 | −4 |
| 3,500 | 658 | −8 |
| 4,000 | 616 | −11 |
| 4,500 | 577 | −14 |
| 5,000 | 540 | −18 |
| 6,000 | 472 | −24 |
| 7,000 | 411 | −30 |
| 8,000 | 356 | −37 |
| 9,000 | 307 | −44 |
| 10,000 | 264 | −50 |
| 12,000 | 193 | −56 |
| 14,000 | 141 | −56 |
| 16,000 | 103 | −56 |
| 18,000 | 75 | −56 |
| 20,000 | 55 | −56 |
| 22,000 | 40 | −54 |
| 24,000 | 29 | −52 |
| 26,000 | 22 | −50 |
| 28,000 | 16 | −48 |

# CHAPTER 3

# Radar Altimeters

Alexander V. Nebylov
*St. Petersburg State University of Aerospace Instrumentation*
*St. Petersburg, Russia*

Felix J. Yanovsky
*National Aviation University*
*Kiev, Ukraine*

## 3.1 INTRODUCTION

### 3.1.1 DEFINITIONS

Altitude is one of the prime parameters for any flying vehicle such as an aircraft, spacecraft, or missile. A crew or a control system needs information about altitude with respect to the ground level over the entire flight. Altimetry is the art of measuring altitude, and an altimeter is a sensor that measures the altitude of a flying vehicle. Normally, an altimeter can also serve as a source of information about the vertical speed of a flying vehicle. Hence, an altimeter is one of the necessary sensors forming the equipment complement of modern aircraft and spacecraft.

### 3.1.2 ALTIMETRY METHODS

Flight altitude can be measured using various different physical phenomena. For example, radar, laser, and acoustic methods are based on the measurement of the time taken by electromagnetic or acoustic waves to travel from an aerospace vehicle to a reflective surface on the Earth or another planet. Alternatively, radioisotope methods may be used to measure radiation backscattering intensity. For aircraft, a barometric altimeter measures the air pressure at the level in which the aircraft is flying and converts that measurement to the height above sea level according to the standard pressure–altitude relationship described in Chapter 2. At stratospheric heights, one method measures a corona-discharge current that depends on air density, which corresponds to altitude. Also, an inertial sensor based on an accelerometer installed on the gyro-stabilized platform of a standard inertial system can measure vertical acceleration and this may be followed by double integration to determine altitude.

The main topics in this chapter are altimeters using active radar principles (Skolnik 1990), that is, radar (including laser) altimeters, though a brief treatment of radioactivity methods is also given.

### 3.1.3 GENERAL PRINCIPLES OF RADAR ALTIMETRY

A radar altimeter is a low-power radar system that measures the height of an aircraft (or other aerospace vehicle) above the ground. Like other radar devices, it measures distance by determining the time required by a radio wave to travel to and from a target, in this case the Earth's surface.

Various classes of waveform can be used in radar altimetry, the most obvious divisions of all possible waveforms being continuous (CW) and pulsed waves. In both contexts, the modulation of a carrier wave is necessary in order to measure the time taken for a signal to reach and return from the ground for conversion to the target range, or altitude. Pulse altimeters are more popular for altitudes above about 5000 feet (1500 metres), whereas CW equivalents tend to be preferred for low altitudes. However, this is not a hard-and-fast rule; 0.4 to 44 GHz (or even higher) carrier waves with various forms of pulse modulation or sawtooth frequency modulation may be used for either low or high altitudes. Frequency choice depends upon regulations, mission objectives, and other constraints as well as technical possibilities and impossibilities.

If the Earth were a perfectly flat horizontal plane or a smooth sphere, the return signal would come only from the closest point and would be a true measure of altitude. However, the Earth is not smooth, and energy is scattered back to the radar receiver from all parts of the surface illuminated by the transmitter. For the radar altimeter to measure distance to the ground accurately, it must distinguish between reflections from points near the vertical and those from points that are more distant. Therefore, a narrow antenna beam pointing vertically down would be desirable. However, aircraft antennas are limited in size, and antenna beam-width is finite and frequently rather wide. Generally, a reflected signal is formed from a large surface area depending on beam width and flight altitude. Such signals contain data not only on the altitude $H_g$ but also on slant ranges $R_i$ within the illuminated area $AB$ as is shown in Figure 3.1, where the shape of the antenna pattern is omitted for simplicity.

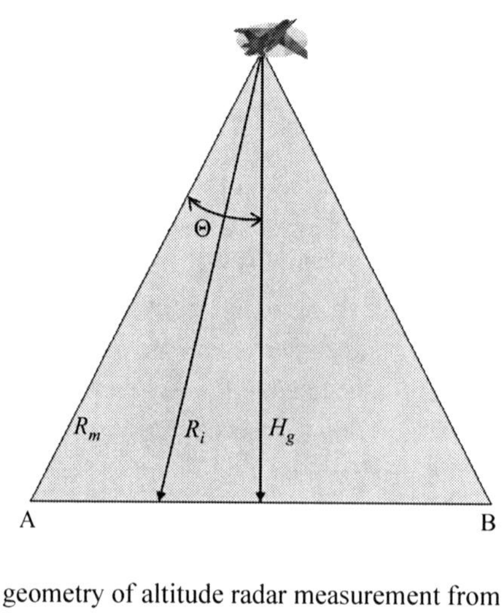


**Figure 3.1.** The geometry of altitude radar measurement from a flying vehicle.

This geometrical feature of radar altimetry using both pulse and CW waveforms can be taken into account during signal processing. Two groups of signal processing methods can be considered: *local methods* and *integral methods*. In the local method, only a part of the received signal that is reflected from the surface area near the perpendicular $H_g$ is processed. However, if the integral method is used, the whole signal reflected within the area *A B* is processed, and different parameters of the complicated received envelope are measured to derive useful information, these depending on which modulation format is employed.

This task becomes more complicated if account is taken of a real antenna pattern that gives different weights to signals from different directions. However, this antenna pattern is normally known so that the measured altitude can be easily biased to give the absolute altitude above the surface. In practice, not only is the antenna pattern important but so is the backscattering diagram of the flown terrain. When flying over different reflecting surfaces (water, forest, ploughed field, buildings, etc.), surface backscattering diagrams continually change, as do the radar signal statistical characteristics (envelope, spectrum, and correlation function). In the case where the illuminated surface within area *A B* (Figure 3.1) is known during the entire flight, the displacements of the time delays corresponding to $R_m > R_i > H_g$ could be balanced out on average. Unfortunately, in most cases *a priori* data on backscattering diagrams are absent. That is why, during flights over heterogeneous terrain, mean values of the indicated time delays become random and cannot be taken into account and corrected.

Biasing of the altitude indication can also be caused by flight vehicle maneuvers such as rolling and pitching. In this case, axis deviations from the vertical in a rigid antenna can result in changes in the measured average time delay of the reflected signal. If data on aircraft roll and/or pitch are not used for altimeter correction, a slant range $R$ measurement instead of a vertical altitude $H_g$ may result.

Another feature of radar altimetry is that the time taken for the transmitted signal to be reflected, received, and successfully processed can be expressed as an altitude gap—the so-called altitude hole or blind zone, which is the smallest altitude that can be measured, typically 0.3 to 5 m. To decrease this altitude hole to a more acceptable level, it is rational to shorten the pulse length in pulse altimeters, or to increase the frequency deviation in CW frequency modulation (FM) altimeters. Furthermore, for a single transmit/receive antenna using very short pulses, the altitude hole can be decreased by reducing the switching time between the two functions.

In practice, both CW altimeters and pulse altimeters are frequently designed with two separate antennas for the transmission and reception functions, which allows for further minimization of the altitude hole by improving the isolation and switching time between them. Such an installation might include a pair of microstrip antennas operating in the C-band, which would typically provide a gain of 10 dB over isotropic. These antennas would be spaced to provide an isolation loss greater than the maximum expected ground return loss at low altitudes: a half-metre spacing between two antennas installed on an aircraft underside would provide about 85 dB isolation (a $10^{8.5}$ reduction in signal power). Normally, an altimeter has an automatic sensitivity range control that limits receiver sensitivity as a function of altitude, especially at low altitudes, in order to detect ground returns but not antenna leakage.

### *3.1.4 CLASSIFICATION BY DIFFERENT FEATURES*

Radar altimeters may be classified on the basis of the *frequency bands* over which they operate. They comprise microwave, millimetre-wave, laser, and radioactive altimeters, though these delineations are rather rough because each waveband is very broad. For example, the frequency band 4.2 to 4.4 GHz in the microwave C-band is assigned to aircraft radar altimeters. This frequency band is high enough to result in reasonably small-sized antennas being able to produce a 40° to 50° beam but is sufficiently low so that rain attenuation and backscatter have no significant range limiting effects. Detailed information on other frequencies of operation will be provided in particular cases.

Another attribute is the *waveform*, as was mentioned in Section 3.1.3, and corresponding details of both pulsed and CW radar altimeters are considered in Section 3.2.

Radar altimeters also differ in their *application* and *functionality*. Obviously, it is possible to distinguish between altimeters for aircraft and spacecraft, military and civil applications, altimeters as sensors measuring motion parameters for navigation purposes, altimeters as sensors for remote Earth surface sensing, low range radio altimeters (LRRA), and high range altimeters.

### *3.1.5 APPLICATION AND PERFORMANCE CHARACTERISTICS*

#### *3.1.5.1 Aircraft Applications*

Radar altimeters are used for pilotage and navigation in airplanes and helicopters in all phases of flight. The basic function of an aircraft radar altimeter is to provide terrain clearance directly beneath the aircraft, particularly in mountainous areas and during bad-weather landings. Additional functions include the measurement of vertical rate of climb or descent and (selectable) low altitude warnings. Radar altimeters are also essential parts of many blind-landing and automatic navigation systems. In civil aviation, they are designed to support automatic landing, flare, and touchdown computations. When landing by category III ILS, once above the runway, the aircraft's bottom-mounted radar altimeter measures altitude, and either the electronics or the pilot accomplishes the subsequent flare maneuver (Kayton 2001).

Another application is for map-matching (Kayton and Fried 1997), also called terrain contour navigation, which is a type of terrain reference navigation (TRN) (Collinson 1996). Here, the profile of the terrain is measured by using the readings of both a baro-inertial altimeter calibrated for altitude above mean sea level (MSL), and a radar altimeter measuring height above the terrain. An on-board computer calculates the autocorrelation function between the measured profile and each of many stored profiles on possible parallel paths that can be taken by the vehicle.

Finally, the aircraft radar altimeter is a key sensor and component of the ground proximity warning system treated in Chapter 4.

#### *3.1.5.2 Spacecraft Applications*

Radar altimeters have been used in various spacecraft, their main tasks being to assist with:

- automatic control in soft landing systems on the surfaces of planets;
- automatic control during the launch of a spacecraft into a ballistic trajectory;
- determining the altitude of a space satellite's current orbit.

#### *3.1.5.3 Military Applications*

Obviously, all the navigational applications of radar altimeters mentioned above are important for military as well as civilian aircraft and are particularly so in the former because of their autonomy. However, in addition there are some purely military applications. For example, radar altimeters are used in bombs, missiles, and shells as proximity fuses to cause detonation or to initiate other functions at set altitudes (Kayton and Fried 1997).

#### *3.1.5.4 Remote Sensing Applications*

Special types of radar altimeter are used in surveying for the rapid determination of profiles. In particular, they have been applied to measuring the shapes of the geoid and the heights of waves and tides over the oceans. Continuous monitoring of sea level and the measurement of terrain relief can be accomplished from orbiting satellites. Spacecraft radar altimeters can also provide topographic information on other planets.

### *3.1.6 PERFORMANCE CHARACTERISTICS*

Performance characteristics are designed to match particular applications and can be very different. One of the main attributes of any altimeter is *measurement accuracy*. Depending on the altimeter function and the class of flying vehicle, measurement accuracy requirements for altitude and vertical speed can be essentially different. Therefore, values for acceptable errors should be set individually for partial cases. For example, high-performance low-flying military aircraft and cruise missile systems require accurate altitude tracking at vertical rates of over 600 $ms^{-1}$ whilst maintaining high degrees of covertness and jam immunity.

Altimeters designed for terrain correlation for navigational purposes must process an extremely small ground illumination spot size at high altitudes in order to provide the required altitude resolution.

*Altitude marking radars* are generally low-altitude altimeters designed specifically to provide mark signals at specific altitudes for the initiation of automatic operations such as fuse triggering at a given distance to a target, or parachute opening upon return from space for an automatic landing.

Performance characteristics and even tasks that are desirable and expedient are, in part, dictated by the level of engineering possible at the time of manufacture. Altimeters built during the early 1960s weighed 7 kg or more and transmitted about 100 W of peak pulse power, whilst at the end of 1990s radar altimeters typically weighed 2–5 kg, exhibited a 0.5 m or 2% altitude accuracy and transmitted 5 W peak pulse or 500 mW average CW power for 1500 m altitude capability. Nowadays, even these figures can be improved upon.

The altimeter is an integral part of an aerospace vehicle navigational system and is used in vehicle control, whilst in remote sensing applications it is a kind of useful load, the aerospace vehicle itself being basically a carrier for remote sensors. The wide class of altimeters designed for remote sensing of terrain is beyond the scope of this book, and will only be touched upon where necessary.

## 3.2 PULSE RADAR ALTIMETERS

### 3.2.1 PRINCIPLE OF OPERATION

The operation of a pulse radar altimeter is based on the principle of reflecting short pulses of electromagnetic radiation from a target, in this case the Earth's surface. Altitude is determined by measuring the time it takes for the pulse waveform (the *sounding pulse*) to travel from the flying vehicle to the Earth's surface and return to the radar transmitter after reflection. Figure 3.2 depicts this principle.

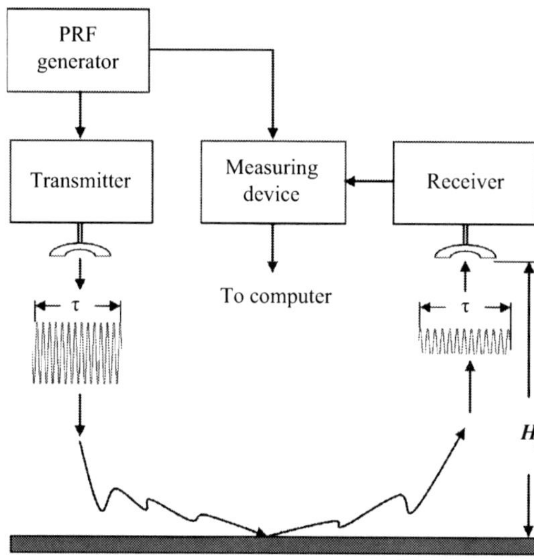


**Figure 3.2.** Simplified block diagram of a pulse-tracking radar altimeter.

The pulse repetition frequency (PRF) generator provides the modulation waveform for the transmitter and a time reference for measuring the time delay $t_d$ that is proportional to the required altitude, $H_g = ct_d/2$. It is obvious that the relationship between the PRF and the maximum altitude $H_{max}$ that can be measured unambiguously is given by formula PRF $\leq c/2H_{max}$. For example, if a spacecraft altimeter is required to measure altitudes up to 4,000 km, its PRF should be less than 37.5 Hz. By contrast, if an aircraft altimeter is designed to measure a maximum altitude of 11,000 m, the PRF can be under 13.6 kHz.

### 3.2.2 PULSE DURATION

Given a sufficient energy level, the measurement accuracy depends upon an adequate signal-to-noise ratio (SNR), and also upon the waveform. Classical radar theory rigorously proves that to achieve high accuracy in measuring target range (altitude in this case), the waveform should be as wideband as possible, so enabling the accurate measurement of the relevant time delay. This actually means that the use of wideband waveforms with time-bandwidth products $\tau B >> 1$ (where $B$ is the spectrum width of the radiated pulse) is necessary. This inequality can be satisfied by modulating inside the sounding pulse ('within-pulse modulation').

Normally, linear frequency modulation (LFM) or phase code manipulation (PCM) is used to expand pulse spectrum width without decreasing pulse duration. A sounding signal that satisfies the latter inequality can be termed a 'complex signal' in the sense that such a waveform is very different from the simplest radar waveform, which is a harmonic curve.

A complex signal is compressed because of processing in the matched filter of the radar receiver providing a very short pulse at the receiver output and therefore high accuracy altitude measurement.

In the case of conventional pulses (no within-pulse modulation), the accuracy becomes better as the pulse duration becomes shorter. In a typical pulse altimeter the radio-frequency carrier is modulated with pulses of duration $\tau < 0.25\ \mu s$. Such short pulses permit measurements, even at low altitudes, of the time delay between the leading edge of the transmitted pulse and that of the pulse returned from the ground.

### 3.2.3 TRACKING ALTIMETERS

Early pulse altimeters displayed the received signal on a cathode-ray tube with a circular sweep, so allowing the pilot to determine the leading-edge position of the echo signal. In a non-tracking altimeter, an echo time-delay reading appears when a pulse edge exceeds the set threshold. Normally, modern pulse altimeters use a *tracking gate* system, and a simplified block diagram of a pulse-tracking altimeter is shown in Figure 3.2 (Zhukovsky, Onoprienko, and Chizhov 1979). In accordance with the generalized simple diagram (Figure 3.3), it consists of a transmitter (T), receiver (R), and a sequence of measuring devices that includes a time discriminator (TD), a time modulator (TM), a smoothing filter (SF), and a display. Such a sequence forms a complete servo system.

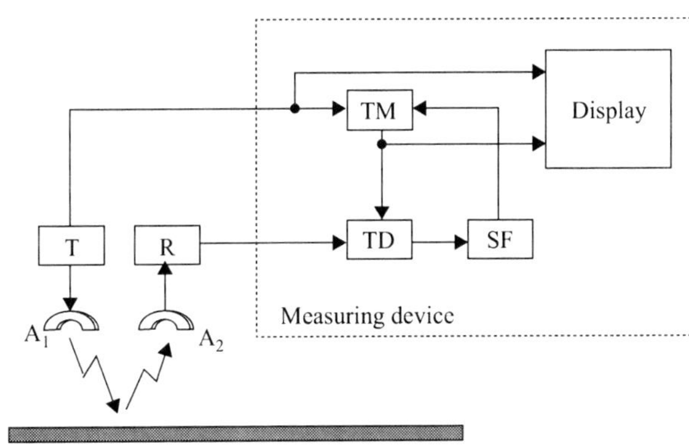


**Figure 3.3.** Block diagram of a pulse-tracking radar altimeter.

A time discriminator compares the time positions of both received and *expected signals*, the latter being generated by electronic gates in the time modulator. The sum of the reflected signal $u(t,t_{\mathrm{H}})$ from the output of the receiver plus additive noise $n(t)$ is applied to the discriminator input as a received signal. The signal at the output of the time discriminator is proportional to (or at least functionally related with) the altitude error $\varepsilon = \tau_{\mathrm{e}} - t_{\mathrm{H}}$, where $\tau_{\mathrm{e}}$ is the expected time delay set by the time modulator and $t_{\mathrm{H}}$ is the real time delay of the reflected signal. After smoothing, the error signal is fed back to the time modulator to change the output of the gates so that the altitude error tends to zero ($\varepsilon \rightarrow 0$). Hence, during continuous operation, the gate pulse is kept close to the leading edge of the reflected signal by the servo system. Altitude information is therefore proportional to the difference between the time positions of the gate and the sounding pulses, and this is displayed.

A measuring device (Figure 3.3) can track the time positions of reflected pulses on wavefronts, maxima, or the main points ("centers of gravity").

Pulse radar altimeters can use integral or local methods of signal processing, as mentioned in Section 3.1.3. If an integral method is used, the resolution characteristic of the pulse signal is not used for selecting a surface section in the vertical direction. Therefore, when measuring the position of the pulse maximum, the pulse duration $\tau$ should exceed the path-length difference in the antenna aperture. That is, $\tau > (1/\cos\Theta - 1)\, 2H_g/c$ where $\Theta$ is the angle between $H_g$ and $R_m$ (Figure 3.1) because under this condition, the signal around the pulse maximum is formed by the full illuminated surface. The converse condition, where $\tau < (1/\cos\Theta - 1)\, 2H_g/c$ provides the local method.

The key element of the system is the discriminator, and this can be implemented in several different ways. Figure 3.4 shows the structure of a discriminator with a single gate. Here, the received signal (after amplification and detection) $u(t - t_H)$ is gated by $v(t - t_e)$, averaged by a filter, and then applied to a subtractor to be compared with a given threshold $u_{th}$. The output voltage $U_D(t, \varepsilon)$is therefore a function of the error $\varepsilon$.

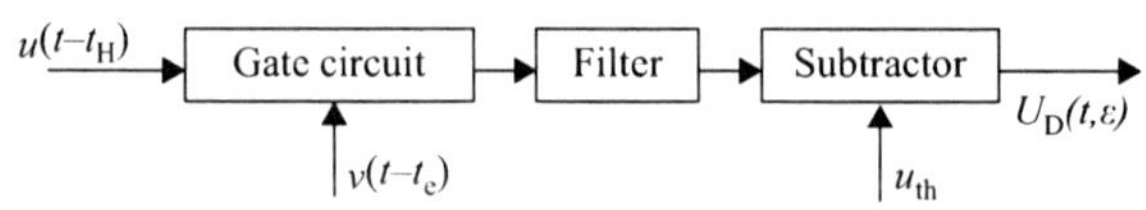


**Figure 3.4.** Structure of a single-gate discriminator.

Although this simple single-gate system can be used, most pulse altimeters use two or three gates to achieve better distance measurement in the presence of noise and fading. Figure 3.5 presents a diagram of a two-channel discriminator (Mityashev 1962). Here, the error signal is formed with help of two gates that are shifted for a certain time $2\Delta t$ relative to one another. The two-channel discriminator can be symmetric or asymmetric depending on the amplitude balance between the channels. The symmetric discriminator allows the measurement of altitude via the time position of the maximum or main point of the reflected signals. For narrow gates $\tau_g << \tau$, when $v(t - t_e)$ can be approximated by the Dirac delta-function $\delta(t - t_e)$, and if the time half-shift $\Delta t$ is less than pulse duration $\Delta t < \tau$, the zero value of discriminator curve corresponds to the maximum of the reflected pulse envelope.

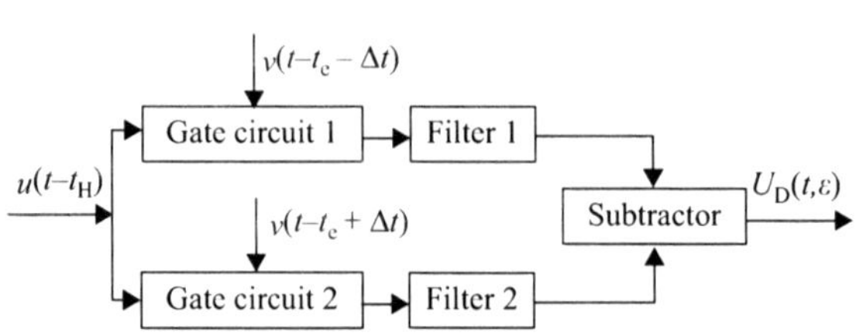


**Figure 3.5.** Structure of a two-channel (two-gate) discriminator.

In the case of wide gates $\tau_g \leq \tau$, and the zero value of the discriminator curve corresponds to the main points or "centers of gravity" of the reflected pulse envelope. The asymmetric discriminator measures the wavefront time-positions of the reflections.

The discriminator curve can be modeled (Zhukovsky, Onoprienko, and Chizhov 1979) as:

$$\alpha(\varepsilon) = f(\varepsilon - \Delta t) - k_{as} f(\varepsilon + \Delta t) \tag{3.1}$$

where $f(\varepsilon \pm \Delta t) = \frac{1}{T}\sum_{n}\int_{0}^{T} u(t - t_H - nT)\, v(t - t_e \pm \Delta t)\, dt$, and $\varepsilon = t_e - t_H$; $T$ is the interpulse period and $k_{as}$ is an asymmetry factor that is unity for the symmetric discriminator. Figure 3.6 shows hypothetical discriminator curves calculated in accordance with this model in the symmetric mode ($k_{as}$=1), for the quadratic $f(\varepsilon \pm \Delta t)$ case, and where $\Delta t_1 > \Delta t_2$.

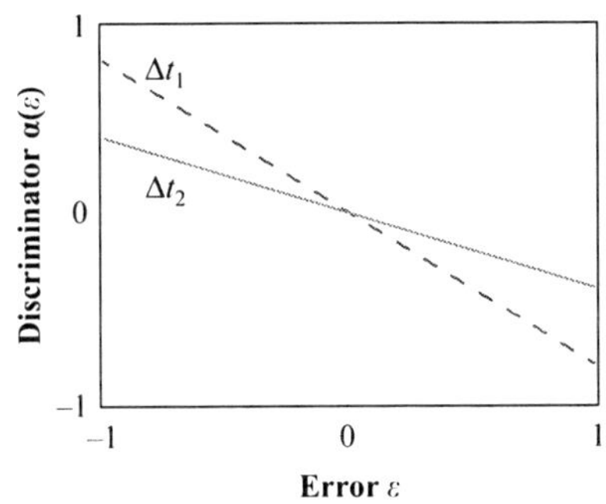


**Figure 3.6.** Example of discriminator curves where $k_{as} = 1$ and $\Delta t_1 = 2\Delta t_2$.

The feedback loop around the measuring devices in a tracking pulse altimeter may be closed within the system or via wave propagation. In the latter case, the PRF is controlled with the help of feedback coupling. A time discriminator in the tracking system measures the altitude according to the time position of the reflected pulse. Such a pulse position can be estimated by the pulse edge, pulse maximum, or weighted pulse center. If an altitude is measured by the position of the pulse edge or pulse maximum, some range selection (that is, gating or time strobing) of the reflected signal is used. However, when an altitude is estimated using the pulse center, range selection is not normally used because the weighted pulse center is determined over the entire reflected pulse.

### 3.2.4 DESIGN PRINCIPLES

The range of altitudes to be measured $H_{min}$ to $H_{max}$ should be specified with regard to the intended application of the proposed altimeter. Consideration must also be given to the expected flight velocity of the vehicle and its maximal angles of roll and pitch. Furthermore, account must be taken of the characteristics of the relevant reflective surfaces, and of various potential errors. The operating wavelength $\lambda$ will be chosen based on a compromise between various contradictory requirements such as propagation properties and the relationships between wavelength and the reflecting properties of the Earth's surface, the antenna size, and resolution, and the required measurement precision, along with technical viability and electromagnetic compatibility. All these factors were taken into consideration when international standards were developed and adopted. As was indicated in Section 3.1.4, the frequency band of 4.2 to 4.4 GHz ($\lambda \approx 7$ cm) is assigned to aircraft radar altimeters. A minimal interpulse period can be derived from the requirement of unambiguous altitude measurement, $T > 2H_{max}/c$. This means that a reflected signal from the Earth's surface should be received earlier than when the next sounding pulse is radiated, whence $T$ is the time interval between two consecutive impulses.

Given a knowledge of possible aircraft roll and pitch excursions, a suitable beam width $\theta_a$ can be determined having regard to an allowable altitude measurement error based on such aircraft maneuvers. Backscattering patterns can be projected using the expected types of reflective surfaces and the wavelength. If it is assumed that the beam width of the backscattered radiation is $\theta_b$, the total effective width $\theta_e$ of the antenna pattern and backscattering can be obtained from the following expression: (Zhukovsky, Onoprienko, and Chizhov 1979)

$$\frac{1}{\theta_e^2} \approx \frac{2}{\theta_a^2} + \frac{1}{\theta_b^2} - \frac{4\pi^2 tg^2\theta_0}{\theta_a^4} \tag{3.2}$$

where $\theta_0$ is the angle of the antenna axis deviation from the vertical. For accurate altimeter design, the limits of the total effective width variation $\theta_{e\,min} - \theta_{e\,max}$ are important.

A method of signal processing can be chosen depending on the ratio between the allowable altitude measurement error $\Delta H_{bias}$ and values $\theta_{e\,max}$ and $H_{max}$. An integral method can be used if $\Delta H_{bias} > H_{max}(\cos^{-1}\theta_{emax} - 1)$. In this case, symmetric discrimination at relatively long pulses and gates can be applied: $\tau + \tau_g \geq t_{H\,max}/\cos\theta_{0max} - t_{H\,max}$. Otherwise, local methods are expedient, and they can be realized based on single-gate or multiple-gate discriminators with short pulses and gates: $\tau + \tau_g < t_{H\,max}/\cos\theta_{0max} - t_{H\,max}$. Local methods are the most applicable for pulse radar altimeters.

Different shapes of pulse envelopes $U(t)$, gates $V(t)$, and receiver frequency responses $F(f)$ may be used in different practical cases. For example, in the case of the Gaussian approximation:

$$U(t) = \exp(-\pi\, t^2/\tau^2) \tag{3.3}$$

$$V(t) = \exp(-\pi\, t^2/\tau_g^2) \tag{3.4}$$

$$F(2\pi f) = \exp(-\pi f^2/2\Delta f^2) \tag{3.5}$$

where $\Delta f$ is the receiver bandwidth, the generalized pulse length parameter is $\tau_\Sigma = \sqrt{\tau^2 + 1/2\Delta f^2 + 2\tau_g^2}$ and the normalized time delay is $\nu_H = t_H\theta_e^2/2\pi\tau_\Sigma$.

The value of the parameter $\tau_\Sigma$ can be determined from the given allowable altitude measurement error $\Delta H_{bias}$. Normally $\tau_\Sigma \approx 2\Delta H_{bias}$, therefore $\nu_{H\,max} = t_H\theta_{e\,max}^2/4\pi\Delta H_{bias}$ and $\nu_{H\,min} = t_H\theta_{e\,min}^2/4\pi\Delta H_{bias}$.

Important parameters that have to be evaluated are SNR, receiver bandwidth $\Delta f$, and radiating power $P_t$ of the sounding pulse. These should be determined based upon an allowable fluctuating error $\sigma_{Ha}$ taking into account the inertia of the measuring device.

### 3.2.5 FEATURES OF ALTIMETERS WITH PULSE COMPRESSION

Pulse compression is an effective method of increasing range resolution, velocity resolution, and maximum range in radar systems (Cook and Bernfield 1967). In pulse altimeters, it makes possible an increase in measurement accuracy. Special kinds of modulation are used within the pulses to generate wideband (WB) signals having time-bandwidth products $\tau B >> 1$ without lessening pulse duration (see Section 3.2.2). Thus, radar designers may choose a pulse duration $\tau$ that provides the necessary radiating energy (and radar range of operation) for a given radiated power limitation, and then achieve the necessary range resolution and altitude accuracy by broadening the spectrum with the help of within-pulse modulation (mostly FM or PSK [Phase Shift Keying]). Special processing of WB signals provides pulse compression at the output of an optimized receiver (Mahafza 2000). The compression ratio is equal to the time-bandwidth product $\tau B$ and may be rather high (up to thousands). The theoretical limits of range resolution and accuracy are defined by the wavelength.

Generally, a received signal is processed in a correlator or matched filter (Turin 1960). The main feature of the pulse compression technique in altimeters is related to distortion in the received signal resulting from statistically non-homogeneous undulating terrain. Therefore, a receiver filter designed to match the sounding waveform is not fully matched with the reflected signal, and the quality of compression decreases. That is why two approaches to signal processing can be considered: (1) matching with the radiated waveform and (2) matching with the reflected signal. The first is a traditional radar approach, whilst the second requires filter parameters that depend on the statistical properties of the surface illuminated during the flight, which complicates the implementation of the relevant equipment. Nevertheless, modern digital techniques allow the memorizing of huge amounts of information on surface properties that, in combination with Global Positioning System (GPS) data on coordinates (Ferguson, Kalisek, and Tucker 1997), opens new possibilities for adaptive signal processing (Grishin 2000).

### *3.2.6 PULSE LASER ALTIMETRY*

The pulse radar principle is used also in laser altimetry. This method relies on measuring the distance from a vehicle in flight to the Earth's surface by precisely timing the round-trip travel time of a brief pulse of laser light. The travel-time is measured from the time the laser pulse is radiated to the time laser light is reflected back from the surface. The reflected laser light is received using a small telescope that focuses the collected laser light onto a detector. Typically, a laser transmitter is used that produces a near-infrared laser pulse that is invisible to humans (Harding 2000).

Airborne laser altimeters can be used to accurately measure the topography of the ground, even where overlying vegetation is quite dense. The data can also be used to determine the height and density of such overlying vegetation, and to characterize the location, shape, and height of buildings and other man-made structures. By scanning the laser pulses across the terrain using a rotating mirror, a dense set of distances to the surface is measured along a narrow corridor. These distance measurements are associated with map coordinates and elevations for each laser pulse by combining the distance data with information on the position of the airplane at the time the laser pulse was fired and the direction in which it was fired. The airplane position along its entire flight path is determined using GPS (Kaplan and Hegarty 2006). The direction of the laser pulse is established using an Inertial Navigation System (INS) that measures the orientation of the airplane along with measurements of the orientation of the scan mirror.

Modern precise laser altimeter systems measure multiple returns for each laser pulse, and typical systems produce laser pulses several feet in diameter (Harding 2000). If such a wide laser pulse reflects off more than one feature, distances to the multiple features can be measured. For example, if part of the laser pulse reflects off tree branches or foliage at several levels and the remainder of that laser pulse reflects off the ground, the elevations of the branches, foliage and ground can be measured. This capability is very important when trying to map ground topography beneath vegetation. The 'last returns' for each pulse are those from the lowest features and thus are more likely to be reflections from the ground (Harding, 2000). Sophisticated algorithms are used to identify these laser returns emanating from the ground itself.

### 3.2.7 SOME EXAMPLES

As historical examples of pulse altimeters, consider the SCR-518-A and SCR-718 altimeters, which were used for measuring absolute altitude in high-altitude bombing, photographic mapping, and terrain clearance. The SCR-718 operates between 50 ft and 40,000 ft, and its accuracy is ±50 ft plus 1/4% of altitude. Such altimeters were used from the 1940s (US Radars 2007). A later example is the Teledyne AN/APN-192 Short-Pulse Radar Altimeter (Parsch 2007) that was used in the Boeing CH-47 Chinook heavy-lift helicopter, which made its maiden flight in 1961.

Typical examples of modern radar altimeters are the Honeywell types (Honeywell Aerospace Products 2007). The AN/APN-194 Radar Altimeter Receiver-Transmitter is a high-resolution device that measures altitude up to 5,000 ft. Its output is fed into an autopilot controlling the altitude of low-flying target drones. This altimeter employs narrow-pulse transmission in the C-band range (4,200–4,400 MHz) with leading edge tracking of the echo pulse. Altitude range information is obtained by comparing the received echo pulse with a timed ramp voltage generated simultaneously with the transmitted pulse. The accuracy is ±3 ft and ±4% of the actual altitude, the allowed vehicle maneuverability is up to ±30° of pitch and roll, and the RF output is 200 W. Its dimensions are 20.3 × 8.9 × 7.6 cm, its weight is 2.0 kg, and the power supply provides 40 W at 104–118 VAC.

Because the AN/APN-194(V) is a range-tracking radar that is reliable over 20–5,000 ft, the height indicator is disabled when the aircraft is flying above 5,000 ft. Also, when the aircraft is on the ground, the system is disabled by the weight-on-wheels switch (Integrated Publishing 2003). Altitudes up to 70,000 ft can be measured by the AN/APN-222 radar altimeter, which has dimensions of 21.3 × 12.7 × 10.2 cm and weighs 2.9 kg for the transmitter-receiver, plus a 0.5 kg antenna, and a 1.1 kg indicator. The pulse repetition frequency is 5 kHz, and the accuracy is ±5 ft +2% of altitude (Aerospace Research Information Center 2005).

Other examples of laser altimeters can be found in Laser Optronix (2006). Accuracies depend on the relevant surface but are less than 1 m for all models and 10–20 cm for the best. These are single-shot values at full speed and if averaged, the error can be decreased down to 5 cm.

### 3.2.8 VALIDATION

An interesting question is how to check the accuracy of high-altitude measurements in order to validate radar altimeters. An evaluation of the tracking-pulse radar altimeter installed in a B-25 aircraft was conducted to determine the feasibility of using it as an altitude and altitude-rate sensor for all-weather landing applications (Zyzys 1964). This evaluation included both flight and laboratory tests. Altitude and altitude-rate accuracies were determined from 186 data runs over runway terrain and over water. Photo-theodolites and ground elevation surveys were used to determine the aircraft altitude-above-terrain. Altitude-rate data were thus obtained from photo-theodolite space-position and time data. Finally, the altimeter outputs were recorded on an airborne oscillograph and correlated in time with the photo-theodolite data. Altimeter performance was also checked over approach lights, grass, trees, dirt, and buildings. It was eventually concluded that it was feasible to use the altimeter as an altitude and altitude-rate sensor for all-weather landing applications. It was recommended that the deficiencies detected during the evaluation be corrected and that the altimeter be operationally evaluated in an aircraft equipped with an all-weather landing system (Zyzys 1964).

### 3.2.9 *FUTURE TRENDS*

The design of radar altimeters is subject to ongoing developments in both technological and methodical areas of radio engineering and electronics. For example, the transition from tubes (or 'valves' in the United Kingdom) to solid-state devices for radio frequency (RF) power generation, as well as in analog-to-digital signal processing, is now complete. An important current trend is the development of conformal and aircraft integrated antennas instead of planar antennas.

Significant progress is very clear in the signal processing area, and a novel tracking and averaging algorithm has been described and analyzed (Ulander 1987). This is called the interpolation tracker and it produces undistorted estimates of mean pulse returns in the presence of frequent gaps in the data stream. The theoretical analysis has been confirmed by calculated first-order statistics using experimental data.

Traditionally, pulsed radar altimeters were used only to measure high altitudes, but new nanosecond technologies make it possible to use the same technique to measure low altitudes. Ultra wideband (UWB) technology shows considerable promise for aviation and Multispectral Solutions Inc. (MSSI), a Germantown, Md., company, is developing a new UWB radar altimeter for use in low-flying vehicles such as helicopters and small unmanned air vehicles (UAVs) (Jensen 2005).

Modern radar altimeters have tracking systems that measure the range to the surface and also produce averages of the radar pulse returns from which geophysical information can be extracted. Such geophysical applications, especially using satellites, are currently undergoing rapid development.

An innovative altimeter concept called D2P has been described (Raney 1998a) that is built around two techniques. The delay/Doppler technique (Raney 1998b) enhances along-track resolution and measurement precision, and reduces transmitter power requirements. The phase monopulse technique (Jensen 1995) measures the across-track angle-of-arrival of the height waveform, which mitigates cross-track slope errors. In future satellite versions, a flight-proven D2P radar altimeter can offer unprecedented measurement accuracy over continental ice sheets and better precision from a smaller instrument over the open ocean. The D2P concept simultaneously offers high signal-to-noise ratios, high signal-to-speckle ratios, and high signal-to-clutter ratios. These characteristics represent a substantial and innovative breakthrough (Jensen and Raney, 1998; Raney 1998c).

Finally, it is expedient to mention that a possible transition from single sensors to multifunctional apertures and sensors located on different platforms with synchronous operation and joint signal processing for remote sensing applications has been described by Zelli et al. (1997).

## 3.3 CONTINUOUS WAVE RADAR ALTIMETERS

### *3.3.1 PRINCIPLES OF CONTINUOUS WAVE RADAR*

As opposed to pulsed radar systems, continuous wave (CW) radar systems emit electromagnetic radiation at all times. In principle, therefore, CW radar can measure the instantaneous rate-of-change of the target distance, and this is accomplished by directly measuring the Doppler shift of the returned signal. The Doppler shift is a change in the frequency of the electromagnetic wave caused by the motion of the radar apparatus itself, the target, or both. However, unmodulated CW radar cannot measure target range itself at all, because there is no basis for the measurement of time delay—the signal is simply continuous. Some sort of modulation is therefore necessary to provide such a basis. For example, frequency modulation (FM) makes it possible to use a CW radar system to measure range by the systematic variation of the transmitted frequency. This gives a unique "time stamp" to the transmitted wave at every instant. By comparing the frequency of the return signal with the frequency of the transmitted waveform in the same time slot, the delay time between transmission and reception can be measured and therefore the range determined as in the case of pulsed radar.

Thus, a range-measuring capability is implemented in FMCW radar, and such radar sensors are widely used as altimeters. Normally they use FM radiation and frequency processing of the sensed signal. However, FMCW radar altimeters with phase processing are also possible (Komarov and Smolskiy 2003); furthermore, FMCW altimeters can also use correlation processing. In these latter two cases, FM is also used for isolating the *sounding waveform* and the received signal too, as explained subsequently. Here, the *sounding frequency* is the frequency of the radiated wave, and in the case of FM it will obviously vary along with the relevant modulation signal. The *sounding waveform* consists of the entire radiated signal, that is, the carrier sinusoid as modified by whatever modulation signal is applied.

The principle of FMCW radar has been described in many books (e.g., Skolnik 1990). A generalized diagram of a CW radar system is shown in Figure 3.7. Here, the transmitter generates a waveform, the carrier frequency of which changes over time and which is radiated by the transmitting antenna.

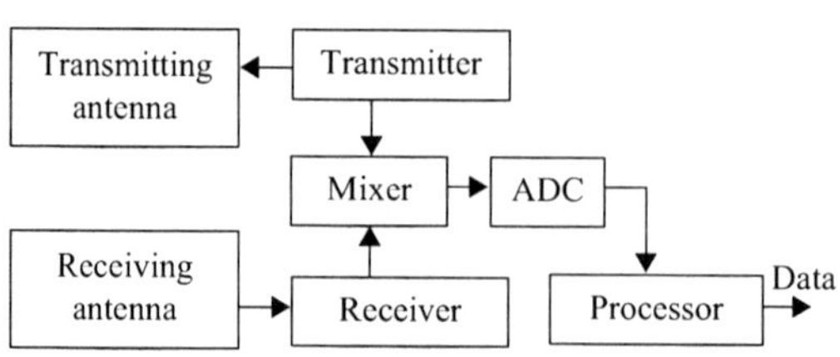


**Figure 3.7.** Generalized diagram for CW radar.

One example is a linear FM waveform that uses a changing sounding wave frequency as a reference for measuring the time delay $t_d = 2H_g/c$. The transmitting frequency changes during the time $\Delta t = t_d$, and the difference between the receiver and transmitter frequencies $\Delta f$ appears as a signal at the output of a mixer. From this, range (altitude) data can be extracted, for example, by digital processing after analog-to-digital conversion (ADC).

### *3.3.2 FMCW RADAR WAVEFORMS*

One of the simplest ways of modulating a wave is to increase the frequency linearly as a function of time $f(t)$ from some initial value $f_{min} = f_0$ to $f_{max} = f_0 + f_{dev}$ during one-half period of modulation $T_M$ and then decrease it back again during the second half-period, $f_{dev}$ being the *frequency deviation*. This is illustrated in Figure 3.8 (upper graph) where the solid line is the frequency of the transmitted waveform and the dashed line is the frequency of the received signal reflected from an immovable object. The lower graph shows the frequency difference $\Delta f$ between transmitted and received frequencies (the beat frequency) that is formed at the output of the mixer (see Figure 3.7). For most of the time, the beat frequency $\Delta f = f_b$ depends on the time difference $\Delta t = t_d$ and hence on the range of the reflective object, that is, the altitude in the case of simple radar altimeters.

From the geometrical considerations of Figure 3.8, it can be seen that the ratio $\Delta t$ to $T_M/2$ equals the ratio of $f_b$ to $f_{dev}$. This means that $t_d = T_M f_b / 2 f_{dev}$ whence, keeping in mind that $t_d = 2H_g/c$, the altitude is proportional to $f_b$:

$$H_g = \frac{cT_M}{4f_{dev}} f_b \tag{3.6}$$

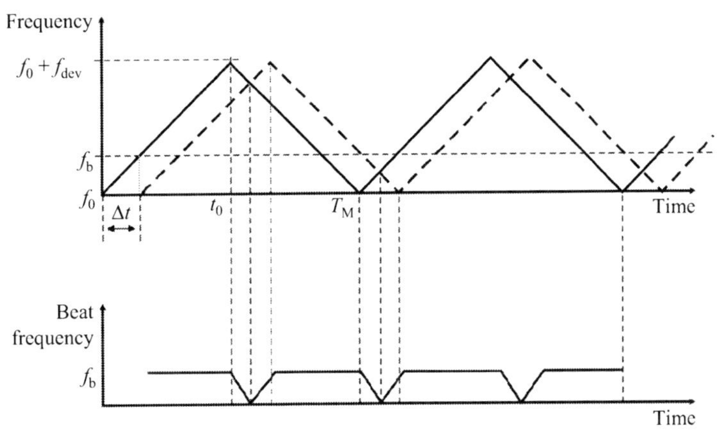


**Figure 3.8.** Frequency–time dependence and beat signal for an immobile target.

The frequency deviation $f_{dev} = f_{max} - f_{min}$, where $f_{max}$ is the maximum frequency during the modulation time $T_M$, and $f_{min}$ is the minimum frequency, as previously noted.

However, there is a slight problem that occurs when the transmitting frequency falls whilst the receiving frequency is still increasing. This is a reason for changing the beat frequency around points of time divisible by $T_M / 2$ where this beat frequency is noninformative. The beat frequency value $\Delta f$ becomes zero if the frequencies indicated in Figure 3.8 by solid and dashed lines become equal. This problem can be solved by a signal processing procedure in which a discriminator is used to clip off the noninformative part of the signal, leaving only the valid part with a beat frequency $\Delta f = f_b$, which is directly proportional to the altitude as given by the formula in Equation (3.6).

However, this sort of modulation is only one of a number of possible FM sounding waveforms, and in many cases the quality of altimeter operation is defined by the properties of these sounding waveforms. Currently, the following types of modulation are used with continuous or quasi-continuous sounding waveform radiation in FM radar altimeters:

1. Harmonic (cosine) FM
2. Linear (sawtooth) FM
3. Linear symmetric FM (triangular—as was shown in Figure 3.8)
4. Coded modulation (Skolnik 1990)
5. Random FM or noise modulation (Skolnik 1990).

Linear and linear symmetric modulations are the most frequently used. Cases of linear FM and harmonic FM are illustrated in Figure 3.9.

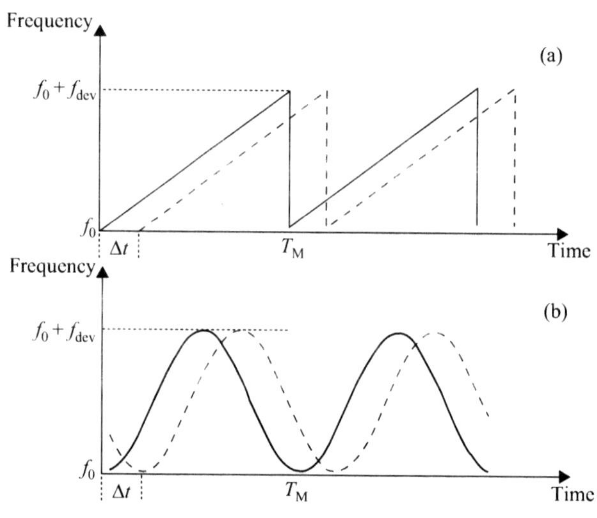


**Figure 3.9.** Frequencies of transmitted (solid line) and reflected (dashed line) signals at linear (a) and harmonic (b) FM sounding waveforms.

For a particular waveform, the FMCW system measures the instantaneous difference between the transmitted and received frequencies $\Delta f$, and this is directly proportional to the distance between the radar and the Earth's surface, as has been mentioned earlier. For example, in the case of linear FM (Figure 3.9[a]), it is easy to obtain the relationship between the altitude and the measured beat frequency as:

$$H_g = \frac{cT_M}{2f_{dev}} f_b \tag{3.7}$$

This expression implies that $f_b$ is 2 times less for a given altitude $H_g$ than for the linear symmetric FM case (Figure 3.8). The harmonic FM case is closer to that of the symmetric FM case.

Generally, for an arbitrary modulating waveform with given frequency deviation and modulation time, a beat frequency is proportional to an altitude:

$$f_b = Kc^{-1}T_M^{-1}f_{dev}H_g \tag{3.8}$$

where $K$ is a proportionality constant that depends on the FMCW radar waveform.

In the case of quasi-continuous waveforms, long pulses with intrapulse modulation are transmitted, as for FM or phase-code signals (Skolnik 1990). Pulse radiation in such cases is necessary only for time separation of transmitted and received waves.

### 3.3.3 DESIGN PRINCIPLES AND STRUCTURAL FEATURES

An antenna for an FMCW altimeter can be essentially similar to that of a pulse altimeter. However, for the same average power, the peak voltage of the antenna and transmission line is much less than in the case of pulse radiation. This allows a significant increase in the average power radiated by the same antenna. Moreover, it facilitates operation at high altitudes.

Transmitters use various methods of creating FM sounding waveforms in the appropriate frequencies, as shown in Figure 3.10. Transmitter (T) consists of a modulating waveform generator (WFG), a frequency modulator (FM), a high-frequency generator (HFG), a frequency multiplier (FMP), and a power amplifier (PA). The transmitter output is connected to a transmitting antenna $A_1$.

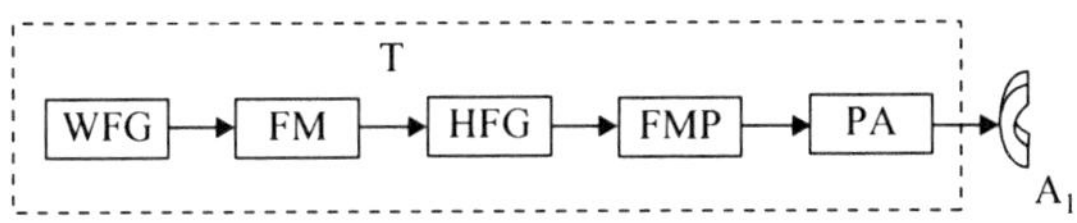


**Figure 3.10.** An FMCW radar transmitter.

The WFG defines the modulating waveform and modulation time $T_M$; the FM unit modulates the HFG frequency to provide the necessary frequency deviation $f_{dev}$; the FMP unit converts the generated frequency up to an operational frequency band (normally 4,200–4,400 MHz for aircraft navigational altimeters); and the PA provides the necessary radiating power.

In reality, the transmitter structure may differ from that of Figure 3.10. For example, a single unit microwave C-band generator can substitute for the HFG and FMP, and even for the PA in special cases. This depends on the basic components and design features, which are outside the scope of this chapter.

The crucial parts of an FMCW altimeter are a receiver and a measuring device, or processing unit. The simplest receiver has a microwave mixer as the first stage to provide a beat signal of frequency $f_b$ that carries the useful information. The structure of such an FMCW altimeter with direct frequency transformation is shown in Figure 3.11. This consists of an FM transmitter (T) with its antenna ($A_1$), a receiver antenna ($A_2$), a receiver that includes a balanced mixer (BM), a low-frequency amplifier (LFA), and a frequency meter (FMT). Such a receiver is simple but disadvantageous because of vibronoise frequencies, power supply ripple, and other interferences in the same frequency band as the useful information. Therefore, it is extremely difficult to eliminate such unwanted frequencies by means of filtering. Moreover, the sensitivity of such a receiver is not sufficient for the majority of applications.

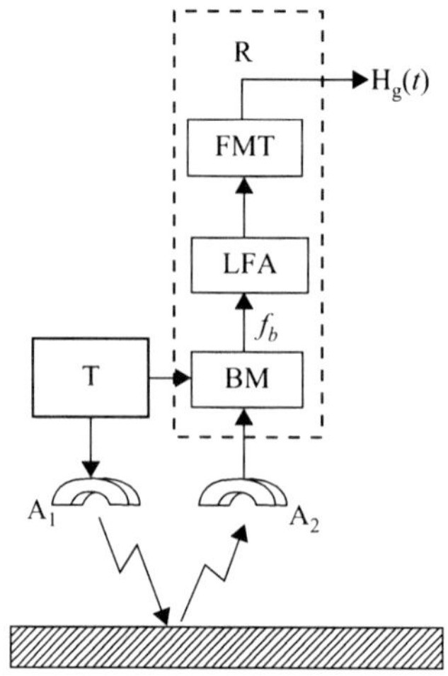


**Figure 3.11.** FMCW altimeter structure with direct beat frequency transformation.

Fortunately, these detrimental interferences can be drastically suppressed if information frequencies are converted into an intermediate frequency (IF) band. An FMCW altimeter with frequency conversion is a radar with nonzero IF, in contrast to the case shown in Figure 3.10 where IF can be considered as equal to zero.

### 3.3.3.1 Local Oscillator Automatic Tuning

An obvious solution is the classical scheme with an autonomous local oscillator (LO) that generates a frequency $(f_{Tx} \pm f_{IF})$ where $f_{Tx}$ is transmitter frequency and $f_{IF}$ is an intermediate frequency. Figure 3.12 illustrates this scheme.

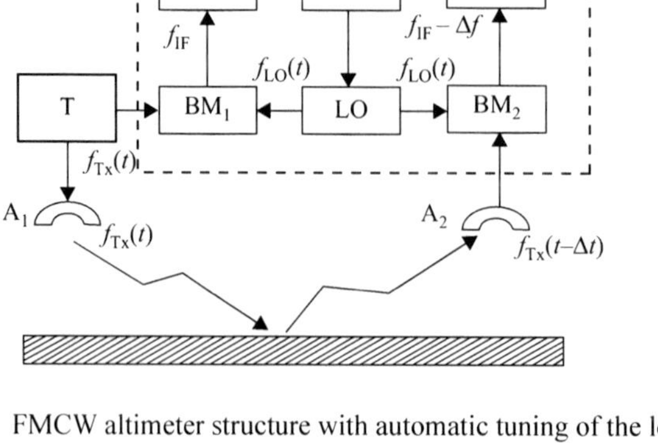


**Figure 3.12.** FMCW altimeter structure with automatic tuning of the local oscillator.

Both the transmitter FM waveform of frequency $f_{TX}(t)$ and the LO signal $f_{LO}(t)$ are applied to the balance mixer $BM_1$, which produces an IF signal of frequency $f_{IF}$ that is eventually used as a reference for the double-balance mixer $DBM_2$. An intermediate frequency amplifier $IFA_1$ passes only a difference-frequency component $|f_{Tx} - f_{LO}|$, which is also applied to the frequency difference detector, FDD. This FDD produces a control signal (in case the difference frequency $|f_{Tx} - f_{LO}|$ is not equal to the nominal value of IF) and tunes the LO to maintain $f_{IF} = |f_{Tx} - f_{LO}|$ constant. $BM_1$ and $BM_2$ are normally balanced mixers to provide additional suppression of any incidental amplitude modulation of the transmitter FM waveform. A reflected signal after frequency conversion by the $BM_2$ mixer is amplified by the bandpass amplifier $IFA_2$ and is applied to the double-balanced mixer DBM, which also receives the IF reference signal $f_{IF} = |f_{Tx} - f_{LO}|$ from the $IFA_1$ at its other input. The output signal of the DBM contains the desired altitude information, $H$.

In this scheme, the difference between the FM transmitter and the tracking LO frequencies should be kept constant and equal to $f_{IF}$. For this reason, the time constant of the LO auto-acquisition circuit should be fast enough to provide quasi-stationary FM reproduction.

Moreover, it is rather difficult to implement a reliable functioning FDD if the $f_{dev} << f_{IF}$ condition is not satisfied. This condition limits the accuracy of altitude measurement, which is why the scheme is applicable only at comparatively small frequency deviations.

### *3.3.3.2 Single-Sideband Receiver Structure*

A single-sideband receiver structure does not require a separate FM LO and has somewhat wider capability. Illustrated in Figure 3.13, this structure uses an LO that generates a given frequency $f_{LO}$ much lower than the transmitter frequency: actually, $f_{LO}$ is of the same order as a normal IF ($f_{LO}$ _$f_{IF}$). At the output of $BM_1$ a signal containing a partially suppressed transmitter frequency and two sidebands is formed, $f_{Tx}(t) \pm f_{LO}$. A single-band filter (SBF) cuts off the carrier frequency and upper sideband component and so passes only the lower sideband component

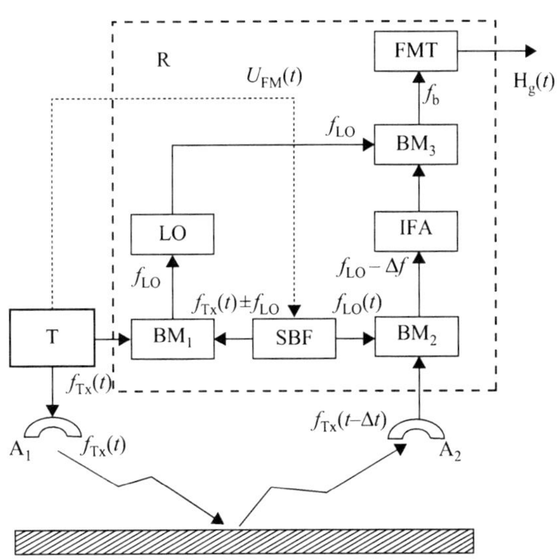


**Figure 3.13.** FMCW altimeter structure with heterodyning by single sideband.

$f_{Tx}(t) - f_{LO}$. This is applied to the input mixer $BM_2$ as an offset heterodyne oscillation. The delay echo frequency $f_{Tx}(t - \Delta t)$ is converted to $f_{LO} - \Delta t$ by $BM_2$ and after amplification in IFA, is applied to $BM_3$ where a beat frequency $f_b = \Delta f$ is formed that is directly related to altitude.

Effective suppression of the carrier frequency and the second sideband frequency by a SBF filter can be accomplished if $f_{LO}$ (that defines a frequency space between the spectral components) is much greater than the frequency deviation $f_{dev}$. This condition requires either a very high IF ($f_{LO}$) or a quite small $f_{dev}$. However, this restriction can be overcome by using a tunable tracking filter as the SBF. This improvement is indicated by the dashed connections in the circuit. The transmitter frequency modulator controls the tracking filter SBF that provides effective suppression of the upper sideband together with a carrier frequency at acceptable values of IF and $f_{dev}$.

### 3.3.4 THE DOPPLER EFFECT

FMCW altimeter principles were described earlier using the tacit assumption that the altitude does not change during the measurement. In reality, the speed of the flying vehicle may have a vertical component, which means that in addition to a range (or altitude) beat frequency, some frequency difference due to the Doppler effect also exists. Of course, the amount of frequency modulation must be significantly greater than the expected Doppler shift or the results will be affected if no special signal processing is implemented.

When the Doppler effect exists, however, the beat frequency depends on both the altitude and the vertical speed, as illustrated in Figure 3.14 for the linear symmetric FM case. The beat frequency during the positive (up) and negative (down) portions of the slope are denoted respectively as $f_u$ (Doppler shift added) and $f_d$ (Doppler shift subtracted) (Mahafza 2000).

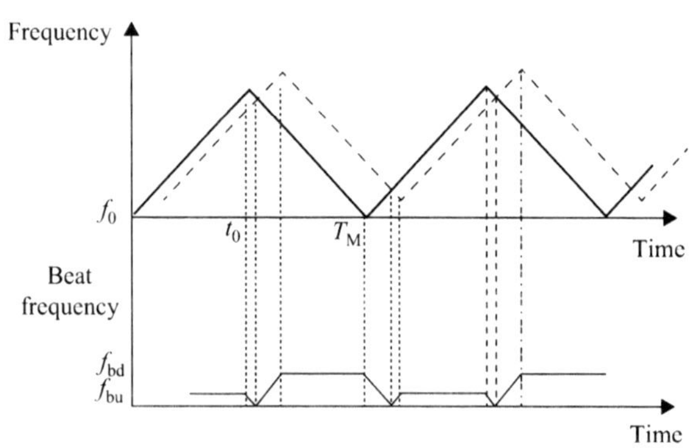


**Figure 3.14.** Frequency–time dependence and beat signal for a moving target.

Both range information and velocity information may be extracted from such a beat signal, particularly $0.5(f_{bu} + f_{bu}) = f_{bA}$ and $0.5(f_{bu} - f_{bu}) = f_{bD}$ with $f_{bA}$ as a component due to the altitude and $f_{bD}$ as a component due to the Doppler effect; and also $f_{bD} = 2V_{ver}/\lambda$ with $\lambda$ as wavelength and $V_{vert}$ as the vertical speed of the aircraft or other flying vehicle.

### 3.3.5 ALTERNATIVE MEASURING DEVICES FOR FMCW ALTIMETERS

Different variants of FMCW radar altimeters, especially the simplest ones, use nontracking measuring devices, for example, beat frequency meters. In this case all the parameters of a sounding waveform are stable, and only the beat frequency is changed as a function of the altitude $f_b = f(H)$ as was shown in Section 3.3.2.

The most widely used FMCW radar altimeters incorporate direct frequency processing of the beat signal and also frequency processing of the converted signal at IF. Such schemes are rather simple and have minute implementation errors because both gain and phase instabilities do not influence the measurement results if altitude information is contained within a frequency parameter.

It is important to note that it is not only the beat frequency that can be used as a measurable parameter in such altimeters. Altitude determination can also be accomplished by measuring the modulation time $T_M$ or the frequency deviation $f_{dev}$ when a frequency difference $\Delta f$ or beat frequency $f_b$ is kept constant in the tracking FMCW altimeter. This follows from the relationship (Section 3.3.2) between altitude and frequency parameters $H = Kc^{-1}T_M^{-1}f_{dev}f_b$, where $T_M^{-1} = F_M$ is the modulation frequency. It is seen that $H$ is similarly related with any parameter $f_b, f_{dev}$, or $F_M$ if the others are stabilized by a tracking system.

Though only FMCW altimeters have been considered with $f_b$ as a measurable variable, similar results can be obtained if altitude information is derived by estimations of $f_{dev}$ or $F_M$ instead of $f_b$. This idea is implemented in the tracking measuring devices that are also widely used in FMCW altimeters. These are designed on the principle of a tracking system using a frequency detector as a discriminator. In this case, the tracking system holds the beat frequency constant. This system forms a loop which is closed via a receiver, a discriminator, a frequency modulator, a transmitter, and the wave propagation medium. Error signals control the parameters of the sounding waveform, namely the frequency deviation $f_{dev}$ or the modulation period $T_M$, and each of them can serve as an analog of the measured altitude. The majority of navigational FMCW altimeters process the completely received signal without range selection.

Radar altimeters that use a quasi-continuous sounding waveform with a complex type of modulation, for example within pulse FM or within pulse PSK (see Section 3.2.5), can form a mutual correlation function between the reflected and reference signals. This then results in a measurement of the time position at its maximum.

Another way of realizing a FMCW radar altimeter is to compare the phase difference between the transmitted and received signals after they have been demodulated to produce sweep information. This system does not have to discriminate the negative values of $\Delta f$. The FMCW altimeter, which uses phase processing, selects only one of several harmonics from the whole signal so that altitude measurements can be made using the phase method.

### 3.3.6 ACCURACY AND UNAMBIGUOUS ALTITUDE

The maximum unambiguous range will be determined by the modulation time, as was also the case for pulse radar altimetry, namely $H_{max} = cT_M/2$. In the case of symmetrical FM determining the frequency difference without a frequency sign, that is, the frequency modulus, the condition of unambiguity is $T_M > 4H_{max}/c$. Normally, $T_M$ is taken as being much greater than $2H_{max}/c$.

The *potential accuracy* of the radar range measurement that can be theoretically achieved is defined by the rms error $\sigma_R = c/2qB$, where $q^2 = 2E/N_0$ is the SNR, $E$ is the reflected signal energy at the input of radar receiver, $N_0$ is the spectral density of the noise power, and $B$ is the effective spectrum width of the sounding waveform (Shirman et al. 1987). Normally the FMCW altimeter has quite a large modulation index $f_{dev}T_M$ and $B \approx f_{dev}$ so that an increase in $f_{dev}$ improves the potential accuracy. Taking into account that the SNR >> 1 at small altitudes, suppose as an example that SNR = 150 and $f_{dev}$ = 100 MHz, so that $\sigma_R = c/2qB \approx 1$ cm.

There are several phenomena that can make the real accuracy much worse, including the step errors due to discrete altitude readings. Other errors may be caused by spectrum asymmetry and aircraft maneuvering, and both random and implementation errors should also be taken into account. The influence of these sources of error is briefly discussed below.

- *Step error.* This is actually a critical distance, that is, a quantization in height $\Delta H = c/4f_{dev}$, which assumes a difference of one zero crossing (Skolnik 1990). This error can vary, but only within the limits of a step $\Delta H$, and it may be assumed to be uniformly distributed within that range, $\Delta H$. Hence, the variance of step error is $\sigma_{SE}^2 = \Delta H^2/12$, and rms $\sigma_{SE} = \Delta H/\sqrt{12} \approx 0.072\,c/f_{dev}$. In the considered example $f_{dev}$ = 100 MHz, so $\sigma_{SE} \approx 21.7$ cm.

  Though the step error is often explained by a difference of one zero crossing (Skolnik 1990), in-depth study of this phenomenon shows that its nature is related to the discrete spectrum structure of the converted signal that is characteristic of an FMCW altimeter. In the case of a discrete spectrum, step error appears using any method of frequency measurement, which means that to decrease such errors it is necessary to destroy the discrete structure of the spectrum and average the error in altimeter smoothing circuitry. The principle of *double modulation* can be used for this purpose, and in particular, double FM is often used to increase the measurement accuracy.
- *Error due to spectrum asymmetry.* A finite antenna pattern width, which extends the land as an area-extensive target, and Doppler broadening are the main factors influencing the spectrum width. In navigational altimeters, the antenna beamwidth can be rather wide—45 degrees, for example. In this case the size of the illuminated area where the slant range can change from $H$ to $R_m = H/\cos\theta$ (see Figure 3.1) is most important. The component of the spectrum width $B_H$ caused by this factor is proportional to altitude:

$$B_H = \frac{Kf_{dev}}{cT_M}\left(\frac{1}{\cos\theta} - 1\right)H \tag{3.9}$$

  and the average spectrum frequency is always more than the main beat frequency, which corresponds to the case where $\theta = 0$. Spectrum spreading causes a bias error which is normally about 3.3–3.6% and this can be taken into account by calibration. However, when in-flight maneuvering takes place (altitude variation, roll, and pitch), the spectral and bias error are changed.
- *Random error.* This is an error due to a fluctuating reflected signal. The noise-like structure of the converted spectrum of such a signal is a source of error in frequency measurement, and hence results in reduced altitude estimation accuracy. This means that even if the SNR is high enough, a random error exists. The wider the effective spectrum of the echo signal, the greater is the rms error in the estimated altitude. This error

can be reduced by increasing the averaging time during the measurement. However, any increase in averaging time means raising the *dynamic error due to aircraft spatial maneuvering*. That is why some optimal averaging time exists, and this is normally between about 0.1 and 1 s.

- *Signal processing*. Different signal processing methods can be used to implement altitude measurement using beat signals, and they also influence the measurement accuracy. In the past, a zero crossing technique and automatic frequency control using a wideband frequency difference detector was widely used (Zhukovsky, Onoprienko, and Chizhov 1979). This was a kind of an integral processing that produced errors due to the biasing of the altitude estimates that depend on the surface type and the changeable spatial orientation of the maneuvering aircraft. Such altimeters process a signal formed by reflections from the entire extended surface. They actually measure some equivalent slant range instead of the altitude, this being defined as a normal with respect to the surface.

  To significantly increase accuracy, other methods of signal processing have been developed, and these are local rather than integral methods. A local method uses only a part of the reflected signal concentrated near the normal to the surface for signal processing. The FMCW principle offers the advantage of using its range resolution ability to select such signals, whilst the smallest value of beat frequency corresponds to the surface patch around the normal.
- *Features of satellite altimetry*. Modern satellite altimeters can produce errors as small as 2–5 cm (Yegorov 2005), though it is not easy to achieve such accuracies. Surface irregularities in the scattering diagram, the carrier velocity, the antenna pattern, the signal processing, and the other factors indicated earlier, are even more important for satellite altimetry. Moreover, though the usual altitude definition assumes invariant measurement conditions, it can be also defined as the moving average of the local altitude. For example, if measurements of local altitudes relative to the carrier are made accurately, the heights of surface irregularities are determined by their density distribution, the size of the effective area, and the correlation window of local surface irregularities such as ocean wave height.

### 3.3.7 AVIATION APPLICATIONS

FMCW systems do not have a minimum range like a pulsed system, which is why they are often used in low-range radar altimeters, or in radar proximity fuses for warheads for example. However, they are normally not suitable for long-range detection because the continuous power level they can transmit is considerably lower than the peak power of a pulsed system.

Historical examples of FM altimeters can be found in US Radars' website (US Radars 2007). Amongst the low-range radio altimeters that were designed for radio navigation, the absolute FM radar altimeter AN/ARN-1 is a typical device. It could be used as landing aid and for use in sea search, torpedo launching, low-altitude bombing, and as a photographic aid. All sets operated between 5 and 400 ft, measuring altitude with an accuracy of ± (5ft + 5%). The AN/APN-1 also operates between 400 and 4,000 ft., but only with an accuracy of ± (60 ft + 5%).

An example of a later dual-purpose altimeter is the AN/APN-232 radar altimeter set. This is a 4.3 GHz, FM, 0–50,000 ft. instrument that was installed in many types of aircraft (AN/APN-Equipment Listing, 2007).

The desired characteristics of a modern radio altimeter intended for installation in all types of commercial transport aircraft are provided by ARINC Characteristic 707-6 (ARINC Incorporated 2009). The primary function of such an altimeter is to determine an aircraft's height above terrain for visual display to the pilot, essentially from ground level to 2,500 ft, and for use by an Automatic Flight Control System (AFCS) during automatically controlled approaches and landings.

Modern airborne electronic equipment should be designed in accordance with DO-254, the Design Assurance Guidance for Airborne Electronic Hardware, which is a standard for complex electronic hardware development published by RTCA, the Radio Technical Commission for Aeronautics (RTCA 2012). This guide is the hardware equivalent of DO-178B for software, also issued by RTCA. One of the latest radio altimeters, designed for the Airbus A380, is described by Hairion et al. (2007). This ERT560 digital radio altimeter meets the stringent requirements for civil aircraft, and the paper presents the application of DO-254 hardware in coordination with DO-178B software. The ERT560 altimeter takes advantage of FPGA (Field Programmable Gate Array) technology (Becker and Manoli 2004) to implement the main features of the equipment. Hairion et al. introduce the main capabilities of the ERT560 product with a focus on the FPGA, this being the key element in the safety-critical analysis of the radio altimeter.

A system suitable for the FMCW altimeter, proposed by Secmen, Demir, and Hizal (2006), utilizes a 3 dB 90-degree hybrid coupler and a dual-polarized T/R antenna. The T/R isolation performance of this system can be made better than that of systems using circulators. There is no power loss either in transmit or in receive modes compared with the 6 dB total loss in a conventional system using a hybrid coupler and a single polarized antenna.

## 3.4 PHASE PRECISE RADAR ALTIMETERS

### 3.4.1 *THE PHASE METHOD OF RANGE MEASUREMENT*

If CW unmodulated oscillation $U_0 \cos 2\pi f_0 t$ is applied to a transmitter antenna and the same oscillation is used as a reference signal, the reflected signal $U_0 \cos[2\pi f_0(t-t_d)]$ has a delayed phase $\phi = 2\pi f_0 t_d = 4\pi f_0 R/c$, where $R$ is a target range. The phase method is based on measuring this phase shift $\phi$. Unambiguous measurement of phase is possible from 0 to $2\pi$, that is, at $\phi \leq 2\pi$, hence the unambiguously measured maximum range is $R_{un} = c\phi/4\pi f_0 = \lambda/2$. In the case of a rather low frequency, $f_0$, such a range may be acceptable, but it is quite small at typical radar operational frequencies. However, the corresponding error is also small in magnitude. For example, for a reasonable phase measurement error $\Delta\phi_m$ of one degree and $\lambda = 1$ m, the range error is 1.4 mm. This means that only the measurement ambiguity is a problem in the phase method.

Two-frequency radars can be used to enhance an unambiguous measuring range, and a major enhancement of performance can be achieved using multiple frequency radars, as will be explained subsequently.

### 3.4.2 *THE TWO-FREQUENCY PHASE METHOD*

Two-frequency CW radar is appropriate in measuring the range of both stable and moving targets. In this case the sounding waveform consists of two sinusoidal components with approximate frequencies $f_0^{I}$ and $f_0^{II}$. The frequency content of both incident and reflected oscillations in the ideal case is shown in Figure 3.15, where $f_D^{I}$ and $f_D^{II}$ is the Doppler shift for each component, and $\Delta f_0$ is the frequency spacing.

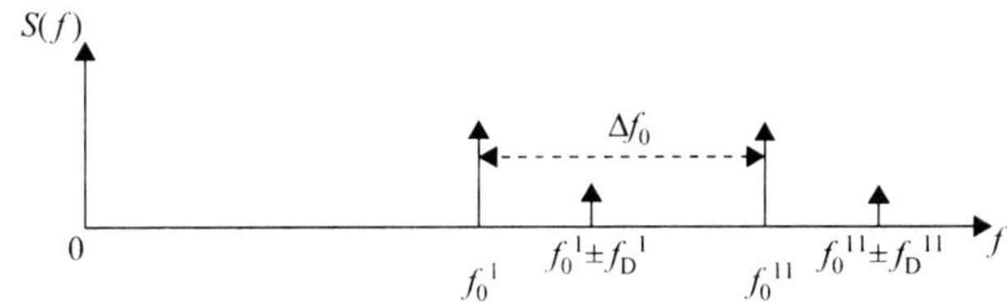


**Figure 3.15.** Spectrum content in the two-frequency phase method.

A simple block diagram of a two-frequency range finder is shown in Figure 3.16, where each receiving channel contains a mixer and a narrow band filter (NBF). The amplifiers are not shown. The components of the reflected signal are separated by the NBF. For effective separation, the maximum Doppler frequency $f_{d\,max}$ should be less than half the frequency spacing $f_{d\,max} < \Delta f_0/2$.

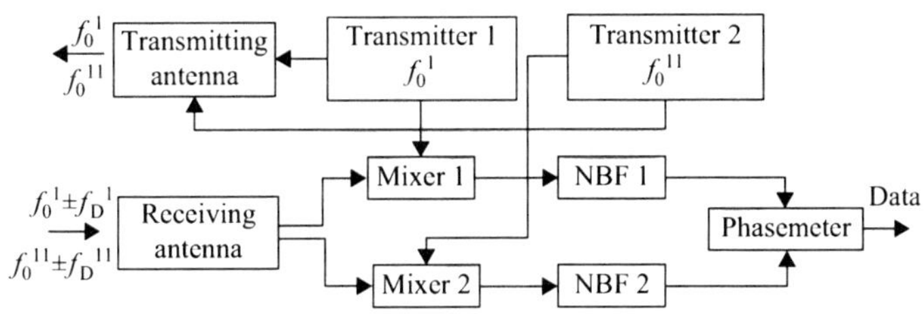


**Figure 3.16.** Simplified block diagram for a two-frequency radar range meter.

The phases of the reflected signals that are applied to mixers 1 and 2 depend on both the Doppler shift and the time delay $t_d = 2R/c$. Because the direct and reflected signals are mixed, the phases $\phi^{I}$ and $\phi^{II}$ at the mixer outputs are correspondingly equal to the phase differences of these signals. The phasemeter produces the phase difference, $\Delta\phi = \phi^{II} - \phi^{I}$, which is proportional to the current range $R$ (Finkelstein 1983):

$$\Delta\phi(t) = \frac{4\pi\,\Delta f_0}{c}(R_0 + v_r t) = \frac{4\pi\,\Delta f_0}{c}R \tag{3.10}$$

where $R_0$ is an initial range and $v_r$ is the range rate of the reflected object.

### 3.4.3 AMBIGUITY AND ACCURACY IN THE TWO-FREQUENCY METHOD

It can be seen that in the last formula $\Delta\phi = 4\pi\,\Delta f_0 R/c$ has the same structure as in the case of the single-frequency method (Section 3.4.1.1) when $\Delta f_0$ is substituted for $f_0$. Thus, the unambiguous measured range can be found from $\Delta\phi = 2\pi$ as $R_{un} = c/2\Delta f_0$. For example, if $\Delta f_0 = 3$ kHz, then $R_{un} = 50$ km. This is quite acceptable for the altimetry process when $H < R_{un}$.

The range accuracy (altitude) measurement is characterized by the error (Finkelstein 1983):

$$\Delta H = \frac{c}{4\pi\Delta f_0}\Delta\phi_m \tag{3.11}$$

where $\Delta\phi_m$ is an error in phase difference measurement.

An important conclusion from this consideration consists of the contradiction between unambiguous range and accuracy of measurement. If the frequency difference $\Delta f_o$ increases, the error becomes smaller, but the unambiguous range decreases.

### 3.4.4 PHASE AMBIGUITY RESOLUTION

The contradiction between accuracy and unambiguity can be solved by using the multiple frequency approach wherein the number of frequencies in the sounding waveform are increased. For example, a radar may have three frequencies: $f_0^{\text{I}}$, $f_0^{\text{II}}$, and $f_0^{\text{III}}$ that fulfill the condition that $f_0^{\text{III}} - f_0^{\text{I}} >> f_0^{\text{II}} - f_0^{\text{I}}$. In this case, $f_0^{\text{III}}$ and $f_0^{\text{I}}$ can provide accurate but ambiguous measurement, and $f_0^{\text{II}}$ and $f_0^{\text{I}}$ fulfill the condition of unambiguity.

There is another aspect that leads to the notion of multifrequency operation: the two-frequency phase method does not possess range resolution, which is an essential disadvantage. In order to provide range resolution, the number of frequencies in the spectrum of the sounding waveform should be increased to make that spectrum wider in accordance with radar theory. This actually provides an approximation of wideband signals, which do possess resolution abilities and are able to measure the ranges of multiple targets. It should also be noted that the limiting case of multiple frequency CW radar is FMCW radar considered in Section 3.2.

A patent (Hager, Petrich, and Almsted 2002) proposed a radar altimeter wherein the phase ambiguity is resolved by placing an additional antenna close to the first antenna so that there are no phase ambiguities between the reflected radar signals received by these two antennas.

### 3.4.5 WAVEFORMS

Conventionally, phase radar altimeters use sounding waveforms that do not provide high range resolution: in essence, they use phase differences between harmonic components of radiated and received signal. Actually, this principle can be implemented not only in CW radar but also in quasi-continuous or even pulse radiation. In reality, phase radar altimeters can use different sounding waveforms: FM and AM in continuous wave radar altimeters, and even pulse modulation with a duty factor of one-half (Zhukovsky, Onoprienko, and Chizhov 1979).

### 3.4.6 MEASURING DEVICES AND SIGNAL PROCESSING

A basic measuring system consists of a tracking loop with a controlled phase changer within an FM generator circuit. In a phase altimeter, though the measuring device must obviously be based on phase comparison, it can be implemented in different ways. For example, the phase of a modulation frequency harmonic of an echo-signal can be compared with the phase of a corresponding modulation frequency harmonic derived from a generator. Another method is to compare the phases of two synchronously modified components of a reflected signal.

Actually, all CW radar altimeter measuring devices, including phase altimeters, process all received signals scattered by area-extensive statistically heterogeneous surfaces. However, the range gating of a signal for measuring the shortest distance between an aircraft and a surface is not usually supported in such altimeters. In other words, phase altimeters normally use integral signal processing. Nevertheless, it is possible to select a small area in the vicinity of the normal to the surface. This can be implemented because of the existence of Fresnel regions in the case of modulating waves with difference frequencies.

### *3.4.7 REMARKS ON THE ACCURACY OF CW AND PULSE RADAR ALTIMETERS*

In contrast to the integral signal processing in CW radar altimeters (both phase and FMCW) described in Sections 3.3.and 3.4, pulse altimeters implement a selection of signals reflected from different parts of a surface. The measuring device in pulse altimeters (Section 3.2) tracks the pulse edge (or maximum) of the received signal. The indicated difference in the processing techniques sometimes leads to the erroneous conclusion that the accuracy of the pulse radar altimeter depends very slightly upon the surface geometry, whereas the accuracy of the CW radar altimeter is dependent mainly upon the properties of the scattering surface. This misunderstanding is because it is not possible to compare techniques of signal processing that may be very different in the cases of different sounding waveform modulations. Both theory and practice show that the type of modulation is not decisive in accuracy assessments of the comparable parameters in sounding waveforms and signal processing techniques.

Accuracy requirements for the measurement of altitude and vertical velocity are different depending on a specific task or application. Some tasks allow for an increase in the relative error when the altitude increases, that is, a permanent absolute error is required. Sometimes, an absolute error should be the same at practically any altitude, that is, the relative error should decrease if altitude increases.

## 3.5 RADIOACTIVE ALTIMETERS FOR SPACE APPLICATION

### *3.5.1 MOTIVATION AND HISTORY*

The first descent vehicle used in the Yuri Gagarin space flight did not have a soft landing control system, so the cosmonaut had to bail out. The next modification of the descent vehicle had a soft landing system equipped with a powder retro-engine that could be ignited at a certain low altitude above the planet's surface. The problem was how to sense this relatively low altitude under the conditions of space flight. It was recognized that known radar altimeters did not satisfy this requirement, so the problem was how to design an appropriate altitude sensor. This situation was exacerbated because the extremely strict technical requirements could not be met by any known device (Yurevich 2004) either. In particular, the sensor was required to be an absolutely all-weather device, and its accuracy should be independent of the surface properties of a planet including water, ice, snow, and local surface irregularities essentially smaller than the bottom of the descent vehicle. It should also be insensitive to the descent vehicle tilting and to the magnitude of any horizontal speed component. Moreover, such an altitude sensor should operate through the skin of the descent spacecraft and through the rocket blast, but be extremely reliable, light, and adequately small.

The solution to this problem was proposed in 1965 with the help of a brand new gamma-ray altimeter (Yurevich 2004). In the spring of 1966, online tests of the soft landing control system of the new Soyuz spacecraft were conducted, and Kaktus, the first operational gamma-ray altimeter was accepted for practical application. The first flight of the Soyuz-1 spacecraft equipped with a Kaktus system in April 1967 failed, though that catastrophe was not related with Kaktus itself; in that tragic flight the cosmonaut, Vladimir Komarov was killed when the parachute of his spacecraft failed during the return to Earth. The first operational mission that included a Kaktus gamma-ray altimeter was on October 30, 1968 when cosmonaut Georgi Beregovoi in the Soyuz-3 spacecraft made a successful soft landing. Since then all Soviet (and then Russian) recovery capsules and spacecraft have been equipped with such systems.

In 1975 a flight of a Soyuz spacecraft equipped with an airborne altitude indicator based on a heat-resistant and shockproof cesium-137 source (Radium Institute 2008) was accomplished in the framework of the Soyuz–Apollo Program. An important new step was the creation of the Kvant system for soft landing control. This system was used in the Luna Program (Encyclopedia Astronautica 2007) for automatic soft landings at unmanned stations on the moon, to retrieve and return lunar surface samples to Earth, and to deploy lunar rovers Lunokhod-1 and Lunokhod-2 on the moon's surface (1970–1976). A new generation of radioactive altimeters, Kaktus-2B for Soyuz MT and other landers, were created (CR&DI RTC 2011) in 1998. A similar system provided a soft landing for the joint Russian and American crew *in extremis* during the ballistic re-entry of the Soyuz TMA-1 spacecraft on May 4, 2003 (Andronova 2008). Currently, the seventh generation of isotopic altimeters is in operation.

The work on gamma-ray altimetry for space applications provided an incentive for the development of a new scientific and technical field called photon engineering, which is nowadays the basic field of one of the State Scientific Centers of the Russian Federation (CR&DI RTC 2008). Applications of similar systems in helicopters and airplanes for landing heavy arms were developed later in addition to unique systems for blind landing, air data measurement, and for formation flight control at extremely low altitude, amongst others (Andronova 2008).

## 3.5.2 PHYSICAL BASES

### 3.5.2.1 Features of Radiation

Photon systems use gamma rays and X-rays. Gamma rays are a form of electromagnetic radiation having frequencies produced by subatomic particle interactions, such as electron–positron annihilation or radioactive decay. Gamma rays are generally characterized as electromagnetic radiation having the highest frequencies and energies, and also the shortest wavelengths (below about $10^{-11}$ m), within the electromagnetic spectrum.

X-rays are a form of electromagnetic radiation with wavelengths longer than Gamma rays but shorter than UV radiation. More precisely, X-ray wavelengths are in the range of $10^{-8}$–$10^{-11}$ m, corresponding to frequencies of $3 \times 10^{7}$ GHz to $3 \times 10^{10}$ GHz. Hard X-rays overlap the range of "long"-wavelength (lower energy) gamma rays, but the distinction between the two terms depends on the source of the radiation, not its wavelength: X-ray photons are generated by energetic electron processes whereas gamma rays result from transitions within atomic nuclei. However, in such super-high frequency bands, emissions are usually considered as a flow of photons as particles rather than as wave oscillations. Engineering systems and devices based on such hard radiation principally exhibit unique qualities, including operability through obstructions, absolute all-weather operation, and interference immunity even under the conditions in atomic explosions.

Photon emission is rather different from electromagnetic radiation at the lower frequencies used in radio engineering, particularly with regard to generation methods, propagation features, and interactions with tangible media. The application of gamma radiation for altimetric purposes required theoretical and experimental research in photon emission, propagation, and the interactions between the radiation and various substances. Also required was an understanding of the transition from wave to corpuscular descriptions, and the development of methods of analysis and design of the relevant engineering systems, including techniques for their mathematical modeling and simulation.

#### *3.5.2.2 Generators of Photon Emission*

Generators of photon emission may use radioactive sources (radioactive isotopes), X-ray sources with photon energies less than 1 MeV, and electron accelerators with photon energies up to tens of MeV (Yurevich 2003).

Radioisotope sources produce discrete gamma emission spectra, and are omnidirectional radiators. However, defined spatial radiation patterns can be formed using arc baffles and other protective screens.

Sources of X-ray photon emission produce radiation with combinations of both continuous and discrete spectra. Spatial radiation patterns are formed by special X-ray tube design and also by collimation of the emission with protective screens. These sources of photon emission are controllable: they can work in both continuous and pulse modes and different kinds of modulation can be applied.

Electron accelerators as sources of photon emission (betatrons and linear electron accelerators) normally produce pulse photon radiation with narrow radiation patterns.

#### *3.5.2.3 Receivers*

Reception in the present context is really the detection and registration of photonic radiation. It is implemented by detectors wherein the interactions of the photons with the detector material are converted into electrical output signals. There are several types of such detectors for the registration of photon radiation including scintillation detectors and semiconductor radiation sensors.

#### *3.5.2.4 Propagation Features*

The spatial propagation of photon radiation is rectilinear, and no physical fields influence photon propagation. The higher the photon energy, the greater is the penetrability, and such photons can pass through materials that constitute obstacles to electromagnetic radiation in lower frequency bands. Nevertheless, when propagating, the radiation does interact with tangible media, which results in some absorption and scattering in addition to the excitation of some secondary radiation. These processes lead to altered spectra, a reduction in intensity, and changes in the scattering diagram, these being defined by the properties of the penetrated medium. Hence, an analysis of the photon radiation after its interaction with a medium allows various parameters of that medium to be estimated. The scattering of photon radiation in a medium defines also the frontier of a possible application of photonic systems to coordinate measurement, information transmission, and for solving other tasks related to radiation traveling some distance in that medium.

### 3.5.3 PRINCIPLES OF OPERATION

Depending on the principle of operation, all photonic systems can belong to one of two classes: (1) systems that register the direct radiation from the transmitter, and (2) systems that register the reflected radiation, more exactly the backscattered signal. As in the case of radars, this backscattered signal is a part of the secondary radiation, or scattering, from the surface layer of a reflecting object, in this case caused by the primary photon radiation. Isotopic altimeters such as the Kaktus instrument belong to systems of the second type as is shown in Figure 3.17.

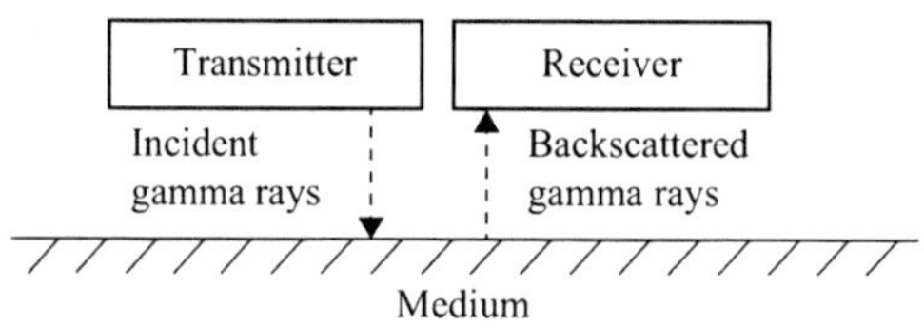


**Figure 3.17.** Layout of a photonic system based on backscattered radiation.

Distance measurement can be implemented by traditional methods using pulse and/or frequency modulation. However, it can also be performed by a device specific to photon technology called a ratemeter or intensity-meter. This can be called the "intensimetric method" and it employs a well-known law of radiation intensity change that depends on the distance to the source (Yurevich 2003). The above-mentioned Kaktus system, designed for the measurement of altitude and the vertical velocity component, belongs to systems of this type. Figure 3.18 taken from

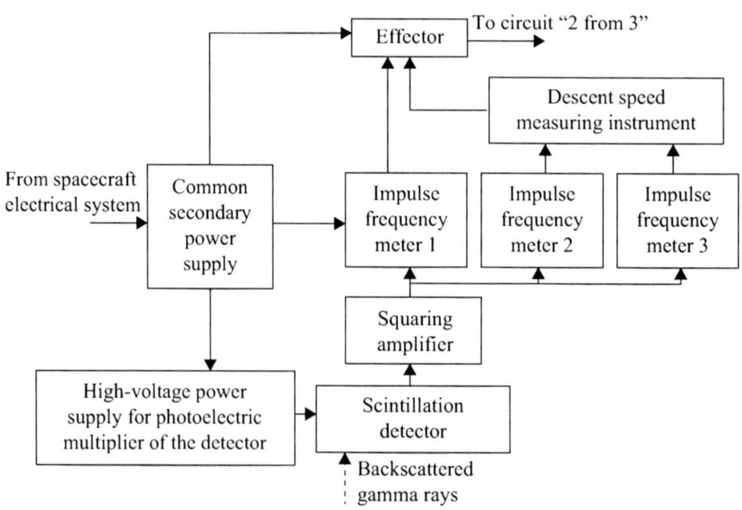


**Figure 3.18.** Functional block diagram of the receiver channel with speed corrector.

Yurevich (2004) presents a receiver layout for such systems. This is a functional block diagram of the receiver channel with a speed corrector, typical of the Kaktus system, which actually gave rise to a new field of photonic engineering. Three channels form discrete signals, and the accepted signal is formed according to majority logic, that is, "two out of three."

In addition to the cesium-137 gamma source, other isotopes can be used in radioactive isotope altimetry, including the cobalt-60 gamma source (Gamma-ray altimeter 2008).

### *3.5.4 RADIATION DOSAGE*

Because of their high-energy content, gamma rays can cause serious damage when absorbed by living cells, and can also influence the operation of electronic systems. In the work by Song et al. (1997), a gamma-ray altimeter containing a gamma-radioactive source of 137Cs with an activity of $2.96 \times 10^{10}$ Bq, was used to detect the height of a spacecraft recovery capsule during landing. To know the gamma-dose distribution near the guidance section of the recovery capsule, especially at a position where the useful load was located, the gamma-dose field was measured. The results showed that the absorbed dose rate at the ground 5 m from the gamma-source was as low as the radiation protection limit when the altimeter was down to 0.16 m from the ground. The ground reflection effect for gamma-rays decreases as the height of the recovery capsule increases, so that the gamma-dose level at the useful load point in the recovery capsule meets the requirements of flight safety.

### *3.5.5 EXAMPLES OF RADIOISOTOPE ALTIMETERS*

Several examples of radioisotope altimeters designed for space applications are given in the book *Avionics of Russia* (Bodrunov 1999), including the Kaktus photonic altimeter for spacecraft soft landing systems (Figure 3.18) which was described earlier. The main performance characteristics of the first Kaktus were as follows:

- Radiant energy of gamma source 661 keV
- Source activity $1.85 \times 10^{10}$ Bq
- Detector type NaJ(TI) crystal detector
- Effective altitude range 0.5–10 m
- Error less than 5%
- Range of approach speeds 4–12 m $s^{-1}$.

Figure 3.19 shows the transmitter and receiver of this Kaktus system.

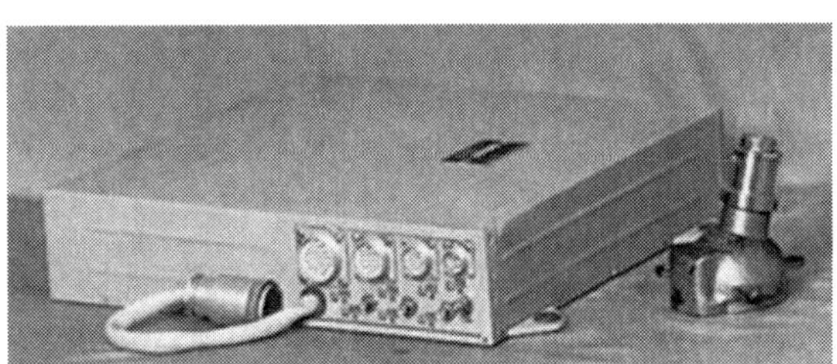

**Figure 3.19.** The Kaktus isotope (photon) altimeter: Transmitter (right) and receiver (left).

Succeeding generations of Kaktus gamma-altimeters exhibited improved measurement accuracies. For example, the Kaktus 5 photon altimeter was designed for producing a signal at a given altitude that switched on the reversing (brake) jets for an automatic soft landing, and independently of the conditions at the landing site. It had an effective altitude range of 0.5–15 m, an rms error of 2–6 %, a no-failure operation probability (during 0.5 hour) of 0.9999, and consumed not more than 25 VA.

The Kvant 2 photon altimeter was designed for the Luna modules to provide soft landings on the surface of the Moon. It had an effective altitude range of 0.5–3.5 m, an rms error of 2–5%, a no-failure probability (over 1 hour) of 0.9999, and consumed not more than 20 VA.

A Louch photon altimeter was designed for measuring airborne vehicle altitudes, particularly for "Wing In Ground-effect" (WIG) vehicle altitudes over water surfaces, and for producing analog signals for motion control systems. The Louch instrument is functional at 100% humidity in sea fog, and at high mechanical loads. Its effective altitude range is 0–11 m, the rms error is 3–5%, the mass is 5 kg, and it consumes not more than 30 VA.

One of the recently developed devices is the automatic control system for the soft landing on Phobos, Mars' satellite (Yurevich 2004).

## REFERENCES

Aerospace Research Information Center. 2005. Retrieved from http://www.aric.or.kr/trend/accessory/content.asp?classify=3&idx=648&search=&page=1

AN/APN Equipment Listing. 2007. Retrieved from http://www.designation-systems.net/usmilav/jetds/an-apn.html

Andronova, O. 2008. CR&DI RTC – 40 years old, website of the Newspaper Computer – inform. Retrieved from http://www.ci.ru/inform03_08/p_04.htm (In Russian.)

ARINC Incorporated. 2007. ARINC Characteristic 707-6. Retrieved from https://www.arinc.com/cf/store/catalog_detail.cfm?item_id=297

Becker, J., and Y. Manoli. 2004. "A continuous-time field programmable analog array (FPAA) consisting of digitally reconfigurable G_m-cells." *Proc. IEEE International Symposium on Circuits and Systems (ISCAS)*, 1: I-1092-5.

Bodrunov, S. D. (Ed.) 1999. *Avionics of Russia* (pp. 102–4). St. Petersburg, Russia: National Association of Aviation Instrument Engineers.

Collinson, R. P. G. 1996. *Introduction to Avionics* (p. 456). London: Chapman & Hall.

Cook, C. E., and M. Bernfield. 1967. *Radar Signals: An Introduction to Theory and Application.* New York: Academic Press.

CR&DI RTC. 2011. Websites of the State Scientific Center "Central Research and Design Institute for Robotics and Technical Cybernetics", Retrieved from http://www.neva.ru/CNII-RTC/; http://www.rtc.ru/index-r.shtml; http://www.rtc.ru/publication/hi-tech2001.shtml

*Encyclopedia Astronautica.* 2007. Luna, Retrieved from http://www.astronautix.com/project/luna.htm

Ferguson, M., R. Kalisek, and L. Tucker. 1997. *GPS Land Navigation: A Complete Guidebook for Backcountry Users of the NAVSTAR Satellite System* (p. 255). Boise, ID: Glassford Publishing.

Finkelstein, M. I. 1983. *Fundamentals of Radar* (p. 536), Radio i Svyaz, Moscow. (In Russian.)

Gamma-ray altimeter. 2008. Answers.com Technology website, Retrieved from http://www.answers.com/; http://www.answers.com/topic/gamma-ray-altimeter#after_ad1

Grishin, Y. P. 2000. Reliable data processing in an integrated GPS-based airborne navigational equipment, EUROCOMM 2000. Information Systems for Enhanced Public Safety and Security. IEEE/AFCEA, pp. 91–4.

Hager, J. R., C. J. Petrich, and L. D. Almsted. 2002. Precision radar altimeter with terrain feature coordinate location capability, US Patent Issued on March 26, 2002, Application No. 498930 filed on 2000-02-04, Retrieved from http://www.patentstorm.us/patents/6362776-claims.html

Hairion, D., S. Emeriau, E. Combot, and M. Sarlotte. 2007. New safety critical radio altimeter for Airbus and related design flow, Design, Automation & Test in Europe Conference & Exhibition, 16–20 April, 2007, ISBN: 978-3-9810801-2-4, INSPEC Accession Number: 9476453, pp. 1–5.

Harding, D. J. 2000. *Principles of Airborne Laser Altimeter Terrain Mapping*, NASA's Goddard Space Flight Center, Retrieved from http://pugetsoundlidar.ess.washington.edu/laser_altimetry_in_brief.pdf

Honeywell Aerospace Products. 2007. Retrieved from http://www.honeywell.com/sites/aero/Avionics_Electronics.htm

Integrated Publishing. 2003. *Absolute (Radar) Altimeter*. Retrieved from http://www.tpub.com/content/aviation/14030/css/14030_46.htm

Jensen, D. 2005, March 1. "Has the time come for UWB radio." *Avionics Magazine*. Retrieved from http://www.aviationtoday.com/av/categories/rotocraft/783.html

Jensen, J. R. 1995. "Design and performance analysis of a phase-monopulse radaraltimeter for continental ice sheet measurement." *International Geoscience and Remote Sensing Symposium, IGARSS'95*, 2: 865–7.

Jensen, J. R., and R. K. Raney. 1998. "Delay/Doppler radar altimeter: Better measurement precision." *Proc. IEEE IGARSS'98*.

Kaplan, E. D., and C. Hegarty. 2006. Understanding GPS: Principles and Applications (2nd edition)., Norwood, MA: Artech House Inc.

Kayton, M. 2001. "Navigation systems." Chapter 13 In *The Avionics Handbook*, edited by C. R. Spitzer, The Avionics Handbook, (Ed. Cary R. Spitzer). Boca Raton, FL: CRC Press LLC.

Kayton, M., and W. R. Fried. 1997. *Avionics Navigation Systems* (2nd edition, p. 773). New York: John Wiley & Sons, Inc. DOI: 10.1002/9780470172704.

Komarov, I., and S. Smolskiy. 2003. *Fundamentals of Short-Range FM Radar* (p. 314). Norwood, MA: Artech House Inc.

Laser Optronix. 2006. Laser Altimeter systems, Airborne altimeters. Retrieved from www.laseroptronix.se

Mahafza, B. R. 2000. *Radar Systems Analysis and Design Using MATLAB*. Boca Raton, FL: Chapman & Hall/CRC. DOI: 10.1201/9781584888543.

Mityashev, B. N. 1962. *Determination of Time Position of Pulses at Presence of Interferences* (p. 232). Moscow: Sovetskoe Radio. (In Russian.)

Parsch, A. 2007. AN/APN - Equipment Listing, Retrieved from http://www.designation-systems.net/usmilav/jetds/an-apn.html

Radium Institute (2008), V. G. Khlopin Radium Institute Website, Retrieved from http://www.khlopin.ru/english/hronology.php

Raney, R. K. 1998a. "The new generation of radar altimeters: Proof of concept." NASA Research Announcements. Proposals Selected Under NRA-98-OES-05. Retrieved from http://esto.nasa.gov/files/solicitations/IIP_98/winners.html

Raney, R. K. 1998b. "Delay compensated Doppler radar altimeter." US Patent No. 5736957, Issued April, 1998.

Raney, R. K. 1998c. "The delay/Doppler radar altimeter." *IEEE Trans. Geoscience and Remote Sensing* 36 (5): 1578–88. DOI: 10.1109/36.718861.

RTCA. 2007. RTCA Incorporated website. Retrieved from http://www.rtca.org/downloads/ListofAvailableDocs_WEB_OCT%202007.htm

Secmen, M., S. Demir, and A. Hizal. 2006. "Dual-polarised T/R antenna system suitable for FMCW altimeter radar applications, microwaves, antennas and propagation." *IEE Proceedings* 153 (5):407–12.

Shirman, Y. D., V. N. Golikov, I. N. Busygin, G. A. Kostin, and V. N. Manshos. 1987. Theoretical bases of radar (selected pages). Transl. into English from Teoreticheskiye Osnovy Radiolokatsii, (Moscow, USSR), 1970 pp. 1–420, 483–541, 550–560.

Skolnik, M. I. (Editor). 1990. *Radar Handbook* (2nd edition). McGraw-Hill.

Song, J., Y. Shi, S. Lin, S. Dong, Z. Zhang, and Z. Shen. 1997. "Measurement and evaluation of radiation dose distribution of gamma-ray altimeter in static test for recovery capsule during simulated landing." *Institute of Atomic Energy, Beijing* 10 (5). Retrieved from http://lib.bioinfo.pl/pmid:11540391

Turin, G. L. 1960. "An introduction to matched filters." *IEEE Transactions on Information Theory, IT-6* 310–29.

Ulander, L. M. H. 1987. "Averaging of radar altimeter pulse returns with the interpolation tracker." *International Journal of Remote Sensing* 8: 705–21. DOI: 10.1080/01431168708948682.

US Radars. 2007. Retrieved from http://www.history.navy.mil/library/online/radar-12.htm

Yegorov, V. V. 2005. "Problems of accuracy and uncertainty in satellite altimetry." Third All-Russian open conference "Modern problems of the Earth remote sensing from the space", Moscow, Russian Federation, RAN Institute for Space Research, 14–17 November, 2005, p. 83. (In Russian.)

Yurevich, E. I. 2003. Photonic Technology. Mechatronics, Automatization, Control, 2003, No5, June, Retrieved from http://www.rtc.ru/publication/foton-teh.shtml

Yurevich, E. I. 2004. The soft landing control system of space vehicle, Aviakosmicheskoe Priborostroenie (Aerospace Instrument Engineering), 2004, No 5, pp. 58–60. (In Russian.)

Zelli, C., M. Martin-Neira, G. Alberti, F. Impagnatiello, and M. Matteoni. 1997. A Bistatic Altimetry Mission for Ocean Topography Mapping, International Astronautical Federation, IAF-97-B.3.07, Retrieved from http://www.corista.unina.it/Docs/bistatic_altimetry.pdf

Zhukovsky, A. P., E. I. Onoprienko, and V. I. Chizhov. 1979. *Theoretical Fundamentals of Radio-Altimetry* (p. 320), Moscow: Sovetskoe Radio (In Russian.)

Zyzys, E. A. 1964. Evaluation of Honeywell Model 7182 Radar Altimeter, National Aviation Facilities Experimental Center, Atlantic City, NJ, Report, JUN 1964, Defense Technical Information Center, Accession Number: AD0608376; Retrieved from http://stinet.dtic.mil/oai/oai?&verb=getRecord&metadataPrefix=html&identifier=AD060837).

CHAPTER 4

# Autonomous Radio Sensors for Motion Parameters

Felix J. Yanovsky
*National Aviation University*
*Kiev, Ukraine*

## 4.1 INTRODUCTION

This chapter describes autonomous avionic devices used to acquire the information necessary for air navigation, except for radar altimeters, which were considered in Chapter 3. Any sensing device that measures motion parameters, or derives additional navigational information without interaction with ground-based or satellite equipment, is considered to be an autonomous sensor in this book.

Autonomous radio sensors are mostly based on radar principles, and such a sensor can actually constitute a rather complicated radar system, namely:

a. An Airborne Weather Radar (AWR) that is a pulse primary radar system (Barton and Leonov 1998) used as a sensor of meteorological information
b. A Doppler Sensor for Ground speed and crab-Angle (DSGA) that is, in most cases, a continuous wave Doppler radar system (Saunders 1990)
c. A Traffic alert and Collision Avoidance System (TCAS) that is a secondary radar system (IEEE standard radar definitions 1998).

This list includes only the basic radar equipment installed aboard an aircraft for solving problems of aeronavigation and flight safety and it can differ for different aircraft. For example, it does not include Stormscopes®, which are passive radar systems (Shirman et al. 1987) for detecting and locating zones of thunderstorm activity (Stormscope® 2007) and are widely applied, especially in small and business aircraft. Also, the listed DSGAs are actually not installed on modern civil airliners but are still important for military aircraft and for helicopters.

In addition to the standard equipment necessary for aeronavigation and flight service, accessory radar devices intended for carrying out some special tasks may also be installed. An example is side-looking equipment incorporating Synthetic Aperture Radar (SAR) that is often used onboard aircraft and satellites for remote sensing (Curlander and McDonough 1991). However these are beyond the remit of this chapter, which describes only the DSGA, airborne weather radar, and collision avoidance systems as sensors providing information considered necessary for aeronavigation and flight safety.

## 4.2 DOPPLER SENSORS FOR GROUND SPEED AND CRAB ANGLE

### *4.2.1 PHYSICAL BASIS AND FUNCTIONS*

Doppler radar installed aboard an aircraft allows a quite accurate measurement of the ground speed of the aircraft by processing a radar signal reflected from the surface. Due to the effect of wind currents, the direction of flight does not exactly coincide with the longitudinal axis of the aircraft. The angle between the heading and the track is called *the drift* or *crab angle*, and it will be shown subsequently that this angle can also be measured using the appropriate radar. These important flight navigational parameters can be autonomously determined during flight using DSGA radars often called simply *Doppler navigators*. For a long time, the DSGA was a part of the standard equipment of any airliner, but in recent years the tendency has been to obtain navigational parameters with the help of other radio-navigational means, though usually with some loss of autonomy. Nevertheless, autonomous sensors like the DSGA will evidently remain essential for some special and military applications.

The physical basis of the DSGA is the Doppler effect, which can be described as the change in observed frequency when there is relative motion between a transmitter and a receiver. This change is called Doppler shift and is directly proportional to the relative speed between the transmitter and the receiver. If the relative velocity is much smaller than the speed of light, as in the case of real aircraft speeds, the Doppler shift can be expressed as $F_D = V_r f/c = V_r/\lambda$, where $V_r$ is the relative velocity between transmitter and receiver, $f$ is the frequency of transmission, $c$ is the speed of light, and $\lambda = c/f$ is the transmission wavelength. In the case of *monostatic* primary radar, where both transmitter and receiver are mounted on the aircraft and electromagnetic energy is radiated toward the Earth's surface, some of the energy is backscattered by the Earth and is received by the radar receiver on the aircraft. If the aircraft is moving with a total velocity directed arbitrarily in 3D space, the Doppler shift is $F_D = 2V_R f/c = 2V_R/\lambda$, where $V_R$ is the component of aircraft velocity along the radar beam centroid. The factor 2 appears in the Doppler shift equation because both the transmitter and the receiver are moving with respect to ground, that is, a target from which the energy is backscattered. When backscattering, the target plays the role of a receiver first and then as a transmitter.

A DSGA is basically a Doppler sensor used for the determination of aircraft velocity components relative to the underlying surface and hence for delivering information on absolute ground speed $V_G$ and crab angle $\beta_c$ to the crew and also to a flight-path computer. In the case of helicopter DSGAs, the longitudinal and transverse speeds are usually defined and sometimes also the vertical speed $V_V$. Information on the ground speed and crab angle is used in dead reckoning systems for the determination of the current position coordinates of an aircraft: airborne Doppler radar measures all velocity components in the Earth-referenced coordinate frame and then

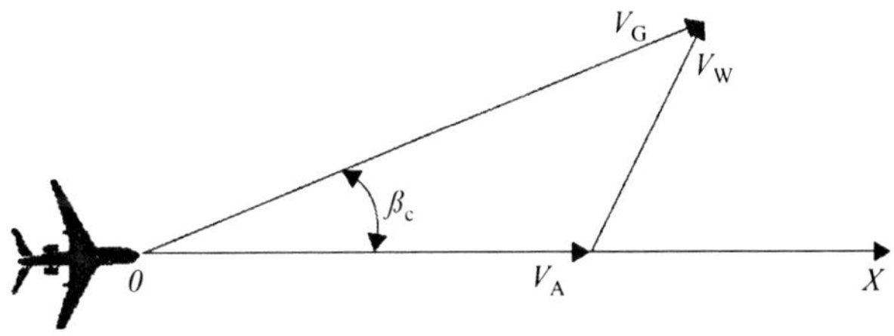


**Figure 4.1.** Crab angle $\beta_c$ caused by wind $V_W$.

integrates them into distance traveled from a known point of departure. From this, the aircraft's geodetic present position and course as well as distance-to-destination can be calculated. Hence, Doppler radar can operate as the primary sensor for a dead reckoning navigation system or as one of the sensors in a multisensor system (Huddle and Brown 1997).

### 4.2.2 PRINCIPLE OF OPERATION

Consider the movement of an aircraft over terrain (Figure 4.1). The airspeed $V_A$ is directed along axis $OX$ and the ground speed $V_G$ is the horizontal component of the true speed relative to that terrain. However, because of wind, this ground speed is not equal to the airspeed in either direction or magnitude: it is actually the resultant of vectors $V_A$ (airspeed) and $V_W$ (wind speed) as illustrated in Figure 4.1. The angle $\beta_c$ between the ground speed and airspeed vectors is the crab angle, and this characterizes the wind-induced drift of the aircraft. More exactly, the crab angle includes the aerodynamic sideslip angle, which is defined by any discrepancy between the longitudinal axis of the aircraft and the direction of the thrust vector. However, this normally small aerodynamic sideslip angle can usually be neglected.

If a Doppler radar installation aboard the aircraft is equipped with a directional antenna that radiates and receives echo signals reflected from the terrain, then the frequency of the reflected signal differs from the frequency of the radiated waveform by the Doppler frequency shift $F_D = 2V_r / \lambda$, where $V_r$ is the projection of the aircraft speed onto the direction of radiation, and $\lambda$ is the wavelength.

The Doppler frequency shift will be maximal at the maximum radial velocity $V_r$, that is, when the antenna beam is directed exactly along the vector of the ground speed $V_G$. Hence, the crab angle can provide a fix for the steering angle of the antenna beam, which should track the maximal magnitude of the Doppler frequency. This is the principle of the *single-beam Doppler crab-angle meter*, and such an instrument obviously requires a scanning antenna beam.

Now suppose that a Doppler radar with two stable antennae is installed aboard the aircraft. These antennae radiate and receive reflected signals at an angle of $\Phi$ degrees to each other and symmetrically relative to a longitudinal axis of the aircraft as is shown in Figure 4.2 using dashed lines.

For initial simplification, any inclination angle of the antenna beams in the vertical plane may be neglected. Then, the signals reflected from the terrain that are received by left-hand and right-hand antennae will have corresponding Doppler frequencies $F_{D1}$ and $F_{D2}$:

$$F_{D1} = \frac{2V_G}{\lambda}\cos\left(\frac{\Phi}{2} - \beta_C\right) \quad \text{and} \quad F_{D2} = \frac{2V_G}{\lambda}\cos\left(\frac{\Phi}{2} + \beta_C\right) \tag{4.1}$$

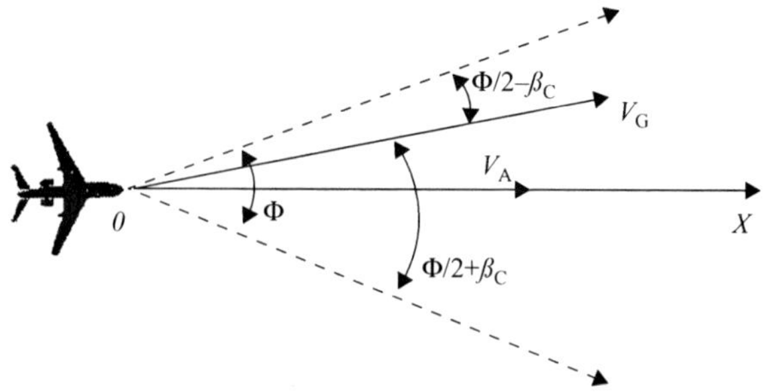


**Figure 4.2.** Two-beam crab angle meter principle.

Adding and subtracting the measured values of Doppler frequencies gives:

$$\begin{cases} F_{D1} + F_{D2} = \dfrac{2V_G}{\lambda} 2\cos\dfrac{\Phi}{2}\cos\beta_C; \\ F_{D1} - F_{D2} = \dfrac{2V_G}{\lambda} 2\sin\dfrac{\Phi}{2}\sin\beta_C. \end{cases} \tag{4.2}$$

The left-hand sides of the equations contain measurable variables and the right-hand sides contain known radar parameters, the only two unknown variables being the ground speed $V_G$ and crab angle $\beta_c$. Solving these simultaneous equations gives:

$$\beta_C = arctg\left(\frac{F_{D1} - F_{D2}}{F_{D1} + F_{D2}}\right) \tag{4.3}$$

$$V_G = \frac{\left(F_{D1} - F_{D1}\right)\lambda}{4\sin\dfrac{\Phi}{2}\sin\beta_C} \tag{4.4}$$

In reality the antenna beams are directed to the land at a known angle $\gamma$ in the vertical plane, which makes the final formulae somewhat more complicated. However, the operational principle of the two-beam Doppler meter remains valid. Thus, a two-beam Doppler meter makes possible the measurement of both crab angle and ground speed.

### *4.2.3 CLASSIFICATION AND FEATURES OF SENSORS FOR GROUND SPEED AND CRAB ANGLE*

Doppler sensors for ground speed and crab angle determination can use continuous or pulse radiation, and in aviation, continuous wave DSGAs are normally used. Waveforms for such DSGAs can simply be coherent oscillations of a single carrier frequency or they can be frequency modulated carrier frequency oscillations. Usually, a sinusoidal voltage is used as a modulating function.

Single-beam, two-beam, three-beam, and four-beam ground speed and crab angle meters all exist because increasing the number of beams corresponds to increasing the number of equations and hence the number of unknowns that can be calculated. The use of three- or four-beam DSGA systems allows for the most complete calculation possibilities. Such DSGAs can define all three components of an aircraft velocity vector including vertical speed. Moreover, three- or four-beam systems make it possible to reduce the influence of aircraft roll and pitch on the accuracy of measurement. Although only three beams are required to provide three components of velocity, most modern Doppler radars employ four beams because planar array antennae naturally generate four such beams. An additional equation results in a more accurate value for a velocity component.

According to Technical Standard (TSO.C65A, 1983), the frequency band recommended for DSGAs is 13.25–13.4 GHz. (Ground speed and crab angle meters developed in the former USSR used an 8.8–9.8 GHz frequency band.) General requirements relating to Airborne Doppler Navigational Radar were developed by the RTCA (formerly the Radio Technical Commission for Aeronautics) in 1975 in *Requirements and Technical Concepts for Aviation, Minimum Performance Standards*. This document recommends standards and test procedures for Airborne Doppler Radar Navigation Equipment, and appendices include conditions of testing and detailed test procedures coordinated with EUROCAE—the European Organization for Civil Aviation Equipment, which is the regulatory agency for certifying aviation equipment in Europe.

### 4.2.4 GENERALIZED STRUCTURAL DIAGRAM FOR THE GROUND SPEED AND CRAB ANGLE METER

A generalized structural diagram for a DSGA is shown in Figure 4.3. It contains the following: an antenna system (AS); a transmitter or generator unit (T) that generates a sounding waveform and also contains a local oscillator (LO) to form an intermediate frequency (IF); a receiver (R) that consists of a balance mixer (BM), an IF amplifier (IFA) and mixer (M); and a frequency meter (FMT). A transmitting antenna A-1 with a three- or four-beam antenna pattern radiates the sounding waveform at a carrier frequency $f_1$ at selected angles in the horizontal and vertical planes. The reflected signal, of frequency $f_2$, is picked up by the receiving antenna A-2, which is identical to antenna A-1 in terms of both antenna pattern and beam spatial orientation.

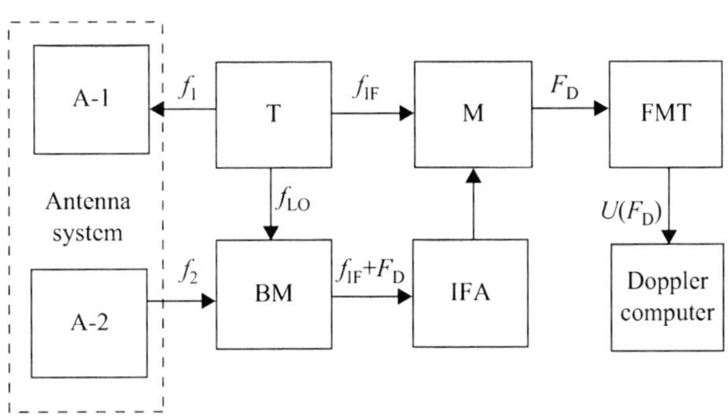


**Figure 4.3.** Generalized DSGA structural diagram.

The balance mixer, BM, produces a converted signal with intermediate frequency $f_{IF}$ as a carrier and a Doppler shift that depends on aircraft speed and beam direction. Since the antenna has a finite beam width, the returned signal associated with the beam comes from a spread of angles concentrated around the beam centroid. Furthermore, the backscattering medium is composed of a multitude of randomly situated scattering centers, each of which produces an individual Doppler shift. In view of the frequency spread and randomness of the amplitude and phase of the reflections from different scattering centers, the Doppler signal associated with a beam is in the form of a noise-type frequency spectrum, the shape of which is roughly Gaussian. More analysis of such signals and spectra can be found in *Airborne Doppler Radar: Applications, Theory and Philosophy* (Schetzen 2006). The Doppler frequency of interest, which is proportional to the velocity component along the beam centroid, is the mean or center of area of the Doppler spectrum.

After amplification by the IFA, the converted signal is applied to the mixer, M. The other input of this mixer is connected to the IF output of the transmitter. An intermediate frequency $f_{IF}$ is formed by the same generators, which specifies the basic frequencies $f_1$ and $f_{LO}$. This helps to improve the accuracy of the speed and drift angle measurement.

Nowadays, DSGA systems are mainly sinusoidally modulated FM/CW Doppler radars employing four separate downward-looking beams aimed at about 15° off the vertical.

### 4.2.5 DESIGN PRINCIPLES

In the past, DSGA equipment consisted of two units: a Doppler radar monoblock installed in the lower part of the fuselage, and a display installed on the flight deck. Modern DSGA configurations dispense with any special display, and speed and drift angle data are transmitted directly to a flight computer.

The antenna system is one of the basic components of DSGA equipment, and the accuracy of the DSGA depends on the design of this antenna and the stability of its parameters. The angles at which the antenna beams are set appear in the basic equations that determine the Doppler frequencies on each of the beams, hence high accuracy must be maintained.

In one of the standard configurations, the antenna system consists of two flat slotted-guide antennae that are fastened to the lower side of the airframe. The transmitter, receiver, and other components of the radar system are situated on the upper side of the same frame. Such a monoblock is mounted on a convertible radio-transparent radome embedded in the lower part of the fuselage. A screen for absorbing electromagnetic energy is mounted between the antennae for better isolation. Modern DSGAs often use a single microstrip antenna, especially for helicopter configurations.

During antenna installation, parallelism between the electrical axis of the monoblock (and hence the antenna system) and the longitudinal aircraft axis should be established with high accuracy (not worse than 15 angular minutes). Moreover, the plane of the antenna system should be parallel to the terrestrial surface at cruise flight (taking into account a static pitch) with an accuracy better than 0.5° (Kolchinsky, Mandurovsky, and Konstantinovsky 1975).

A Doppler radar frequency metre usually provides the search, detection, and capture of a Doppler signal as well as center-frequency tracking. It is implemented as a kind of follow-up system. Usually, however, this frequency metre does not measure Doppler frequency directly. It is a narrowband automatic frequency control system that forms a pulse sequence with the help of its own generator, and the pulse repetition frequency of that sequence tracks the Doppler frequency $F_D$. The filter pass-band of such a system is close to the signal spectrum width, which makes possible an instrumental error as small as 0.1–0.2% at an SNR of few decibels. Pulse signals that are formed at the output of the frequency metre are easily converted to a code for further digital processing in a flight computer.

### 4.2.6 SOURCES OF DOPPLER RADAR ERRORS

Airborne Doppler radar errors are analyzed in detail by Fried, Buel, and Hager (1997). Among them the following kinds of errors are considered:

1. Doppler fluctuation error, which is caused by the noise-like nature of the Doppler signal spectrum due to the backscattering properties of terrain (usually less than 0.05%)
2. Errors due to inexact knowledge of beam direction (some estimates give this as about 0.08%)
3. Transmission frequency fluctuation error (typically negligible)
4. Frequency tracker error (less than 0.05 knots at 6 dB or higher SNR)
5. "Altitude hole" error due to the effect of modulation (normally less than 0.02%)
6. Land-terrain error, which is normally very small unless a very small antenna with a large beam-width is used
7. Three types of overwater errors, namely non-accurate calibration error, sea-current error, and surface-wind-induced water motion error. If no special measures are used this can be up to 0.6% or more (1-sigma), however known compensation techniques (Fried, Buel, and Hager 1997) reduces it radically
8. Maneuver-induced errors caused by acceleration and turning that can be reduced by using smoothing times of about 0.1 s
9. There are also errors of attitude stabilization or conversion from vehicle to ground coordinates, calibration errors, installation errors, and errors in data conversion and readout. The estimates given by Fried, Buel, and Hager (1997) for high performance Doppler radar velocity error (over land) gives a typical total orthogonal velocity component error of 0.10% + 0.05 knots (1-sigma).

### 4.2.7 EXAMPLES

One of the first examples of Doppler radar equipment developed, the AN/APN-81, which became operational in the mid-1950s, radiated an average power of 50 W and consumed 1700 W. The weight of the navigation system together with its computer was about 320 kg. Doppler velocity and drift indicator DISS-013, developed in the USSR in the 1960s, was an FMCW Doppler radar that measured ground speeds of 180–1300 km per hour with an error 0.25% (2-sigma) and a drift angle of –30° to +30° with an error of 15 angular minutes. The weight of the double set with interface and damper strut was 62 kg, and the total consumed power was 290 W.

Doppler systems performing the equivalent functions in 1996 weighed approximately 5 kg including the antenna, all the electronics, and an MIL-STD-1553 data bus interface (Spitzer 1997). These radiated an average power of 20 mW and consumed 20 W.

Because commercial airlines have shifted to non-radar forms of navigation, current DSGAs are designed principally for helicopters and military aviation. Particular examples of commercial Doppler navigation radars operating at 13.3 GHz are described by Saunders (1990). One system employs a Gunn oscillator as the transmitter, with an output power of 50 mW, and utilizing a 30-kHz modulation frequency. A single microstrip antenna is employed. A low-altitude apparatus (below 15,000 ft), the unit weighs less than 12 lbs (5.4 kg). A second radar cited has an output power of 300 mW, dual antennae, dual modulating frequencies, and an altitude capability of 40,000 ft.

## 4.3 AIRBORNE WEATHER SENSORS

### *4.3.1 WEATHER RADAR AS MANDATORY EQUIPMENT OF AIRLINERS AND TRANSPORT AIRCRAFT*

For flight safety, any aircrew needs opportune and reliable information about the location of dangerous weather zones. Airborne Weather Radar (AWR) is a radar system used in commercial and general aviation to detect Dangerous Weather Phenomena (DWP). It can also be used for ground mapping to aid in navigation. Airborne radar is the basic source of information on meteorological conditions *en route*. It provides the crew with information about the presence of cumulonimbus clouds and such DWP as thunderstorms, severe turbulence, wind shear, hailstones, downpours, and other phenomena important for flight safety. Thus, airborne radar allows the pilot to select a reasonable trajectory for the avoidance of dangerous zones in bad weather conditions.

Decreasing the probability of encountering zones of DWP during flight is one of the major ways of increasing flight safety. Nowadays, with air traffic volumes rapidly increasing, the role and financial viability of reliable weather sensors for airplanes flying under bad weather conditions is also increasing. It is also important to note that the number of delayed and canceled flights increases considerably faster than air traffic density for a fixed airport capability. This tendency is connected with increases in flight timetable density, when even in good weather it becomes more difficult to find a slot needed for the departure of a delayed flight (Baranov and Yanovsky 1976). Weather radar, both airborne and ground based, helps to solve this problem.

Summarizing, AWR is the most important safety-related item of equipment in an aircraft. The necessity of using a weather sensor in any aircraft follows from the need to provide a high level of safety and good flight regularity. In the United States it is required by the Master Minimum Equipment List (MMEL) and is fully harmonized with the Federal Aviation Regulations (FAR) and also the Joint Airworthiness Requirement (JAR) including JAR-AWO, that is, the JAR for All-Weather Operations. It is incorporated into the documents of corresponding bodies in many countries, for example in the airworthiness requirements NLGS-3 in Russia, and in international recommendations as document ARINC-708A, the relevant ICAO documents, and standard MOPS (MOPS 1993).

### *4.3.2 MULTIFUNCTIONALITY OF AIRBORNE WEATHER RADAR*

Historically, the first function of airborne radar (in civil aviation) was related to autonomous navigation using characteristic landmarks like cities, lakes, rivers, coastal strips, and so forth.

Until recently, such airborne radar systems provided pilots with navigational information, or more exactly, Earth surface mapping. Nowadays, this assignment can be considered as an auxiliary one because air navigation can be accomplished more accurately using global systems. However, the use of a global system also implies a loss of autonomy that is inadmissible in some cases. Moreover, there are still vast regions in the world that are inadequately covered by navigation equipment. In such cases it is necessary to have access to navigational information that allows the determination of an aircraft's position with respect to the geographic map. Landmark coordinates relative to the airplane that are measured by airborne radar make it possible to update a flight computer for more exact and efficient *en-route* flight.

Weather information is also necessary for successful navigation. Principally, an AWR is used in modern navigation systems as a sensor for meteorological information. Nowadays, the development of AWR is mainly associated with providing growing functionality for the detection of different kinds of dangerous weather during flight. There are actually several meteorological functions of AWR, and these functions gave the modern name of the AWR, which is a *multifunctional system*.

As a multifunctional system, weather radar provides the pilot with visual information on the meteorological situation in the forward semisphere during the flight, and also provides an autonomous means of Earth surface observation (Belkin, Dzubenko, and Yanovsky 2001). In addition, when flying in a complex environment, AWRs are used to avoid obstacles like mountains, hills, and towers. They also assist in some navigational tasks in flight, for example the delivering and dropping of loads, and for search-and-rescue operations. Thus, though the name "weather radar" is given because of the main class of AWR meteorological functions, such equipment normally has several operational modes, not only those related with weather observation. This capability improves tactical possibilities for airliners, transport aircraft, airplanes for search-and-rescue service, local airways, and military transport aircraft. Another specific functional mode of AWR is interaction with ground-based responder beacons. New functions for advanced AWR include the detection and visualization of runways on landing approaches as well as the visualization of taxiways and obstacles thereon.

Though not all of the abovementioned functions are implemented in a particular airborne radar system, AWR is nevertheless always a multifunctional system that provides both Earth surface surveillance and weather observation. Weather radar should at least be able to detect clouds and precipitation, select zones of meteorological danger, and show radar images of the surface in the map mode. In accordance with ARINC-708A and DO-220 (MOPS 1993) recommendations, at least the following operational modes of an airborne weather radar system should be implemented:

- *Weather*—reflectivity of meteorological objects should be displayed
- *Map*—reflectivity of the Earth's surface should be displayed
- *Wind shear*—wind shear zones should be detected and displayed
- *Turbulence*—turbulent zones should be detected and displayed

Different modes can be combined: for example, weather and turbulence, weather and wind shear, and turbulence and wind shear. Normally, the wind shear detection mode is automatically enabled on takeoff and landing.

A major aspect of the operational efficiency of AWR is the reliability of dangerous meteorological phenomena detection.

### *4.3.3 METEOROLOGICAL FUNCTIONS OF AWR*

Sometimes, people erroneously believe that the air traffic control service ground-based radar detects DWP better than AWR. However, pilots are responsible for avoiding DWP in flight (Melvin 1987) and this approach is very well coordinated with the modern "free flight" concept (Maracich 2005). As was explained in Section 4.3.2, at least three modern AWR modes perform functions related to weather information. Analysis of weather radar functionality and the latest AWR requirements show that AWR can be designed to solve the following tasks for extracting meteorological information (Yanovsky 2006):

1. The visualization of clouds and precipitation in the area of coverage, and the determination of levels of hazard in its different sectors (or resolution elements) based on the reflectivity of these sectors
2. The detection of dangerous turbulence zones in clouds and precipitation
3. The detection of wind-shear zones at takeoff and on the landing glide path
4. Representation of the vertical structure of a weather formation by imaging its vertical profile at the selected azimuth
5. The detection of strong weather formations (like high-intensity rain) that are located behind zones of weak weather formations (like slight rain)
6. Compensation for signal attenuation in weather formations
7. The prevention of pilot error during decision making about the absence of danger behind a zone of weak intensity masking a zone of strong intensity (the display of ambiguity zones in such sectors)
8. The detection of a dangerous weather formation situated on the aircraft heading by generating a warning signal when switching off the display imaging mode (automatic weather detection modes)
9. Determination of the coordinates of a detected weather formation and danger zones within it and
10. Clutter (unwanted reflection from the Earth's surface) suppression during the detection and risk analysis of weather formations.

No real AWR can completely solve all the listed tasks, and furthermore, this list does not claim to be exhaustive. However, it can be supplemented with further modes such as those for hail zone and probable inflight icing detection.

### *4.3.4 PRINCIPLES OF DWP DETECTION WITH AWR*

#### *4.3.4.1 Developing Methods of DWP Detection*

The first AWRs could just detect precipitation and clouds that reflected sufficiently powerful signals. However, danger still existed because of problems associated with the distances of weather phenomena because only a qualitative integrated evaluation of meteorological situations was available. Methods for the quantitative evaluation of the *radar reflectivity* (RR) of weather formations were therefore developed, and various schemes were proposed for signal correction for distance, and for the compensation of signal attenuation in precipitation. The next step was to evaluate the degree of danger, and a working principle was derived from past experience: a higher RR corresponds to a higher probability of danger. Data presentation in the form of contour indication (equal RR lines) was developed. As before, no separation of types of danger was possible at that stage of AWR development. However, it was followed by obtaining some relationships between RR parameters and turbulence (Kessler, Lee, and Wilk 1965). On this basis the technique of reliability estimation of dangerous turbulent zone detection was developed, and the first quantitative estimates of turbulent zone detection by RR were obtained. However, at a detection probability $D$ of 0.9, the false alarm probability $F$ was too high at 0.4 (Yanovsky and Belkin 1977).

Studies have been carried out (Lhermitte 1973; Yanovsky 1974) on the application of new information parameters, particularly including Doppler spectrum width or correlation time, and correlation coefficient at zero shift, instead of, or in addition to, the RR. During the 1980s significant advances were made in airborne pulse Doppler radar, greatly enhancing its ability to distinguish various features in the returns. Digital technology underwent tremendous advances, which made practical the signal and data processing required for modern radar. This resulted in the implementation of methods for Doppler spectrum width evaluation in both frequency and time domains. Also, new modes of turbulence detection, and later wind shear detection, were applied and introduced to commercial Doppler AWRs (Commercial Avionics Systems 1996). Pulse-Doppler methods for ground-based observation are described in detail by Doviak and Zrnic (1993).

The detection of hail zones during *en-route* flight is possible using polarization techniques in AWR (Shupiatsky and Yanovsky 1994). This capability is sufficiently desirable to be implemented in future AWR in combination with another Doppler characteristic comprising the Doppler-polarimetric technique (Yanovsky, Unal, and Russchenberg 2005).

Another important advance is in the development of passive devices for thunderstorm activity detection based on the detection and analysis of atmospheric electric discharges or lightning radiation (Seymour and Baum 1978). Such lightning sensor systems can be used separately or they can be successfully combined with AWR.

Thus, the identification of different kinds of danger such as turbulence, wind shear, hail, and lightning has become a reality. However, whereas on the one hand such further information about the meteorological situation can result in improved weather-oriented knowledge for the pilot and hence better danger-avoidance accuracy, on the other hand, pilots, who must make quick decisions, are overloaded. In the majority of cases they need only those parts of the available information that are really necessary at the relevant moment. It is therefore much better if the pilot is able to improve his knowledge by accessing more details in dialog mode when he considers it necessary. The design of integrated presentations of danger on the basis of comprehensive detection and evaluation of the level of each source of danger separately is predicted as a further development in airborne equipment for DWP detection. Summarizing, information should be presented to a pilot in a convenient generalized form along with the capability of transforming it to more detailed form in dialog mode (Yanovsky 1991).

Regarding the future evolution of airborne sensors and systems for weather surveillance in flight, it is possible to conclude that advances in the various DWP detection techniques have arrived at a point where integrated danger evaluation is necessary, as indeed it was in the very beginning when RR was the only measured value. However, because a number of information parameters can now be measured, it should be done at new and significantly higher levels (Yanovsky 2006).

Methods for the radar detection and identification of each component of weather danger are briefly described below.

### 4.3.4.2 Cumulonimbus Clouds and Heavy Rain

Heavy rain is the first and the simplest component of dangerous weather from the point of view of radar method complexity. The detection of heavy rain, other kinds of precipitation, and the presence of cumulonimbus clouds is done by RR measurement: the higher the RR the higher is the equivalent precipitation intensity $Z$ (Atlas 1964). The well-known empirical formula $Z = \alpha I^{\beta}$ relates the RR (in mm$^6$ m$^{-3}$) with rain intensity $I$ (in mm per hour), where $\alpha$ and $\beta$ are coefficients that can be considered as more-or-less stable (one version of these parameters gives $\alpha = 200$ and $\beta = 1.6$). AWR can estimate the RR of each resolution volume by measuring the average reflected power $P_r$ that is received by the radar receiver in accordance with a simplified equation:

$$P_r = \frac{CZ}{R^2}|K|^2 \tag{4.5}$$

where $R$ is the radar range, dimensionless factor $|K|^2$ depends on the dielectric qualities of radar scatterers or *hydrometeors* (0.93 for water), and dimension factor $C$ characterizes the energetic potential of the radar. Compensation should be made for range action and attenuation action (not shown in the simplified equation), and a calibration should be made for accurate measurements. In the 1970s, a so-called multicontour indication was developed for reflectivity analysis and information representation as is shown in Figure 4.4 (Baranov et al. 1976). Nowadays color coding is used, normally green, yellow and finally red as the highest level of RR.

More sophisticated and potentially more accurate polarization methods for rain intensity measurement have also been developed (Bringi and Chandrasecar 2001), but these still await application in AWR.

### 4.3.4.3 Turbulence Detection

Atmospheric turbulence is one of the crucial meteorological factors affecting aircraft behavior in flight. The majority (97%) of dangerous turbulent zones (DTZs) in the troposphere is associated

**Figure 4.4.** Images of a horizontal section of the same cumulonimbus on the AWR screen. (Right-hand image shows the cloud when the multicontour indicator is switched on).

with clouds and precipitation, which are detectable by radar.[1] Nevertheless, it does not mean that all detected clouds contain DTZ, so the pilot should not avoid clouds, but find a safe passage through them. A major aspect of the operational efficiency of AWR is the reliability of DTZ detection. Detecting turbulence is difficult, because the RR involved can be rather weak. Hence, noise and interference can essentially influence the reliability of the wanted information, resulting in rather short detection ranges. At the same time, one of the major applications of DTZ detection is during cruise flight at maximum airspeed. These factors require large detection ranges if the turbulence detector is to be of any practical use. If the AWR does not accurately detect DTZ, it is necessary to use an excessively cautious decision-making strategy to ensure an acceptable flight safety level. This leads to a decrease in flight regularity, an increase in flight time, unproductive fuel expenditure, and deterioration of other air transport economic parameters.

The Doppler spectrum (DS) parameters of the reflected signal are the most obvious indicators of DTZ. These depend on the distribution of the radial velocities of particles weighted on their radar cross sections (RCSs). The relevant theory (Doviak and Zrnic 1993) relates the spectral parameters of signals reflected from weather objects with moving scatterers. Air eddies involve radar scatterers (hydrometeors) in the motion of the turbulence. This is why turbulence increases the velocity dispersion of such scatterers and leads to widening of the DS. However, at least three circumstances can decrease the magnitude of the relationship of turbulence intensity with DS width and can therefore hamper the interpretation of data. Firstly, the captivity of drops by air whirls is not absolute; secondly, not only turbulence but also other factors influence particle motion; and thirdly, the particle RCS distribution is unknown in advance. Therefore, the connection of the DS width with the turbulence intensity in the radar resolution volume has a statistical character. In the case of noncoherent radar, the width of the intensity fluctuation spectrum (satisfying the Nyquist condition) is unambiguously connected to the DS width. Hence, it also contains information about DTZ. The DS width is inversely related to the interperiod correlation factor, which makes possible the use of correlation parameters in the envelope of the echo signal in radar meteorology and avionics (Yanovsky 1974). A similar idea is widely used in the so-called *pulse-pair algorithm* (Doviak and Zrnic 1993). Pulse-pair algorithms detect the decrease in the correlation factor of the signal and noise mix, which is why it works rather well but only when the SNR is high enough. It can be shown that the SNR should be above some minimum value determined by the pulse repetition period, wavelength, and noise power. On the basis of the synthesis theory, a two-sample adaptive algorithm was developed by Ligthart, Yanovsky, and Prokopenko (2003), which is invariant with the intensity of background scattering. The advantages of the new algorithm are especially apparent at low SNR levels.

Other approaches for turbulence intensity measuring were proposed by Yanovsky, Unal, and Russchenberg (2005). One of them is related with the estimation of the drop fall velocity contribution to the DS and subtracting this contribution from the measured value of DS width. This increases the accuracy of turbulence estimation. Another possibility is based on the promising Doppler-polarimetric approach (Unal Moisseev, Yanovsky, and Russchenberg 2001).

Turbulence detection is successfully implemented in modern AWR, for example, by Honeywell Aerospace (RDR–4B 2007; RDR–4000 2007) and Rockwell Collins (FMR–200X 2008).

[1] Clear air turbulence should be detected by other means, for example, by lidar.

#### *4.3.4.4 Wind Shear Detection*

Wind shear is one of the most serious threats to flight safety. It refers to a wind speed or direction change affecting an airplane over a specific distance or period of time. Whereas wind shear is usually not strong enough to be hazardous to an airplane in *en-route* flight, a certain subset of wind shear may be critical to flight safety during low-altitude, low-speed flight, especially during landing and takeoff. A hazard to flight safety exists if energy-reducing wind shear removes airplane energy faster than engine thrust can restore it. Under such conditions, the airplane is forced to either lose airspeed or descend. A weather condition known as a *microburst* is a major cause of hazardous low-altitude wind shear, and is formed when a column of air at high altitude quickly cools due to evaporation of ice, snow, or rain. This cooling air becomes denser than the surrounding atmosphere and falls rapidly to the ground. Upon nearing the ground, the downward moving air spreads rapidly in all directions away from the descending core. Inadvertent encounters with low-altitude wind shear have been a major cause of transport airplane accidents and passenger injuries and fatalities (Wolfson et al. 1994).

At the end of the 1980s, NASA and the FAA signed a Memorandum of Agreement establishing the NASA/FAA Airborne Wind Shear Program to investigate the feasibility of remote airborne wind shear detection. Terminal Doppler Weather Radar (TDWR) systems were developed to determine the location and severity of low-altitude wind shear phenomena and other weather hazards (Evans and Turnbull 1989). However, whereas the principles of ground-based wind shear detection systems cannot be simply applied onboard, AWR can benefit from radar airborne location methods. Wind shear phenomena, including microbursts, influence the flying aircraft via (a) horizontal wind shear, (b) vertical wind shear, and (c) downflow. Each of these components individually or some combination thereof can cause a critical altitude loss at landing or a corresponding problem at takeoff, and depend also on aircraft properties in accordance with flight mechanics. That is why a dimensionless parameter called the $F$-factor was developed (Bowles 1990) and this is described by the following formula:

$$F = \frac{1}{g} \cdot \frac{dV_{wh}}{dt} - \frac{V_{wv}}{|\bar{V}|} \tag{4.6}$$

where $g$ is gravitational acceleration;

$V_{wh}$ is the horizontal component of wind velocity along the flight path;

$V_{wv}$ is the vertical component of wind velocity; and

$\bar{V}$ is the airspeed vector relative to the air in which the aircraft moves.

The Bowles $F$-factor is now used by onboard sensors for the detection of hazardous wind shear, having been developed, tested, and accorded FAA certification (Proctor, Hinton, and Bowles 2000). It is recommended in an updated version of ARINC 708A as the basis for decision making via wind shear detection with AWR. Airborne equipment designed for wind shear detection should detect zones of changing wind in both horizontal and vertical planes and generate appropriate alarm signals. Such signals should be clear, automatic, short, and expressive, that is, suitable for immediate and correct interpretation by the pilot. Transition to wind shear detection mode should be made automatically without pilot intervention during takeoff and landing. The AWR designer should establish that the pilot be informed about wind shear hazard

10 to 40 seconds before entering the relevant zone. A time less than 10 seconds is not normally sufficient for an appropriate reaction in the "pilot-aircraft" system; and a time of more than 40 seconds is considered too long to guarantee that a new significant change in the atmosphere has actually occurred.

Technological problems in *F*-factor measurement as well as the satisfying of other requirements have been solved, and the wind shear mode has been implemented in airborne pulse-coherent (pulse-Doppler) radars (RDR - 4B 2007; FMR-200X 2008).

The possibility of measuring wind shear by using non-coherent or quasi-coherent radar is discussed by Pokrovsky, Belkin, and Yanovsky (2005). This paper proposes applying wind shear detection capability to a noncoherent AWR with a magnetron as a transmitter, and the decision on wind shear detection is to be made on the basis of the *F*-factor. The problem of remote estimation of the necessary wind components by means of non-Doppler airborne radar is solved by digital intrinsic coherence, taking into account any real instability in the radar system.

#### *4.3.4.5 Hail Zone Detection*

A wide variety of different hailstones can be found at high levels in the troposphere. These are safety hazards for aircraft and should be detected in a timely manner during flight. The problem of hail zone recognition with AWR was raised long ago, and a polarimetric method was proposed for this purpose (Shupiatsky and Yanovsky 1990a). This principle was implemented in a polarimetric research AWR (Yanovsky and Panits 1996) designed for an air laboratory based on an Ilyushin Il-18 aircraft, and successfully tested. However, industrial AWR still does not have polarimetric capability, though this approach is promising. Mathematical models and appropriate software were developed to calculate scattering parameters, RCS and polarimetric characteristics of hydrometeors (Braun and Yanovsky 2004), particularly for single hailstones and ensembles of hailstones. Radar parameters and sounding modes (wavelength, polarization, antenna pattern, and elevation) as well as scatterer properties (size distribution, shape, permittivity, and distribution of orientation parameters in orthogonal planes) should be taken into account. This is a complicated problem that can never be solved exactly. However, for remote sensing purposes, and particularly for hail detection, it is necessary to have the capability of estimating different radar measurables in different situations. RCS at different polarizations, including cross-polarized components, differential reflectivity (DR), and linear depolarization ratio (LDR) plus other polarimetric measurements, have been modeled. Peculiarities of the polarization parameters in returns from hail stones are rather different compared with those from raindrops. Model verification using experimental data in specific cases, as well as the application of results from hail detection and the remote estimation of hailstone maximum size, have been discussed by Braun and Yanovsky (2004).

Several algorithms for hail zone detection have been developed on the basis of different statistical characteristics using definite models (for example, Yanovsky, Shupiatsky, and Kapitalchuk 1995). A more sophisticated nonparametric method of hail detection was proposed by Yanovsky, Sinitsyn, and Braun (2002). Fuzzy logic and neural network algorithms developed for hydrometeor type recognition are very effective in the case of increasing numbers of measurable variables (Liu and Chandrasekar 2000; Ostrovsky, Yanovsky, and Rohling 2007).

#### 4.3.4.6 Probable Icing-in-flight Zone Detection

Aircraft icing is one of the most frequent and dangerous phenomena that can be encountered by any type of airplane or helicopter in cloud and precipitation. Strong icing not only makes flight and the aerodynamic quality of an aircraft worse, but can also lead to serious accidents. Standard AWRs can only associate zones of probable icing with the presence of clouds and precipitation *en route* when flying above zero isotherms. Furthermore, the reliability of such methods of localizing probable icing zones also appears inadequate because in such cases practically all detected clouds should be considered as dangerous. Polarization methods for the detection of probable aircraft icing was proposed by Shupiatsky and Yanovsky (1990b). Analysis and further development of the method and its application to polarimetric AWR for the remote detection of supercooled water in clouds and precipitation is presented by Yanovsky (2004). This method uses a difference in polarization properties between ice and water clouds. (The combination of supercooled liquid water and negative cloud temperature causes significant aircraft icing.) Several algorithms have been proposed, specifically (a) a simple algorithm based on measuring a linear depolarization ratio (Kropfli et al. 2002; Shupiatsky and Yanovsky 1994), (b) a heuristic-logical polarimetric algorithm (Shupiatsky and Yanovsky 1990b; Yanovsky 2004), (c) a statistical polarimetric algorithm (Yanovsky 2004), (d) a fuzzy logic algorithm, and (e) a neural network algorithm (Pitertsev and Yanovsky 2006).

Modes of hail detection and icing-in-flight zone detection based on the polarimetric approach are still awaiting implementation in AWRs, none of which are currently polarimetric.

### 4.3.5 SURFACE MAPPING

#### 4.3.5.1 Comparison of Radar and Visual Orientation

Surface mapping with AWR locates the orientation of an aircraft in flight by observing an imaged radar map of the surrounding terrain on a radar display. Such a method resembles visual orientation, which involves the observing of characteristic ground-based landmarks or the specificities of terrain relief that can serve as flight reference points (this is actually a traditional method of navigation known as *pilotage*). However, the range to the useful landmarks is different in the two cases. Landmarks located at 15–20 km from an aircraft (roads, lakes, parks, boroughs, rivers, etc.) can be resolved visually by human eyesight, whereas resolution by AWR is much poorer, and AWR cannot distinguish between numerous visual landmarks. However, AWR compensates for this by being able to observe objects located significantly farther away. Objects such as towns, railways, arterial highways, lakes, dams, bridges, high-tension transmission lines, coasts, islands, woodland belts, forest plantations, large rivers, hills, and cities, can all be detected at distances of 100–600 km. These can serve as good radar landmarks provided their positions are indicated on geographic maps. Airborne radar is also advantageous due to the possibility of measuring the coordinates of such landmarks. Moreover, it keeps its functionality in zero visibility, for example, at night and inside clouds. Radar landmark coordinates measured with AWR in map-mode allow for correcting the flight computer and hence dropping a load accurately at a given point. Thus, airborne radar improves aircraft performance capabilities in transport aviation, search and rescue services, and over local routes.

### *4.3.5.2 The Surface-Mapping Principle*

Assume that the antenna beam is directed from the aircraft to the Earth's surface at an angle $\gamma$ as shown in Figure 4.5. Differing segmental surfaces within the irradiated area can be seen in the upper panel of the picture, in particular meadows, forests, bodies of water, and buildings. Such different segmental surfaces possess different reflecting properties because of different dielectric behavior, surface properties, and configuration peculiarities. Portions of the Earth's surface can be regarded as distributed target areas. The reflection properties of such targets are described by $\eta_0$, the specific radar cross section, that is, the RCS per unit area. This is often called the *differential scattering cross section* or *scattering coefficient.*

Under given conditions of illumination, the reflection power of a grassy meadow $\eta_0^{(1)}$ is less than that of a forest region $\eta_0^{(2)}$; whereas that of a water body $\eta_0^{(3)}$ is less than that of a grassy meadow. Normally, the best reflection power is characteristic of buildings, constructions, and industrial objects $\eta_0^{(4)}$. That is, $\eta_0^{(4)} > \eta_0^{(2)} > \eta_0^{(3)} > \eta_0^{(1)}$. Given a more powerful reflected signal resulting from a higher $\eta_0$, the signal at the output of a receiver (after an envelope detector) can be represented as is shown in the lower panel of Figure 4.5. The sounding waveform is indicated at the beginning ($t = 0$) of the pulse repetition period, $T$. The first reflected signal comes from the surface point located directly under the aircraft ($\gamma = 0$), and its delay time is $2H/c$, where $H$ is the flight altitude. Later on the time axis, signals appear that are reflected from surface points situated further and further away with increasing angles of $\gamma$, and each time delay is $t_d = 2H/c\sin\gamma$.

Thus, the output voltage of an AWR receiver for Earth surface illumination carries information about the character of the surface (received reflected power, $P_r$), the target range (time delay, $t_d$), and the target azimuth (current antenna position). During antenna scanning, a radar beam sequentially illuminates narrow sectors in radial directions, and finally a composite image of the surveyed surface is formed. The power of the received signal $P_r$ serves as another informative parameter and is a function of the coordinates of the resolution element on the observed

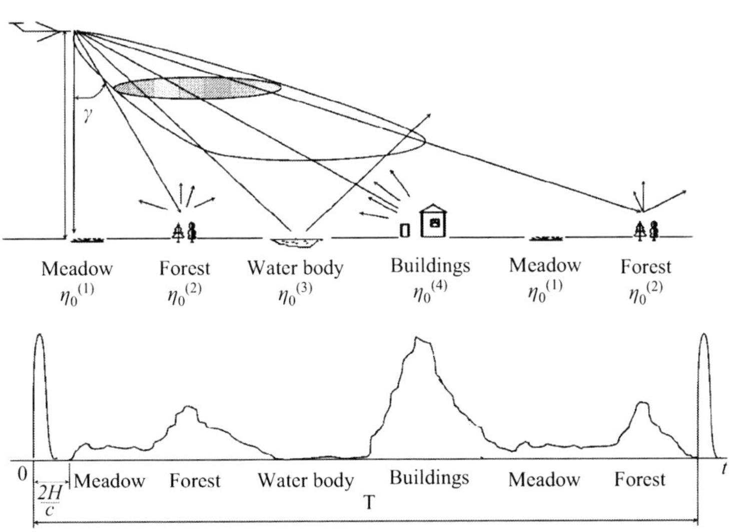


**Figure 4.5.** Forming radar reflections from the Earth's surface.

surface (range $R$ and azimuth $\alpha$), the aggregate of the radar parameters $\vec{C}_{\text{RADAR}}$, and the conditions of observation $\vec{Y}_{\text{conditions}}$:

$$P_r = f\left(\underbrace{R, \alpha}_{\eta_0}, \vec{C}_{\text{RADAR}}, \underbrace{\vec{Y}_{\text{conditions}}}_{\gamma}\right) \tag{4.7}$$

Resolution elements that have different coordinates $R$ and $\alpha$ are characterized by different scattering coefficients, $\eta_0(R,\alpha)$, and the basic parameter of the conditions is the viewing angle $\gamma$ that appears in the formula inside the brackets.

All the information on the range $R$, the azimuth $\alpha$, and the power $P_r$ is displayed as a conditional map of the surface. If the brightness increases when $P_r$ increases, then the darkest elements of the screen correspond to water bodies and the brightest elements correspond to artificial constructions. Instead of relying on such traditional black-and-white images however, modern AWRs use a coded version of the received power $P_r$ leading to a multicolor display.

#### 4.3.5.3 *Reflecting Behavior of the Earth's Surface*

Values of the scattering coefficient are often expressed in decibels (dB) relative to 1 $m^2\,m^{-2}$ and the relationship between absolute value of $\eta_0$ and the relative value $\eta_{0\text{dB}}$ is:

$$\eta_{0\text{dB}} = 10 \lg \eta_0; \; \eta_0 = 10^{0.1\eta_{0\text{dB}}} \tag{4.8}$$

Scattering coefficients for the surface of the Earth depend on wavelength, polarization, weather, season, and so forth. Statistical data on the scattering coefficients for different surfaces under various conditions are known and can be found in handbooks (Moore 1990). Table 4.1 presents a range of scattering coefficient magnitudes $\eta_{0dB}$ for different area distributed targets at a wavelength $\lambda = 3.2$ cm, horizontal polarization, and viewing angle $\gamma = 80°$. Note that $\eta_{0\text{dB}}$ takes negative values because $\eta_0 < 1$ for natural conditions. However, in conversations with experts and sometimes even in the literature, the minus sign may be omitted. This should be taken into account to avoid misunderstandings during calculations.

**Table 4.1.** Scattering coefficients for different surfaces at $\lambda = 3.2$ cm, $\gamma = 80°$, and horizontal polarization

| **Type of Surface** | $\eta_{0dB}$ **Range of Scattering Coefficient, dB** |
|---|---|
| City | –13 ... –25 |
| Cultivated land | –19 ... –33 |
| Asphalt | –45 ... –49 |
| Concrete | –52 ... –54 |
| Calm sea | –52 ... –56 |
| Sea (3-point wind) | –34 ... –38 |

### 4.3.5.4 *The Radar Equation and Signal Correction*

In the case of surface-mapping with AWR radar, an equation can be written as follows (Yanovsky 2003):

$$P_r = \frac{C_R \eta_0}{R^3} \sin \gamma \tag{4.9}$$

where $C_R$ is a dimension factor that depends on transmitter power, antenna pattern, wavelength, and pulse duration; $R$ is the range-to-current resolution element.

This equation shows that the receiving power is inversely proportional to the cube of the range. However, when flying at constant altitude, the viewing angle and range are interrelated, so that the dependence $P_r(R)$ can be complicated. Consequently, in order to adequately represent a surface map on the screen, it is necessary to exclude the dependence of the receiving power on the range $P_r(R)$. This can be done by digitally enabled automatic receiver gain-control. Also, a method of correction based on special fan antenna patterns (cosecant-squared, or barrel antenna) is well known and widely used in airborne radars with reflector antennae. Unfortunately, this method is not applicable in the case of AWR with passive slot arrays.

### 4.3.5.5 *Automatic Classification of Navigational Landmarks*

When flying at a constant altitude, a cosecant-squared antenna will provide a balance in the brightness of radar images from equally reflective surface cells located at different distances. However, if the altitude is changed, the rationality of the cosecant-squared pattern becomes worse. This results in the need for additional tuning of the radar during flight. The transition to passive slotted antenna arrays with symmetrical needle-like antenna patterns not only has improved weather modes but has also worsened the quality of the radar map. For distinguishing navigational landmarks, tuning on a particular landmark is often required. The application of widened antenna beams and temporal gain controls partially mitigates these deficiencies, though adaptive gain control is required depending on the flight mode and changeable radar parameters.

Terrain complexity and the presence of a number of reflecting objects on that terrain, neither of which are important for navigation in most cases, nevertheless clutter up the radar map, so complicating the separating out of useful information.

To solve this problem, a classification of landmarks and principles of indicating them by automatically forming a radar map of the Earth's surface are proposed by (Belkin, Dzubenko, and Yanovsky 2001). The idea is simple: (1) detecting and distinguishing typical classes of landmark by using special algorithms for every class; and (2) synthesizing a map without unwanted objects. The proposed classification includes aqueous, area, and point landmarks, in addition to the background of an underlying surface. The radar map of the terrain is then synthesized by the integration of the selected landmarks.

The developed method and subsequent device involve rather simple engineering implementation (Buran A-140 2007), which improves the quality and self-descriptiveness of the radar map. Also important is the imaging of the surface map, which can be significantly simplified. Simple algorithms for the relevant signal processing are described by (Belkin, Dzubenko, and Yanovsky 2001). However, there are possibilities for further improvement, and more sophisticated radar signal parameters can be used to improve the recognition quality. Specifically, it is proposed to use polarization parameters for improving the identification of landmarks.

### 4.3.6 AWR DESIGN PRINCIPLES

#### 4.3.6.1 The Operating Principle and Typical Structure of AWR

Typically, AWR is an active monostatic pulse primary radar system. Its operating principle is based on the use of secondary radiation, specifically the backscattering of electromagnetic waves by objects like hydrometeors, clouds, precipitation, and the Earth's surface (Shirman et al. 1987). A simplified functional diagram that explains this operating principle is shown in Figure 4.6, which represents a version of the classical scheme for a pulse radar that also takes into account features related to the airborne installation of the system.

Another feature of AWR is related to its airborne location. This results in the possible influence of inflight maneuvers on the quality of the radar data. Incorporating an Antenna Stabilization System as a part of the AWR allows compensation for the influence of aircraft roll and pitch on the radar images. Signals from the aircraft attitude sensors (see Chapter 6) are used as initial information for such a system, as is shown in Figure 4.6. A Control Panel enables remote operational control of the AWR during both flight and maintenance.

This functional diagram does not reflect a real AWR block structure, but simply illustrates the operating principle. Here, a clock driver unit coordinates the work of all other units, and a transmitter generates the microwave electromagnetic sounding waveform, usually consisting of short pulses of duration $\tau$ with interpulse intervals $T$. A duplexer automatically connects the antenna to the output of the transmitter (during the pulse generation time $\tau$) and to the input of the receiver (in the remaining time). Hence, the duplexer switch frequency is equal to the pulse repetition frequency. The antenna is designed to form the most desirable antenna pattern, which is normally symmetric and rather narrow in order to radiate sounding pulses of electromagnetic energy and to receive reflected energy scattered by the objects under observation. A receiver detects the signal over the background of interference and noise by filtering and amplifying the initial useful information. Further signal processing is implemented in the Processing Unit, which extracts information including data on hazardous weather phenomena.

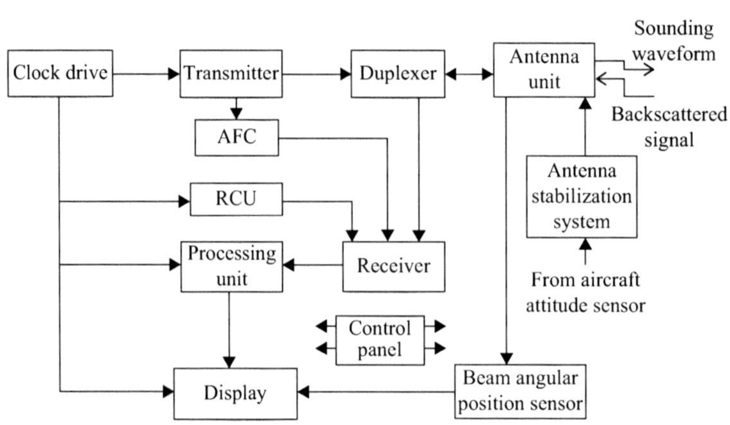


**Figure 4.6.** Simplified functional diagram of an airborne weather radar.

The display unit provides the pilot with radar information in polar coordinates (*azimuth range*) as well as auxiliary information. The Beam Angular Position Sensor provides the display with data on the current antenna beam direction and hence the angular coordinates of the current resolved volume of the object under study. Range information is obtained from the time delay.

An Automatic Frequency Control (AFC) system implements automatic receiver tuning in accordance with the transmitter frequency. In the case of coherent radar, this link—which connects the transmitter and receiver—provides a reference signal for Doppler measurements.

An important feature of AWR is the provision of qualitative reflectivity measurements to determine the danger level of clouds and precipitation. For this purpose, the Range Compensation Unit (RCU) provides a sensitivity time control to remove any receiver power dependency on range, that is, to obtain range correction of signal magnitude as the inverse square of range according to simplified Equation 4.5 in Section 4.3.4.2. The receiver amplification factor is set to its minimum at the beginning of each repetition interval, when a target range is minimal, and then to increase it step by step up to a maximal value when the target range is great enough.

### *4.3.6.2 AWR Structures*

Airborne radars are block-designed systems. In the modern digital AWR, the signal—at least after the envelope or phase detector—is digitized. Signal processing and radar data forming are implemented digitally and all interblock electrical connections are in digital form apart from the microwave connection between the antenna and the transceiver. Several configurations of AWR structure are recommended by ARINC-708, where the main blocks are the antenna and the transmitter–receiver that are present in any structure. One such configuration is shown in Figure 4.7.

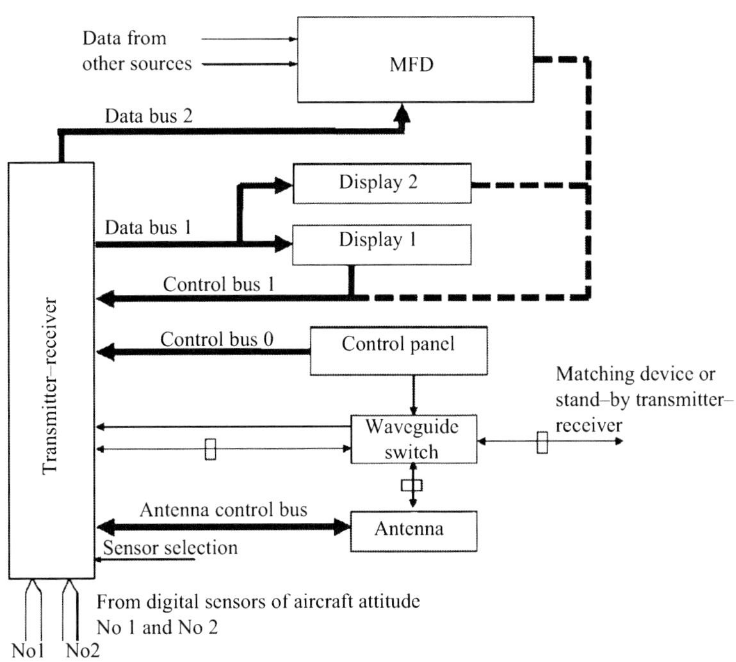


**Figure 4.7.** A possible structure for a digital AWR.

This structure contains all the functional elements of Figure 4.6. In particular, the transmitter, receiver, duplexer, AFC, RCU, and antenna stabilization system are situated in a single unit, which is called the *transmitter–receiver* or *transceiver*. However, there are some peculiarities related to the system's digital control. For example, changes in receiver amplification to accord with the required temporal gain control law (provided by the RCU in Figure 4.6) are made with regard to a changeable code derived from the control panel or basic display in specific bins of the digital control word. All signal processing is carried out entirely digitally.

Digital AWRs that accord with ARINC-708 can be produced and used in different configurations, of which three are common, namely: (1) a three-unit AWR consisting of antenna, transceiver, and a combined control panel with a display that is assigned only for the tasks of weather object detection and analysis; (2) a three-unit AWR consisting of antenna, transceiver, and control panel, all display functions being carried out by a multifunctional display at progressively higher hierarchy levels; and (3) a four-unit AWR that includes antenna, transceiver, control panel, and a specialized display for weather object detection and analysis.

All the units in a modern AWR are designed to be installed within a system that can contain either single or double transceivers, control panels, and displays though a single antenna is used. The second set is used as a backup to increase the reliability of the system. The example shown in Figure 4.7 represents a single system with two data buses. The AWR units are interfaced in digital form except for the microwave connection between the antenna and the transceiver, this being normally made using a waveguide. Digital interfaces connect antenna with transceiver, and transceiver with control items and displays. Data exchange between the units and also control signal transmission are normally implemented in sequential digital form using a serial digital data interface, that is, all data and commands are transmitted without time overlap. Thus, it can be seen from Figure 4.7 that information from the control panel, or from a display combined with the control panel, are applied to the transceiver via successive control buses 0 and 1. A code control word contains 32 bits and can have two formats for transmitting all the necessary control information, totaling 64 bits. Via the transceiver, the control signals are applied to displays using data buses designed to transmit data about targets, including the coordinates and features of targets in every resolved volume, such as target reflectivity, turbulence, wind shear, and so forth. All data are represented as code words of 1,600 bits containing 1,536 bits for the data itself (3 bits for each range bin by 512 range bins), and 64 bits for control signals. This word has a well-defined structure according to ARINC-708A.

In the configuration of Figure 4.7, a system with two data buses is shown, that is, displays 1 and 2 and the MFD may receive different data. Control signals and data for each respective range and azimuth corresponding to every radar unit cell arrive at the digital memory of the display unit and are transferred to the screen on demand by the pilot. Information from aircraft attitude sensors such as gyros arrives at the antenna stabilization system located in the transceiver via low-speed buses 1 and 2.

A waveguide serves to connect one or the other of the two transceivers to the antenna. During switchover, the waveguide switch applies a blocking command to the transceiver.

#### *4.3.6.3 Performance Characteristics: Basic Requirements*

The performance characteristics of AWR should reflect the requirements in terms of flight safety, flight regularity, airworthiness, ease of technical maintenance, reliability, and other AWR operating benefits. The maximum range, coverage area, scanning interval, composition, and structure of information along with its pertinence, accuracy, resolution, reliability, and interference immunity are among such characteristics. The characteristics indicative of the technology used to achieve the desired performance include operational wavelength, transmitter power, pulse duration, pulse repetition frequency, receiver noise figure, and so forth. An overall parameter for AWR is the performance index (PI), which is calculated according to a procedure developed by ARINC.

Airborne weather radars are basically X-band systems, though the C-band has also been used. For X-band AWR applications, two carrier frequencies have been designated as 9.375 and 9.345 GHz.

Many users consider that the lowest maximum range should be 150 nautical miles, and this corresponds to a PI of 204 dB in the frequency band 9.3–9.5 GHz. Many AWRs exhibit a PI > 220 dB, which corresponds to a weather-formation detection range of more than 300 nautical miles, or 550 km. Normally, developers try to maintain the unification trend contained in the ARINC-708A recommendations. The operating conditions are also important for AWR—for example, the operating temperature range should be at least –60°C and up to 100% relative humidity at 35°C.

Information on the meteorological situation along with an indication of DWP zones should be presented to the crew in obvious and user-friendly form using polar (range, azimuth) coordinates, that is, as sections of the coverage area in the horizontal plane at flight altitude. A capability for beam tilt in the vertical plane is also required, and in recent years, automatic scanning in the vertical plane has often been provided to display vertical cloud sections at chosen azimuths.

The main mode for obtaining weather information is still horizontal scanning at the flight altitude. The coordinates of selected DWP zones are determined with help of range rings and azimuth marks. A majority of crew members believe (Yanovsky, Golubchik, and Fishman 1985) that in addition to coordinates, the displayed information should also include the type of danger, for example, turbulence, hail, lightning, wind shear, and icing. Moreover, information on the level of danger is also desirable, and 80% of crew members think that three gradations of danger are sufficient, these corresponding to a three-alternative decision strategy designed into the DWP airborne sensor, for example, "safe," "potentially hazardous," and "dangerous," or "negligible," "weak," and "strong."

The necessary updating times are related to both the flight speed and the velocity of weather object process behavior. The optimal updating time is 2 to 10 seconds depending on the flight regime. Excessive and increasing information content leads to growth in the pilot's workload and may increase the probability of error during data interpretation and decision making. In this context, the creation of effective information filters is needed, and these can be implemented as special software and hardware used interactively (Yanovsky and Fishman 1986). Such smart support tools should select and compose only those parts of the sensor data that are directly intended to define decision-making on evasive action and/or airspeed.

### *4.3.7 AWR EXAMPLES*

Leading producers of globally marketed airborne avionics equipment such as Honeywell and Rockwell Collins deal in a variety of AWRs. In Russia AWRs are produced by Kontur-NIIRS Ltd (Kontur-NIIRS 2008). Ukrainian AWRs are developed at the Kiev Buran Research Institute and are produced by the JSC Kiev Radar Plant (JSC Kiev Radar Plant 2008). Modern AWRs solve the many difficult challenges related to the airborne detection of DWP including wind shear prior to its encounter. Sophisticated digital signal processing is implemented in modern AWRs, which are really multifunctional, hazardous weather detection and avoidance systems.

An example of a Honeywell AWR is the RDR-4B Weather Radar with Forward-Looking Wind shear Detection (Weather Radars 2008a). The RDR-4B provides essential weather detection up to 320 nautical miles over ±90° of azimuth adjustment about the aircraft centerline, turbulence detection up to 40 nautical miles over ±90°, and forward-looking wind shear detection up to 5 nautical miles over ±40°. The RDR-4B, along with its cockpit display, offers crucial information during all flight phases and especially addresses the challenges of forward-looking wind shear detection during takeoff and landing. The RDR-4B system offers advanced capability with Terrain-Based AutoTilt and a true dual redundant antenna drive system. The basic configuration of the RDR-4B system completely fulfills all ARINC requirements and has Antenna Drive with either a 30″ or 24″ diameter antenna and an optional MFRD (Multifunction Radar Display) for non-EFIS (Electronic Flight Information System) aircraft. Another Honeywell AWR, the Primus 880 (P-880), combined with an optional Lightning Sensor System (see Section 4.3.8 below), offers an additional dimension in severe weather avoidance.

Rockwell Collins AWRs are named WXR or TWR (Weather radars 2008b). The family of solid-state weather radars TWR-850/WXR-840/WXR-800 achieves full performance using a lightweight 30 W transmitter with weather detection ranges of up to 320 miles and available Doppler turbulence detection ranges of up to 50 miles. Another Collins airborne radar, the WXR-2100, is marketed as a Multiscan Fully Automatic Weather Radar that emulates an "ideal" radar beam by taking information from different radar scans and merging the information into a total weather picture. This allows the elimination of a manual operation, which is often a compromise between observing the most reflective part of the thunderstorm and reducing ground clutter returns. The curvature of the Earth, taken into account along with ground clutter suppression algorithms, results in flight crews being able to view all significant weather over 0–320 nautical miles on a single display that is essentially clutter free.

The AWR antenna is located in the aircraft nose inside a radio-transparent radome as shown in Figure 4.8 (Wallace 1994).

**Figure 4.8.** AWR antenna installed aboard (from Wallace 1994).
(http://fr.wikipedia.org/wiki/Fichier:Airborne_weather_radar_NASA.jpg,
http://trendypicture.com/ifa/40276/ifa-buenaventura-hotel-playa-del-ingles-gran-canaria-canary-islands.html)

Transceivers are usually located in the base of the nose section behind the bulkhead where the antenna is mounted. In classical designs, the transceiver is connected to the antenna by a waveguide. However, monoblock designs with the antenna and the transceiver structurally formed as a single block are also available. This latter design allows the elimination of a waveguide transmission line with a rotary joint, which is a source of noise and energy loss.

The monoblock design is used, for example, in the Ukrainian AWR Buran A-140 (Nosich, Poplavko, Vavriv, and Yanovsky, 2002; JSC Kiev Radar Plant 2008); in Honeywell's Primus-880 weather radar system (Weather Radars 2008a); and in the new Russian Kontur-10SV AWR tested and certified in 2008 (Kontur-NIIRS 2008). In the case of such single-unit designs, the antenna-transceiver monoblock is located in the nose of the aircraft and the displays and control panels are installed in the cockpit front panel.

Figure 4.9 shows the AWR Buran A-140 developed by the Kiev Buran Research Institute for the AN-140 aircraft and, in various modifications, also installed in the AN-148, AN-38, AN-4TK-300, IL-114 aircraft, and the BE-200 seaplane.

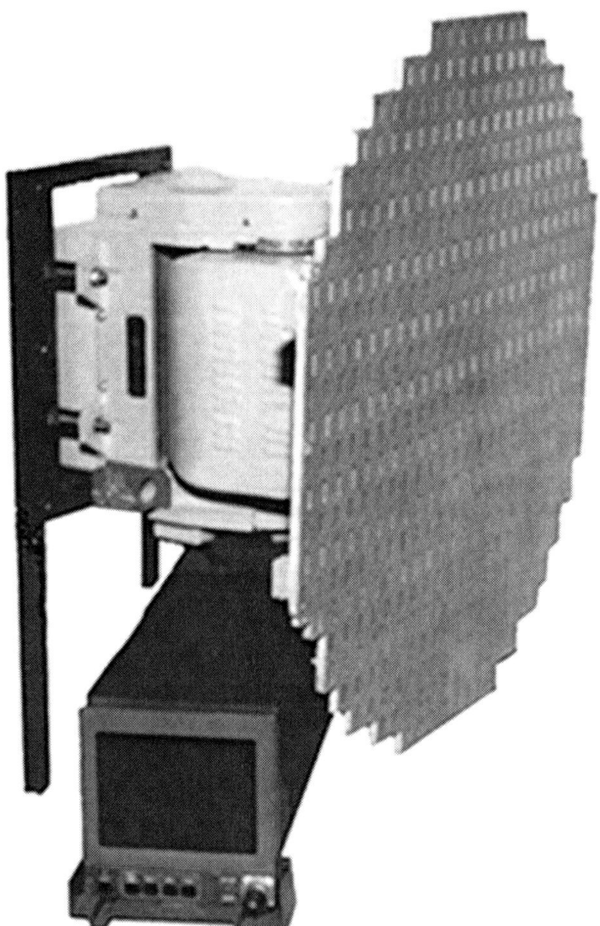

**Figure 4.9.** AWR Buran A-140. Antenna–transceiver single-unit and specialized display combined with control panel (Picture courtesy of F. J. Yankovsky).

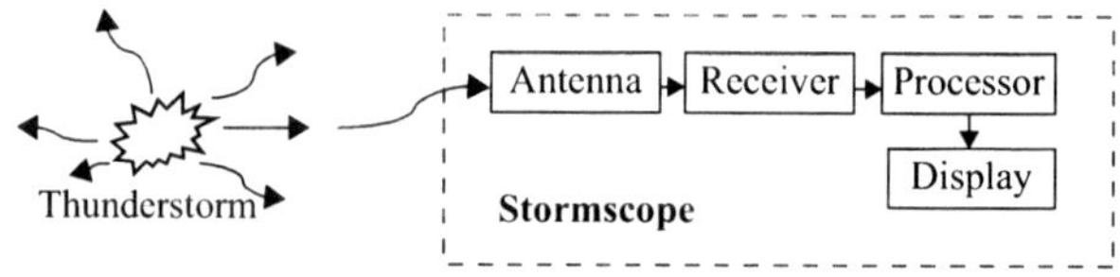


**Figure 4.10.** Generalized block diagram of a Stormscope®.

### 4.3.8 *LIGHTNING SENSOR SYSTEMS: STORMSCOPES®*

The passive detection of thunderstorm activity forms an essential complement to the radar detection of deep clouds and thunderstorms using AWR. However, there are different ways of achieving this (Yanovsky 1997) and nowadays the most common are methods based on the detection and analysis of electromagnetic radiation from electrical discharges in the atmosphere. Stormscopes®, the devices that can remotely detect lightning discharges, register the direction of the lightning and—in what is more difficult—estimate the distance to the sources of such discharges. These are widely used together with AWR by airliners and, as a simple and cheap alternative to AWR, in small airplanes (Stormscope® 2007).

The principle of the passive Stormscope® is illustrated in Figure 4.10. However, the main problem of such airborne passive sensors is still associated with the need to measure the target range from a single location with the required accuracy.

The original Stormscope® was first invented, produced, and used for aircraft safety by Paul Ryan (Ryan and Spitzer 1977), who sold his StormScope company to the 3M Corporation in 1981. Since that time several, and more accurate, methods of estimating the distance from an aircraft to lightning have been developed by Ryan himself and by other authors (Coleman 1984; Fishman, Golubchik, Ignatov, and Yanovsky 1989; Korablev, Yanovski, and Ligthart 2000; Markson, Warwick, and Uhlir 1990; Yanovsky, and Korablev 2000).

Lightning sensor systems produced by L-3 Avionics Systems (Stormscope® 2007), Honeywell (Weather Detection & Avoidance 2008), and Goodrich (FlightMax 2007) are designed as devices for both standalone applications and combined with AWR. For example, Honeywell's Lightning Sensor System LSZ-860 is combined with their AWR P-880.

Stormscopes® are much cheaper than AWRs and are therefore common in small, business aircraft where AWR is physically impossible or unreasonably expensive.

### 4.3.9 *OPTICAL RADAR*

Microwave weather radar works well in conditions of rain and clouds but much worse in dry weather when hydrometeors, which serve as scatterers, are absent. Optical (light) radar has the opposite characteristics, therefore it can serve as a desirable addition to AWR. Optical quantum radar, or *lidar* (Light Detecting and Ranging) is used for measuring important atmospheric parameters. The basic principles of the interaction of laser emission with the atmosphere and its application in the remote sensing of atmospheric phenomena are discussed by Weitkamp (2005).

Measurements of temperature, humidity, and the chemical composition of the atmosphere can also be provided with help of lidar. Moreover, the application of Doppler radar principles to light radar allows wind parameter measurements using Doppler lidar. Taking into account that electromagnetic waves in the optical band are scattered by aerosols (which are always present in the atmosphere, especially in air routes), Doppler lidar makes possible the detection of clear air turbulence. However, in the case of on-board lidar, measures for the stabilization of a narrow optical radar beam are necessary to take account of changes in the spatial orientation of the aircraft.

#### *4.3.9.1 Doppler Lidar*

Doppler lidar detects backscattered signals from aerosols and the very small particles of moisture that are moving within the air mass; and the system depends on the measurement and processing of the subsequent Doppler frequency shifts in the optical band. The narrow lidar beam makes possible the avoidance of interference reflections from the Earth's surface that are characteristic of microwave band systems.

The use of airborne lidar to measure wind velocities and to detect turbulence in front of an aircraft in real time was described by Targ et al. (1996), where 10.6-μm and 2.02-μm laser systems were used. Wind measurement accuracies of better than 1 m $s^{-1}$ were observed for both lidars, and it was shown that such sensors can significantly increase fuel efficiency, flight safety, and terminal area capacity. Results of atmosphere observations with Doppler lidar are also described and thoroughly analyzed by Weissmann et al. (2005).

A disadvantage of lidar systems relates to problems of operation in the case of heavy rain, which is why they cannot be considered as alternatives to AWR, but only as useful additions.

#### *4.3.9.2 Infrared Locators and Radiometers*

Infrared instruments can be used for measuring the temperature along an aircraft route and, in particular, may detect temperature changes characteristic of the cold air flow that can be an indicator of microbursts. By comparison with microwave radar and lidar, the infrared radiometer is cheap.

An infrared system can also be considered as providing an indirect detection of wind shear zones (Kuhn, Kurkowski, Caracena 1982), though a radiometer cannot measure the wind speed as does infrared lidar (Schwiesow and Lightsey, 1986). However, a radiometer's ability to measure temperature allows its use as an element in a system for the remote detection of zones of probable inflight icing on the basis of polarimetric AWR as was indicated in Section 4.3.4.6.

### *4.3.10 THE INTEGRATED LOCALIZATION OF DANGEROUS PHENOMENA*

The concept of integrated localization of DWP was developed in the late 1980s and is described, for example, by Yanovsky (1991). The first level of integration, based on the composite nature of different meteorological threats, concerns integration within the framework of airborne capabilities. This level covers the following aspects:

- The complex of sources of danger such as turbulence, wind shear, lightning, hail, and so forth

- The complex of informative parameters, which may have different natures both physically and mathematically, such as amplitude, Doppler, polarization, one-point, double-point, and so forth
- The complex of information channels that may have different physical natures, including active and passive radar, different frequency bands, lidars, radiometers, and so forth etc.

The second level of integration also includes information exchange on air-to-air, air-to-ground, air-to-satellite, and ground-to-satellite communication lines with addressed transmission of selective meteorological information to the aircraft. At this level the system is no longer autonomous. Some aspects of such integration related with ADS-B surveillance are considered by Yanovsky (2008). Here, it is reasonable to simply note that the prognoses and proposals on complex DWP localization performed in the 1980s are increasingly becoming realities. Moreover, a tendency to integrate different safety-related on-board surveillance means into a united complex, at least on the level of information display, but also on the level of decision making, has come to the fore in recent years. For example, AWRs that provide necessary information on probable meteorological danger are integrated not only with Stormscopes® but also with the airborne transponders, Ground Proximity Warning Systems (GPWS), and Traffic Alert and Collision Avoidance System (TCAS) considered below.

## 4.4 COLLISION AVOIDANCE SENSORS

### *4.4.1 TRAFFIC ALERT AND COLLISION AVOIDANCE SYSTEMS (TCAS)*

#### *4.4.1.1 The Purpose*

Automatic airborne TCAS is the information means intended for collision avoidance between aircraft, and such systems have become standard equipment comparatively recently. A TCAS should detect any aircraft that may constitute a collision threat and provide the pilot with information to facilitate appropriate avoidance actions. According to ICAO recommendations, such systems should be compatible with air traffic control (ATC) discrete address beacon systems, so the TCAS concept makes use of the radar beacon transponders carried by aircraft for ground ATC purposes. Such TCAS assignments implement the principle "See and Avoid," as accepted by the ICAO. Functioning independently of ground-based ATC to provide collision avoidance information, the TCAS undoubtedly reduces the risk of midair collisions between aircraft. However it provides no protection against aircraft that do not have operating transponders.

According to the official definition from PANS—Procedures for Air Navigation Services (PANS-ATM 2007)—TCAS is an aircraft system based on secondary surveillance radar (SSR) transponder signals that operate independently of ground-based equipment to provide advice to the pilot on potential conflicting aircraft that are also equipped with SSR transponders. In modern glass cockpit aircraft, the TCAS display may be integrated into the Navigation Display; in older glass cockpit aircraft and those with mechanical instrumentation, a TCAS display replaces the mechanical Instantaneous Vertical Speed Indicator that indicates the rate at which the aircraft is climbing or descending.

#### *4.4.1.2 A Short History*

In the past, airborne radar has been used to obtain information about oncoming aircraft, the sensitivity of an AWR being quite adequate to detect an aircraft at the required range. At low traffic densities this was acceptable, but as traffic densities increased, aircraft collisions in the 1950s became one of the main causes of aeronautical catastrophes. The efficient working of autonomous systems based on multifunctional airborne radars was found to be limited by their inability to distinguish between staggered flight levels because of low angular resolution in the vertical plane. Hence, it was difficult to estimate the altitude of a conflicting aircraft. However, this altitude is accurately known on board that aircraft, which led to an airborne system based on secondary radar (Stevens 1988), that is, a kind of radar using an active reply from the object under observation. This technique was chosen despite its obvious disadvantage—the necessity of equipping all participating airplanes with additional equipment.

Between 1955 and 1975 several concepts were developed and tested in both the United States and the USSR, some of which became operational, for example, Echelon (Bychkov, Pakholkov, and Yakovlev 1977). Nevertheless, only in the mid-1970s did a viable concept for a practical collision avoidance system technology arise. It was based on Mode C transponder replies, which contain altitude information for ATC systems, and is known as the Beacon Collision Avoidance System. Though test results showed that such a system was promising, they also highlighted some problems, the main one being an inability to distinguish and decode replies from several aircraft simultaneously.

ATC systems also suffered for the same reasons. Modes A and C secondary radar for the ATC radar beacon system (ATCRBS) were developed many years ago for aircraft identification and altitude reporting, and it was, and still is, an important component of ATC. However, as more and more aircraft appeared in zones of observation, this basic form of surveillance increasingly came to restrict the capacity of the ATCRBS. This technology is also associated with such problems as False Reply Uncorrelated In Time (FRUIT), which refers to the reception of a reply from another interrogation leading to the garbling of one reply with the other.

The next step was taken when Mode S discrete interrogation technology was developed and the Mode S transponder was introduced in the 1980s. Mode S—or Mode "Select"—was a new way of interrogating an aircraft such that only that particular aircraft responded. This was done by assigning a unique address to every aircraft. This discrete-address interrogation and data exchange method provided a solution to the main problems of information exchange between aircraft. By 1981 it was already recognized that a viable and very useful collision avoidance system could be implemented, so a concerted effort was made by the FAA, the Radio Technical Commission for Aeronautics, the Air Transport Association, private research organizations, airframe and electronic equipment manufacturers, and others, to develop the requirements of TCAS and its application rules. The first version of the Minimum Operating Performance Specifications (MOPS) for TCAS was developed in 1983 (US Congress 1989) and then followed by a series of changes.

#### *4.4.1.3 TCAS Levels of Capability*

The TCAS MOPS specify three levels of capability. TCAS-I is aimed at aiding the pilot in the visual acquisition of aircraft posing potential threats. It provides a traffic advisory (TA) display only, showing range and bearing. Altitude will also be displayed if the intruder has a Mode C or S transponder. Thus, the sole reason for TCAS-I is to assist the pilot in making visual contact with other traffic in the vicinity. The minimum requirements and elements of TCAS-I are described in (DO-184 1983). This document discusses both passive and active TCAS-I applications and provides the minimum performance requirements for electromagnetic compatibility and test procedures for both active and passive systems.

TCAS-II provides Traffic Alert and Resolution Advisories (recommended evasive maneuvers), and operates in conjunction with a Mode S transponder. It provides a traffic display and voice alert for Traffic Advisories (TAs) in addition to vertical speed commands and voice alerts for Resolution Advisories (RAs). TAs are intended to expedite visual contact with an intruder and RAs provide avoidance maneuvers in the vertical plane, that is, climbing or descending.

TCAS-III was initially defined as an enhanced TCAS-II, which provides resolution advisories in both the vertical and horizontal planes, so improving the Resolution Advisory maneuver options (Williamson and Spencer 1989). However, the development of TCAS-III was discontinued in favor of emerging systems such as Automatic Dependent Surveillance-Broadcast (ADS-B). This uses the Global Positioning System (GPS) and a radio frequency link to broadcast information between aircraft equipped with ADS-B as well as between aircraft and ground-based ADS-B receivers. An aircraft equipped with ADS-B would broadcast its identification along with its position, velocity, and other time-sensitive surveillance information to other aircraft and would receive the similar information from those other aircraft. Clearly, these capabilities will be fully realized only when every aircraft in the system has an operating ADS-B.

Sometimes, future systems that recommend avoidance maneuvers in both vertical and horizontal planes are called TCAS-IV.

Secondary radar systems can work efficiently only if the full and reliable compatibility of equipment can be developed and produced by different manufacturers for on-board installation. Therefore a system such as TCAS must be subject to international law. So, in parallel with the development of TCAS equipment in the United States, the ICAO has worked since the early 1980s to develop standards for Aircraft Alert and Collision Avoidance Systems (ACAS). The ICAO officially recognized ACAS on November 11, 1993. Its descriptive definition appears in Annex 2 of the Convention on International Civil Aviation and its use is regulated in Procedures for Air Navigation Services-Aircraft Operations (PANS-OPS) and Procedures for Air Navigation Services-Rules of the Air and Air Traffic Services (PANS-RAC). In November 1995, the Standards and Recommended Practices (SARPs) and Guidance Material for ACAS-II were approved, and these appear in Annex 10 of the Convention on International Civil Aviation. ACAS has therefore become subject to a current international standard since the ICAO approved the mandatory requirements. Though based on the TCAS system developed in the United States, European documentation frequently uses the name ACAS (Hawkes 1998). For example, ACAS-II is the same as the latest TCAS-II, which is called Version 7. The Version 7 requirements were completed in 1997 and published in RTCA document DO-185A (DO-185A 1997).

Based on a congressional mandate (Public Law 100-223), the FAA issued a rule effective on February 9, 1989 that required the installation of TCAS-II on airline aircraft with more than 30 seats by December 30, 1991 (Stead, Gambarani, and Tillotson 1995). Public Law 100-223 was later amended (Public Law 101-236) to permit the FAA to extend the deadline for TCAS-II fleet-wide implementation to December 30, 1993. In December of 1998 the FAA released a Technical Standard Order (TSO) that approved Change 7, resulting in the DO-185A

TCAS-II requirement. Change 7 incorporates software enhancements to reduce the number of false alerts. By the year 2000, TCAS equipage on aircraft with 30 or more seats was mandated in India, Argentina, Germany, Australia, and Hong Kong (Henely 2001). ACAS-II was mandated by the ICAO in 2003 and, based on varying criteria, throughout much of the world.

However, only TCAS-I and TCAS-II are currently available, so the following material will focus on TCAS-II, Version 7.

#### *4.4.1.4 TCAS Concepts and Principles of Operation*

A fundamental principle underlying TCAS is that the system is designed solely to prevent mid-air collisions and near mid-air collisions, and is autonomous and independent of any external systems such as aircraft navigation equipment, ground systems, or satellites. The basic concepts behind the collision avoidance system relate to (a) the Tau criterion, (b) sensitivity levels, and (c) items relevant to the protected volume, all of which are considered below.

- *The Tau criterion.* One of the core concepts of TCAS development was that the flight time to the threat aircraft (usually called the intruder) is more important than the distance. Hence, the collision avoidance criterion—the Tau criterion—is based on estimating the flight time (not distance) to the Closest Point of Approach (CPA). The CPA is that point ahead that the processor predicts will be the area of conflict with the intruder. Figure 4.11 illustrates the conflict geometry according to (Henely 2001). If $R$ is the distance between aircraft 2 and aircraft 1 at the initial time $t_0$, the vector $V_R$ is the relative velocity, which is the geometric sum of $V_1$ and $V_2$. $V_\theta$ is the tangential velocity of aircraft 2 relative to aircraft 1, and $m$ is the miss distance, that is, the distance between the aircraft at the CPA.

Though aircraft velocities $V_1$ and $V_2$ are constant, the distance $R(t)$ and the bearing $\theta(t)$ vary with time. From Figure 4.11, the time taken by aircraft 2 to reach the CPA is:

$$\tau = \frac{\left(R^2 - m^2\right)^{1/2}}{V_R} \text{ seconds} \tag{4.10}$$

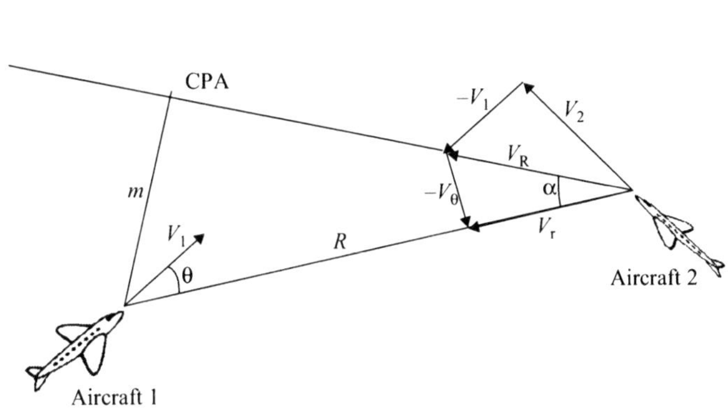


**Figure 4.11.** Conflict geometry when the velocities of two aircraft are constant.

The bearing rate is $|V_\theta|/R = \theta'$ rad sec$^{-1}$, and $\sin\alpha = m/R = |V_\theta|/V_R$. Combining these equations gives:

$$\theta' = \frac{mV_R}{R^2} = \frac{m/V_R}{\tau^2 + m^2/V_R^2} \text{ rad sec}^{-1} \tag{4.11}$$

Equation 4.11 illustrates the fact that the bearing rate is zero if and only if the aircraft are on a true collision course, $m = 0$ (Henely 2001). Thus, the simple Tau is an approximation of the time in seconds to the CPA and is equal to the slant range divided by the closing speed. A decision on any action is made if Tau is less than a particular threshold level.

- *Sensitivity level.* Another important notion related to the TCAS concept is the Sensitivity Level (SL). Effective collision avoidance logic operation requires a tradeoff between necessary protection and unnecessary advisories. This tradeoff is accomplished by controlling the SL, which controls the Tau, and therefore the dimensions of the protected airspace around each TCAS-equipped aircraft. The greater the SL, the more protection is provided, but the higher is the incidence of unnecessary alerts. The pilot can select three modes of operation, which are converted to SL by the logic. These include STANDBY, TA-ONLY, and AUTOMATIC. When the STANDBY mode is selected, the TCAS equipment does not transmit interrogations, and this level is normally used when the aircraft is on the ground. In the TA-ONLY mode, the equipment performs all the surveillance functions and provides TAs but not RAs. This level is generally used by pilots to avoid unnecessary distractions whilst at low altitude on final approach to an airport. When the Pilot selects the AUTOMATIC mode, the TCAS is free to select its SL based on the current altitude of its own aircraft.
- *Range test.* The joint application of Tau and SL concepts leads to three range criteria based on Tau estimation:
  (1) The simple Tau criterion as considered above: In this case $\tau = \tau_u = -R/R'$ with $R$ as slant range and $R'$ as closing speed. In reality $\tau_u$ is equal to time-to-CPA only if the miss distance of the aircraft $m$ at the CPA is zero. This is not true in most cases—the difference increases with miss distance. Therefore this criterion does not suit either low rate closures or accelerating aircraft.
  (2) The Modified Tau criterion: This was introduced for the cases in (1) and is calculated as $\tau = \tau_m = -(R - D_m)/R'$ with $D_m$ as distance-to-threshold, which depends on the SL.
  (3) The Bramson criterion: The modified Tau criterion leads to unnecessarily large protected volumes, and for this reason the Bramson criterion was implemented as the range test for the TCAS/ACAS standard (Consistent Values 1987). The difference from the Modified Tau criterion is that if the distance at the CPA equals the distance threshold, then the alert time should be equal to time-to-CPA. The Distance Modified (DMOD) Bramson Tau is defined as $\tau = \tau_b = -[R - D_m^2 / R]/R'$.
- *Altitude test.* In addition to the range test, the TCAS-II logic includes an altitude test that estimates whether the appropriate altitude difference is below the vertical separation at the CPA. The altitude information is derived from the transponder Mode S and C readouts, and the vertical speed is calculated from the altitude changes. However, because

the transponder altitude information appears in discrete steps of 100 ft, and hence the readout is not conducted at a constant rate, the vertical speed can only be estimated.

Any TCAS-II range or altitude test is based on the Tau concept for all alerting functions. The boundaries can be calculated for any combination of range and closure rate as well as the vertical separation that would trigger a TA (for example, with a 40 second Tau) and an RA (for example, with a 25 second Tau).

- *Protected airspace.* The Tau concepts considered above together with SL determine the dimensions of the protected airspace (see below) around each TCAS-equipped aircraft. Table 4.2 shows the altitude thresholds at which the TCAS automatically changes its sensitivity level selection and the associated Tau values for altitude-reporting aircraft.

Each TCAS-equipped aircraft is surrounded by a protected volume of airspace. The boundaries of this volume are shaped by the Tau and the DMOD—the distance modification criteria described above. The sensitivity level is used in defining the size of the protected volume around the protected aircraft. For encounter geometries involving low vertical closure rates (the Bramson criterion), the vertical dimensions of the protected volume for TAs are 1,200 ft above and below the altitude of the protected aircraft. The vertical dimensions for RAs vary from 750 to 950 ft depending on the protected aircraft's altitude regime. For high vertical closure rates, a TA or RA would be triggered when the predicted time to co-altitude drops below the Tau values for the sensitivity levels.

Because horizontal dimensions are not based on actual distances, but on the time to the CPA, the size of the protected volume depends on the speeds and headings of the aircraft involved and also on the SL. The protected volume is in general a truncated ellipse, with the long axis equal to the distance the faster aircraft would travel during the TA time. TCAS-II is designed to provide collision avoidance protection in the case of any two aircraft which are closing horizontally at any rate up to 1,200 knots and vertically up to 10,000 ft per minute.

Figure 4.12 shows the protected area based on the range and altitude criteria used by TCAS for determining an aircraft's threat status. In case an unsafe separation distance is found, TCAS will issue a Traffic Advisory (TA) and if necessary, TCAS-II will issue a Resolution Advisory (RA), which is in essence a recommended escape maneuver. If both aircraft have TCAS-II systems with Mode-S capability, they are able to issue nonconflicting (coordinated) RAs. It is

**Table 4.2.** Sensitivity level selection based on altitude

| | | **Tau values (in seconds)** | |
|---|---|---|---|
| **Altitude (in Feet)** | **Sensitivity level** | **TA** | **RA** |
| 0–1,000 AGL | 2 | 20 | — |
| 1,000–2,350 AGL | 3 | 25 | 15 |
| 2,350 –5,000 MSL | 4 | 30 | 20 |
| 5,000–10,000 MSL | 5 | 40 | 25 |
| 10,000–20,000 MSL | 6 | 45 | 30 |
| 20,000–42,000 MSL | 7 | 48 | 35 |
| Greater than 42,000 MSL | 7 | 48 | 35 |

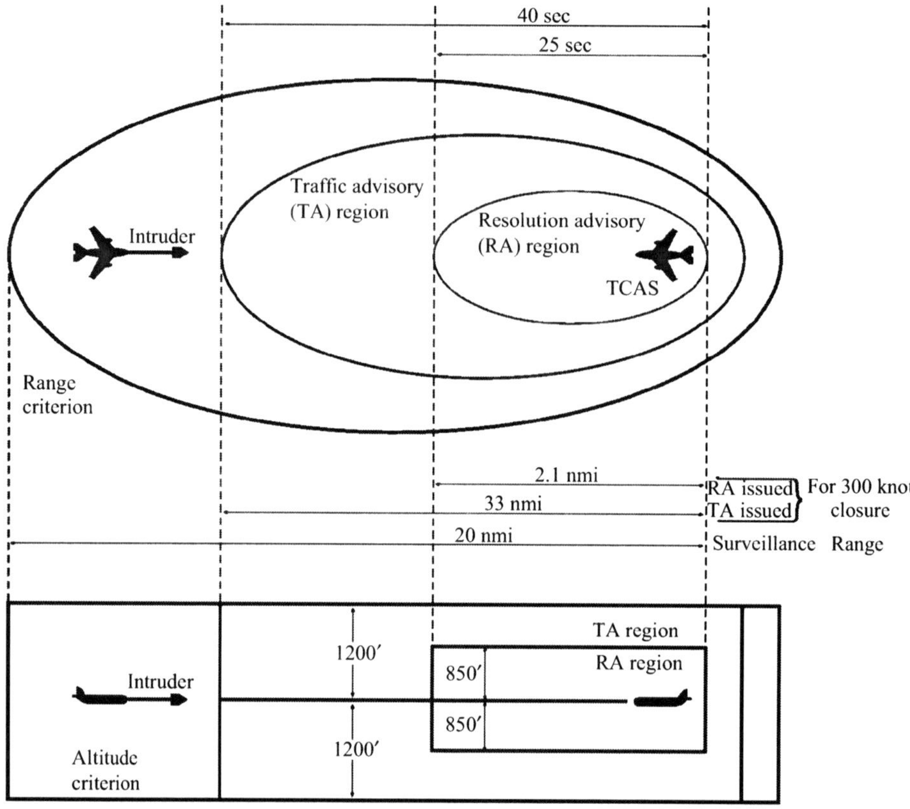


**Figure 4.12.** TCAS protected airspace at SL 5.

important to note however, that TCAS cannot issue advisories with regard to aircraft that do not report their altitude.

It is also important to note that the TCAS protection thresholds vary with altitude. In particular, the numerical data shown in Figure 4.12 are valid at SL5, that is, 5,000–10,000 ft MSL, according to Table 4.2. Under this condition, a TA with a 40 sec Tau and an RA with a 25 sec Tau are triggered if the combinations of range and range rate satisfy range and altitude tests.

#### *4.4.1.5 Basic Components*

Normally TCAS-II consists of a Mode S/TCAS Control Panel, a Mode S transponder, a TCAS computer (processor), antennae, traffic and resolution advisory displays, and an aural annunciator. Figure 4.13 is a block diagram of a typical TCAS-II taken from Henely (2001).

Control information from the Mode S/TCAS Control Panel is provided to the TCAS computer via the Mode S Transponder. TCAS-II uses a directional antenna mounted on top of the aircraft. In addition to receiving range and altitude data on targets above the aircraft, this directional antenna is used to transmit interrogations at varying power levels in each of four 90° azimuth segments. An omnidirectional transmitting and receiving antenna is mounted at

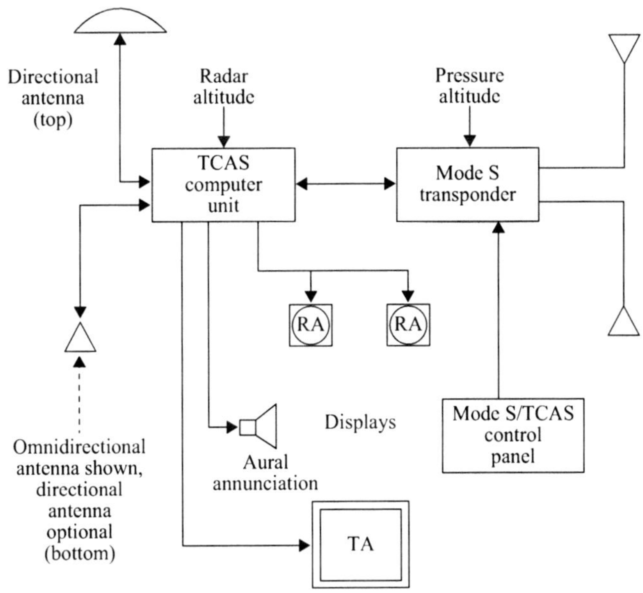


**Figure 4.13.** Typical block diagram of TCAS-II.

the bottom of the aircraft to provide the TCAS with range and altitude data from traffic below. TCAS-II transmits transponder interrogations and receives transponder replies. Provided that the TCAS uses the radar beacon transponders installed on aircraft to operate with ATC ground-based radars, the interrogation carrier frequency is 1030 MHz, whilst the reply carrier frequency is 1090 MHz.

The Traffic Advisory (TA) display depicts the position of traffic relative to the TCAS aircraft to assist the pilot in visually acquiring intruder positions. The Resolution Advisory (RA) can be displayed on a standard Vertical Speed Indicator (VSI) modified to indicate the vertical rate that must be achieved to maintain safe separation from such intruders. When an RA is generated, the TCAS-II computer lights up the appropriate display segments, and RA compliance is accomplished by flying to keep the VSI needle out of the red segments. On newer aircraft, the RA display function is integrated into the Primary Flight Display (PFD). Displayed traffic and resolution advisories are supplemented by synthetic voice advisories generated by the TCAS-II computer.

#### *4.4.1.6 Operation*

1. *Squitter, interrogation, and reply.* In a one-second cycle, TCAS-II should monitor other aircraft by interrogating their SSR transponders, select an advisory when a range test and an altitude test are satisfied, and present the information to the pilot. The system processor constantly evaluates a specific volume of airspace and the geometry around it to solve any possible conflicts that may arise within it. The Mode S transponder pseudo-randomly radiates (squitters) its unique Mode S address omnidirectionally once a second so as to let its presence be known to other similarly equipped aircraft. TCAS equipment listens

for this broadcast squitter, and upon receipt of a valid message the transmitting aircraft identification is added to a list of aircraft the TCAS aircraft will interrogate. Following receipt of a squitter, the TCAS then sends a Mode S interrogation to the specific Mode S address contained in the message. The replies then received by the TCAS are used to determine bearing, range, and altitude.

2. *Whisper–Shout method.* When transmitting, the processor has the ability to control the effective radiated power by utilizing a scheme called Whisper–Shout. This method helps to manage the reach of the system. Intruder range is determined by the time delay between the interrogation and the reply sequence and can also be supplemented by the results of the Whisper–Shout routine. The aircraft is now an active interrogator in the system and as a result has increased the 1030/1090 MHz traffic. In order to limit this TCAS-generated traffic, especially in high density areas, a TCAS Presence Message is broadcast. The TCAS is designed to limit its own transmissions when a specific traffic threshold is met, thus helping to minimize that traffic. The results of the Whisper–Shout routine will also allow certain aircraft to be de-selected as primary threats as the ranging and altitude data are gathered.
3. *Tracking and predicting.* Once an interrogation is recognized, the receiving aircraft logs in that data and can then start the tracking process. The constant interrogation, acquisition, tracking, and predicting of the crossing geometries for all intruders keeps the system quite busy. Once established, this bidirectional data-link between each TCAS-II equipped aircraft is crucial for a successful resolution.
4. *Operation with ATCRBS transponders.* TCAS handles Mode A/C transponders differently: It does not interrogate Mode A because this does not provide altitude information, but it does perform a Mode C-only all-call. A Mode C transponder sends a reply that contains critical target altitude data and also the objects for tracking, but these may be not synchronous and so special algorithms are employed to provide the proper filtering.
5. *Conflict resolution* starts with a TA that informs the pilot about nearby traffic. This traffic advisory only later indicates a threat if conditions change. The next step is a Resolution Advisory (RA), and a Preventive RA basically advises the pilot not to deviate from the current vertical flight profile. This indicates to the crew that the situation is resolving itself as long as the current flight path is maintained. The Corrective RA is the last step in conflict resolution, and this command is given to advise the crew to take action vertically in order to avoid the developing threat. Vertical changes in flight paths have been deemed the quickest resolutions to possible conflicts. All these actions must be performed before the aircraft reaches the CPA point (see Section 4.4.1.4) as computed by the processor.

#### *4.4.1.7 TCAS Logistics*

Collision avoidance logic functions are shown in Figure 4.14 (Henely 2001; Introduction to TCAS II 2000). Here, the collision avoidance logic tracks the slant range $R$ and closing speed $R'$ of each target to determine the time in seconds until the CPA, as was explained in Section 4.4.1.4. A range test must be met and the vertical separation at the CPA must be within 850 ft for an altitude-reporting target to be declared a potential threat and a traffic advisory to be generated.

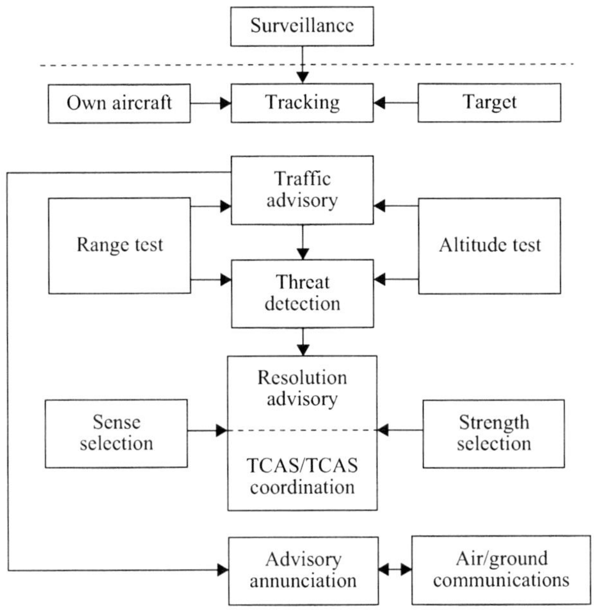


**Figure 4.14.** Functions of collision avoidance logistics.

A two-step process is used to determine the type of RA to be selected when a threat is declared. The first step is to select the sense (upward or downward) of the RA. Based on the range and altitude tracks of the potential threat, the collision avoidance logic models the potential intruder's path to the CPA and selects the RA sense that provides the greater vertical separation. The second RA step is to select the strength of the resolution advisory. The least disruptive vertical rate maneuver that will achieve safe separation is then selected. If both conflicting aircraft are equipped with TCAS-II, each aircraft transmits Mode S coordination interrogations to the other to ensure the selection of complementary resolution advisories. Coordination interrogations contain information about an aircraft's intended vertical maneuver.

One of the first operational software schemes for TCAS-II was version 6.04 developed by the Lincoln Lab at MIT, and was accepted by the FAA as the obligatory version at the end of 1993. Later, it became clear that this version was not fully coordinated with ICAO SARPS for TCAS-II, and also it was not completely adopted for the operation in RVSM (Reduced Vertical Separation Minimum) Airspace. RVSM means 1000 ft vertical separation between aircraft at Flight Levels (FL) 290-410. (HBAT_03-06-EAA 2003). That is why TCAS-II logic and software entered further development. The MOPS for Version 7 was approved in December 1997 and Version 7 units became available for installation in late 1999. Version 7 significantly improved TCAS compatibility with air traffic control systems throughout the world and is currently the main operational TCAS version.

During a joint session in March 2008, the RTCA Special Committee (SC) 147 and EUROCAE Working Group (WG) 75 agreed on the final version of the TCAS-II MOPS, to be known as TCAS-II version 7.1. The MOPS were approved by the RTCA Program Management Committee in June 2008 (document DO-185B) and by the EUROCAE Council in September 2008 (document ED-143). The MOPS were revised following the identification by

EUROCONTROL of two safety issues in the existing TCAS logic: one relating to the performance of the RA-reversal logic, and the other involving incorrect responses to Adjust Vertical Speed RAs. The FAA has already published the TCAS-II version 7.1 Technical Standard Order (TSO), and the European Aviation Safety Agency (EASA TSO) followed before the end of 2009 (TCAS-II version 7.1 2008). This version is being implemented step by step starting from 2010.

#### *4.4.1.8 Cockpit Presentation*

The traffic advisory display can be either a dedicated TCAS display or a joint-use weather radar and traffic display. In some aircraft, the traffic advisory display will be an electronic flight instrument system (EFIS) or a flat panel display that combines traffic and resolution advisory information on the same display. Targets of interest on the traffic advisory display are depicted in various shapes and colors. More details and the images can be found in numerous sources (DO-185A 1997; Henely 2001; Introduction to TCAS II 2000). ACSS. 2012.

#### *4.4.1.9 Examples of System Implementation*

Most recent TCAS implementations that meet the ACAS-II standards set by the ICAO correspond to Version 7.0 of TCAS-II. Several examples of TCAS are produced by three famous avionics manufacturers: Honeywell (Traffic Collision Alerting System 2008), Rockwell Collins (Traffic surveillance systems 2008), and ACSS (Aviation Communication & Surveillance Systems—An L-3 Communications and Thales Avionics company; ACSS 2008). These systems can be considered as second generation TCAS-II, also known as TCAS 2000.

A novel design concept of TCAS-II was implemented by the Kiev Buran Research Institute (Ukraine) in the SPS-2000 system, which combines a Mode S transponder and a TCAS-II processor in a single multifunctional unit (Belkin and Yanovsky 2005). This concept makes it possible to halve the number of necessary aircraft antennas as well as to decrease the number of units and the equipment cost. All functions and requirements of both TCAS-II and Mode S transponder systems correspond to ACAS-II standards. The SPS-2000 system and the results of its flight tests are described by Belkin and Panofsky (2007).

The third generation of TCAS-II has been recently developed as ACSS TCAS 3000 according to ACSS (2008).

Although real TCAS systems occasionally suffer from false alarms, pilots are now under strict instructions to regard all TCAS messages as genuine alerts demanding an immediate, high-priority response. Only stall warnings and Ground Proximity Warning System warnings have higher priority than the TCAS. FAA rules (and those of aviation authorities in most other countries) state that in the case of a conflict between TCAS RA and ATC instructions, the TCAS RA always takes precedence. If one aircraft follows a TCAS RA and the other follows conflicting ATC instructions, a collision can occur, such as the July 1, 2002 Überlingen disaster (Mid-air collision 2004). In this mid-air collision, both airplanes were fitted with TCAS-II systems which functioned properly, but one obeyed the TCAS advisory while the other ignored the TCAS and obeyed the controller. Both aircraft descended into a fatal collision. Hopefully, such situations will be impossible after the implementation of logic improvements in accordance with upcoming TCAS-II version 7.1.

## 4.4.2 THE GROUND PROXIMITY WARNING SYSTEM (GPWS)

### 4.4.2.1 Purpose and Necessity

The GPWS is a system designed to alert pilots if their aircraft is in immediate danger of flying into the ground. Also called the Ground Collision Warning System (GCWS), the GPWS is nowadays defined as equipment installed in an airplane for the purpose of automatically providing a timely and distinctive warning to the flight crew when the aeroplane is in potentially hazardous proximity to the Earth's surface.

Controlled Flight Into Terrain (CFIT) is the act of flying a perfectly operating aircraft into the ground, water, or a man-made obstruction. Historically, CFIT is the most common type of fatal accident in worldwide flying operations (Breen 2001). Prior to the development of GPWS, large passenger aircraft in the United States were involved in three or four fatal accidents per year due to ground-collision. However, since 1974, when the FAA required large airplanes to carry GPWS equipment, there has not been a single passenger fatality in a CFIT crash by a large jet in US airspace. In 2000, the FAA extended this requirement to smaller commuter planes as well. The ICAO introduced GPWS carriage requirements in 1978 to alleviate the CFIT problem. Statistics show that introduction of the GPWS into the scheduled air carrier turbojet fleet has been accompanied by a dramatic drop in the frequency of CFIT accidents (Breen 1999). Requirements for GPWS equipment are defined by Terrain Awareness and Warning System (TAWT) regulations (Regulations 2005; TSO C151a 1990).

### 4.4.2.2 GPWS History, Principles, and Evolution

#### HISTORY

Don Bateman is credited with the invention of GPWS (Wikipedia 2008). He spearheaded the development in the late 1960s after a series of CFIT accidents killed hundreds of people. Tragic airline crashes during this period prompted airline owners to take steps to minimize crashes caused when pilots failed to recognize that they were flying too low or approaching a mountain. The advent of the GPWS made possible the automatic warning for pilots if their aircraft were approaching the ground or water.

#### PRINCIPLE

The GPWS principle is simple: the system monitors an aircraft's height above ground level as determined by the Low Range Radar Altimeter (LRRA) described in Chapter 3. It also uses some other on-board sensors for air data and attitude. A computer then keeps track of these readings to assess the aircraft's current state compared to known hazardous situations. GPWS will warn the captain if the aircraft is in certain states of undesirable behavior. These undesirable states define flying configurations called "modes" (see Section 4.4.2.3).

The heart of the GPWS is a computer that receives inputs from several sensors on the aircraft, calculates trends, and issues warnings to the pilot through visual and aural alerting devices. The primary sensors are the LRRA, the barometric altimeter, the electronic glide slope indicator of the Instrument Landing System (ILS), and sensors that indicate aircraft control surface configurations such as the flap position and also the position of the landing gear.

## EVOLUTION

As technology improved, a series of advanced devices were developed that made the warning systems more effective and reliable. These added more sophisticated ways of determining the distance from the aircraft to threatening terrain, provided wind shear warnings, integrated other avionics systems, and included computerized colored pictures of topographical data.

The evolution of GPWS from the Mode 4 ground proximity warning of 1970 to the Mode 7 Enhanced GPWS (EGPWS) of 1996 is described by Breen (1999). Modes 1 through 4 are the original classic GPWS modes, first developed to alert pilots to unsafe trajectories with respect to the terrain. The original analog computer model had a single red visual lamp and a continuous siren tone as an aural alert for all modes. Aircraft manufacturer requirements caused refinement to the original modes and added the voiced "Pull Up" for Modes 1 through 4 and a new Mode 5 "Glideslope." Mode 6 was added with the first digital computer models about the time of Boeing 757/767 aircraft introduction; and Mode 7 was added when wind shear detection became a requirement in about 1985. Further improvements are related to the utilization of GPS data and electronic databases.

### *4.4.2.3 GPWS Modes*

These systems are designed to detect and warn the pilot of an excessive descent rate near the ground, an excessive terrain closure rate, a ground approach with landing gear or flaps not in the landing configuration, and a descent significantly below the ILS electronic glideslope when on approach to landing. Also, during takeoff and immediately after initiating a missed-approach go-around, the system warns the pilot if the aircraft is descending when it should normally be climbing. These features of different undesirable aircraft behavior underlie the different GPWS modes.

The "classic" GPWS, according to Breen (1999), produces warnings in five modes:

- *Mode 1*—Excessive barometric sink rate with respect to terrain clearance ("PULL UP," "SINK RATE")
- *Mode 2*—Excessive rate of terrain closure with respect to terrain clearance ("TERRAIN," "PULL UP")
- *Mode 3*—Excessive altitude loss after takeoff ("DON'T SINK")
- *Mode 4*—Unsafe terrain clearance with respect to phase of flight, airspeed, and/or aircraft configuration ("TOO LOW – TERRAIN," "TOO LOW – GEAR," "TOO LOW – FLAPS")
- *Mode 5*—Excessive descent below ILS glideslope ("GLIDESLOPE").

Newer models of GPWS also have a Mode 6—Bank angle protection, which produces altitude awareness callouts and warning of excessive roll attitude ("BANK ANGLE") and a Mode 7—Reactive wind shear detection ("WIND SHEAR").

Spitzer (2007) provides detailed descriptions and discussions of all modes.

#### 4.4.2.4 Shortcomings of Classical GPWS

Despite the rather high efficiency of the GPWS in reducing the incidence of CFIT accidents, it does have limitations. In particular, late warnings can occur as a result of flying into precipitous terrain, as the standard GPWS depends upon a downward looking radar altimeter to detect rising terrain. That is, traditional GPWS sensors cannot detect hazards that may be ahead of the aircraft such as steeply rising terrain or artificial obstacles. In addition, the effectiveness of GPWS is largely dependent on the pilot's prompt reaction to the system's warnings. Also, no warnings may be given if the aircraft is flown in the landing configuration where there is no runway, because standard GPWS algorithms are desensitized when gear and flaps are down.

In the case of classical GPWS, imminent danger of ground collision is inferred by the relationship of other aircraft performance data relative to a safe height above the ground. With this type of system, level flight toward terrain can only be implied by detecting rising terrain under the aircraft; for flight toward steeply rising terrain, this may not allow enough time for corrective action by the flight crew.

#### 4.4.2.5 Enhanced GPWSs

As described above, the traditional GPWS can only gather data from directly below the aircraft so it must predict future terrain features. If there is a dramatic change in terrain, such as a steep slope, GPWS will not detect the aircraft closure rate until it is too late for evasive action.

In 2002 a new technology, the "Enhanced Ground Proximity Warning System" (EGPWS/TAWS) solved this problem by combining a worldwide Digital Terrain Elevation Database (DTED) with a Global Positioning System (GPS). The DTED can also include man-made obstacles. The addition of the DTED allows the EGPWS to display terrain in proximity to the aircraft, so providing enhanced situational awareness to the pilot when maneuvering near the terrain.

On-board computers compare the current location of an aircraft with a database of the Earth's terrain so that pilots may receive much more timely cautions and warnings of any obstructions in that aircraft's path. Analysis of the conditions surrounding CFIT accidents, as evidenced by early flight recorder data, ATC records, and experiences of pilots in CFIT incidents, have identified common conditions that tend to precede this type of accident. Utilizing various onboard sensor determinations of the aircraft current state and projecting that state dynamically into the near future, the EGPWS makes comparisons to the hazardous conditions known to precede a CFIT accident. If the conditions exceed the boundaries of safe operation, an aural and/or visual warning/advisory is given to alert the flight crew to take corrective action (Breen 2001).

Thus, the EGPWS includes all traditional GPWS functions, but also utilizes a proprietary worldwide terrain database. Referencing the host aircraft location from the main navigation system, the EGPWS can display nearby terrain and provide aural warnings approximately 60 sec in advance of a terrain encounter, compared with 10 sec for a traditional GPWS. The EGPWS provides the following improvements over the GPWS: look-ahead alerting algorithms, multiple radio altimeter inputs, significant reductions in unwanted warnings, and landing-short alerting algorithms. Optionally, an embedded 12-channel GPS receiver can be incorporated to provide an upgrade capability to aircraft that have no existing GPS or FMS output available (Jane's 2008).

In addition to the seven basic function modes described in Section 4.4.2.3, the enhanced functions are:

- Terrain clearance floor
- Terrain look-ahead alerting, and
- Predictive wind shear

The development of statistical models of terrain derived from an actual terrain database allows estimation of the probability of a CFIT accident following a GPWS alert (Kuchar 2001).

#### *4.4.2.6 Look-Ahead Warnings*

The introduction of DTED provides the EGPWS look-ahead warning function. The system continuously computes alerting envelopes ahead of the aircraft, these envelopes being functions of current aircraft position, performance, and direction, including looking into turns (Campbell 2005). If the computed boundary is found to be in conflict with the terrain as indicated by comparison with the DTED, alerts are issued. There are two levels of alerts, based upon the approximate "time to impact." The longer look-ahead distance is computed to give approximately one minute of advanced alert, and if this distance conflicts with the DTED, a cautionary vocal alert is issued—"CAUTION, TERRAIN" or "TERRAIN AHEAD." If the flight path of the aircraft is not corrected, the alert is reissued every seven seconds. The shorter look-ahead distance is computed to give approximately 30 sec of warning, and if this alerting envelope reaches into the DTED, a vocal warning "TERRAIN - TERRAIN - PULL UP - PULL UP," or "TERRAIN AHEAD - PULL UP" is issued. In the case of alerts for man-made obstructions, the word "Terrain" is replaced by "Obstacle," giving "CAUTION-OBSTACLE" (or "OBSTACLE AHEAD"), and "OBSTACLE - OBSTACLE - PULL UP - PULL UP." This provides additional situational awareness of the flight path problem, especially in flat areas with tall towers, where a "terrain" warning might not seem credible (Breen 1999).

#### *4.4.2.7 Implementation Examples*

Now that GPWS has become standard equipment, such systems are produced by Honeywell which has had some 30 years of ground proximity warning experience. According to Honeywell Aerospace EGPWS (2008), it offers cost-effective solutions to TAWS regulations. Because EGPWS uses aircraft inputs such as position, attitude, air speed, and glideslope along with internal terrain, obstacle, and airport databases, it can predict a potential conflict between an aircraft's flight path and any terrain or obstacle. The result is a visual and audible caution or warning alert. When coupled with display, the surrounding terrain can be viewed relative to the aircraft position, so providing strategic terrain information up to 30 min before a potential terrain conflict.

Allied Signal Aerospace has announced an EGPWS about the size of a paperback novel designed specifically for the general aviation market, so providing terrain protection for private planes and other GA aircraft (AlliedSignal 1999).

South East Aerospace has developed the KGP-860, an EGPWS that meets TSO C151a Class B TAWS requirements and is specifically designed for light turbine and piston aircraft (SoutheastAerospace 2004). It combines early alert/warning capability with optional terrain display (when used with a compatible display), and incorporates GPS position and terrain/obstacle databases to indicate terrain above and below the aircraft. Terrain display can be presented on multifunction display KMD-550 or KMD-850, weather indicator (with compatible terrain input), or a dedicated display such as the Honeywell TRA-45A. The system uses an updateable worldwide terrain database and accepts GPS data from an external sensor with RS-232/422 and ARINC 743/743A formats. Six-color terrain alerting is available as well as the usual display output, and the system implements all the basic and enhanced functions described earlier. A typical system weighing less than 1.5 pounds consists of KGP-860 EGPWS computer, a database card (Americas, Atlantic, or Pacific), a configuration module, and appropriate installation kits.

According to Jane's (2008), the following variants of the EGPWS are available:

- The *Mk V air transport version* for aircraft with digital data interfaces. This has a worldwide database including runways longer than 3,500 ft and all man-made obstacles in North America.
- The *Mk VI regional aircraft version*, which is similar to the Mk V but smaller and lighter, is intended for installation in turboprop aircraft with appropriate analog avionics. It has a regional terrain database (rather than global) and lacks wind shear capability.
- The *Mk VII air transport version* is similar to the Mk V, but for aircraft that have analogue data interfaces. Its worldwide database includes runways longer than 3,500 ft and all man-made obstacles in North America.
- The *Mk VIII* also uses analog inputs and includes an integrated 75 × 75 mm Terrain Awareness and Display System (TADS), which can be retrofitted into the space occupied by an existing altimeter. It has a worldwide terrain database including runways.

However, traditional (nonenhanced) systems are also used. The Boeing 727 GPWS described and explained with diagrams in Boeing (2001) is an example of a five-mode GPWS, which alerts the flight crew when one of the thresholds corresponding to the five classical modes is exceeded between 50 and 2450 ft radio altitudes. Inputs to the ground proximity computer are altitude from the radio altimeter, the barometric altitude rate change, the Mach number from an air data computer, glide slope deviation signals, and landing gear and flap positions. The loss of one of these inputs will deactivate only the affected mode or modes. Aural alerts and warnings for modes 1 through 4 are accompanied by red "PULL UP" lights.

## REFERENCES

ACSS. 2008. An L-3 Communications and Thales Company. Retrieved from http://www.acssonboard.com/

Allied Signal. 1999. New EGPWS. Retrieved from http://www.aeroworldnet.com/3tw05179.htm

Atlas, D. 1964. "Advances in radar meteorology." In *Advances in Geophysics* (Volume 10, pp. 317–488), edited by H. E. Landsberg and J. van Mieghem. New York: Academic Press.
Baranov, I. M., V. V. Belkin, V. T. Bogatyr, E. M. Mescheryakov, P. M. Sokolov, A. A. Tereschuk, and F. J. Yanovsky. 1976. "Device for operational measurement and representation of radar reflectivity on the section of a weather target." VINITI (All-Union Research Institute for Science and Technical Information), No 1819-76 DEP, pp. 1–23. (In Russian.)
Baranov, I. M., and F. J. Yanovsky. 1976. "About the necessity of solving the problem of finding an optimum flight trajectory in difficult meteorological conditions." In *Computer Science and Simulation of Complex Systems* (Volume 2, pp. 56–62), edited by G. E. Pukhov. Kiev, Ukraine: KIIGA Press. (In Russian.)
Barton, D. K., and S. A. Leonov. 1998. *Radar Technology Encyclopedia*. Norwood, MA: Artech House, Inc.
Bateman, C. D. 1974. "Aircraft landing approach ground proximity warning system." U.S. Patent No 3922637, Priority 3 Oct 1974.
Belkin, V. V., V. P. Dzubenko, and F. J. Yanovsky. 2001. *Automatic Forming of the Earth Surface Map with Airborne Weather Radars, Proceedings of the International Conference on Land Use / Cover Change Dynamics LUCCD-2001*, Beijing, China, August 26–30, 2001, pp. 105–16.
Belkin, V. V., and F. J. Yanovsky. 2005. "Aircraft traffic collision avoidance system." *Proceedings of the 2nd International Workshop on Intelligent Transportation* (WIT 2005), Hamburg, Germany, March, 2005, pp. 195–200.
Belkin, V. V., and F. J. Yanovsky. 2007. "Aircraft collision avoidance system." *Proceedings IEEE Aerospace and Electronic Systems Symposium*, Big Sky, MO, USA, 1-4244-0525-4/07/$20.00 ©2007 IEEE, p. 9.
Boeing. 2001. Boeing 727 GPWS, Retrieved from http://www.boeing-727.com/Data/systems/infogpws.html#mode2
Bowles, R. L. 1990. "Windshear detection and avoidance—Airborne systems survey." *Proceedings of the 29th IEEE Conference on Decision and Control* (Volume 2, pp.708–736). DOI: 10.1109/CDC.1990.203685.
Braun, I. M., and F. J. Yanovsky. 2004. "Models of scattering on hailstones in X-band." *Proceedings European Radar Conference*, Amsterdam, pp. 229–232.
Breen, B. C. 1999. "Controlled flight into terrain and the enhanced ground proximity warning system." *IEEE AES Systems Magazine*, January, 14 (1): 19–24. DOI: 10.1109/62.738350.
Breen, B. C. 2001. "Enhanced situation awareness." Chapter 18 In *The Avionics Handbook* (p. 12), edited by Cary R. Spitzer. Boca Raton, FL: CRC Press LLC.
Bringi, V. N., and V. Chandrasecar. 2001. *Polarimetric Doppler Weather Radar*. Cambridge: Cambridge University Press. DOI: 10.1017/CBO9780511541094.
Bychkov, S. N., G. A. Pakholkov, and V. N. Yakovlev. 1977. *Radio-Engineering Systems for Avoidance of Collisions between Aircraft* (p. 268). Moscow: Sovetskoe Radio. (In Russian.)
Campbell, N. A. H. 2005. The Use of Enhanced Ground Proximity Warning System (EGPWS) Data for Aviation Safety Investigation, Retrieved from www.asasi.org/papers/2005/Use%20of%20EGPWS.pdf
Coleman, E. W. 1984. "Storm warning method and apparatus." US Patent 4672305, December 14, 1984.
Commercial Avionics Systems. 1996. AlliedSignal Inc., ACS-6026D, 6 p.
Consistent Values. 1987. "Consistent values of DMOD and TAU for The Bramson Range Test." Lincoln Laboratory Report, Nov. 1987.
Curlander, J. C., and R. N. McDonough. 1991. *Synthetic Aperture Radar: Systems and Signal Processing* (p. 648). Hoboken, NJ: John Wiley & Sons Inc.
DO-184. 1983. Traffic Alert and Collision Avoidance System (TCAS) I Functional Guidelines, 5-13-83. Retrieved from http://www.rtca.org/downloads/List%20of%20Available%20Docs%20-%20Jun%202012.pdf
DO-185A. 1997. Minimum Operational Performance Standards for Traffic Alert and Collision Avoidance System II (TCAS II), Airborne Equipment, 12-16-97. Retrieved from http://www.rtca.org/downloads/List%20of%20Available%20Docs%20-%20Jun%202012.pdfv

Doviak, R. J., and D. S. Zrnic. 1993. *Doppler Radar and Weather Observations*, Academic Press, Inc.

Evans, J., and D. Turnbull. 1989. "Development of an automated windshear detection system using Doppler weather radar." *Proceedings of the IEEE* 77 (11): 1661–73. DOI: 10.1109/5.47729.

Fishman, B., V. Golubchik, V. Ignatov, and F. Yanovsky. 1989. Airborne device for operative determination of hazardous lightning, USSR Patent No 1692261, Priority 07 Feb 1989, Reg. 1991.

FlightMax 2007. Goodrich Avionics Systems Stormscope®, WX-1000E weather mapping system, Goodrich Corporation website. Retrieved from http://www.avidyne.com/techpubs2/FlightMax%20 Lightning.pdf

FMR-200X. 2008. Multimode Weather Radar, Rockwell Collins website. Retrieved from http://articles.janes.com/articles/Janes-Avionics/FMR-200X-multimode-weather-radar-United-States.html

Fried, W. R., H. Buel, and J. R. Hager. 1997. "Doppler and altimeter radars." In *Avionics Navigation Systems* (2nd edition, pp. 449–502) edited by M. Kayton and W.R. Fried. New York: John Wiley & Sons, Inc. DOI: 10.1002/9780470172704.ch10.

Hawkes, D. 1998. Airborne Collision Avoidance System II (ACAS II), Joint Aviation Authorities, CNS/ATM Steering Group Position Paper, No. and Revision pp. 17–12, R. Lundberg (coordinator). Retrieved from http://www.eurocontrol.int/msa/gallery/content/public/documents/PP017_12.pdf

HBAT_03-06-EAA 2003. Retrieved from http://www.faa.gov/about/office_org/headquarters_offices/ato/service_units/enroute/rvsm/documents/HBAT_03-06.doc

Henely, S. 2001. "TCAS II." Chapter 18 In *The Avionics Handbook* (p. 10), edited by Cary R. Spitzer. Boca Raton: CRC Press LLC.

Honeywell Aerospace EGPWS. 2008. Retrieved from http://www.honeywell.com/sites/aero/Egpws-Home.htm

Huddle, J. R., and R. G. Brown. 1997. "Multisensor navigation systems." In *Avionics Navigation Systems* (2nd edition, pp. 55–98), edited by Myron Kayton and Walter R. Fried. John Wiley & Sons, Inc.

IEEE standard radar definitions. 1998. Radar Systems Panel of the IEEE Aerospace and Electronic Systems Society, USA.

Introduction to TCAS II. 2000. "Version 7, U.S. Department of Transportation." FAA, Nov 2000, 45 pp.

Jane's (2008), Enhanced Ground Proximity Warning System (United States). Retrieved from http://www.janes.com/extracts/extract/jav/jav_0561.html

JSC Kiev Radar Plant. 2008. For civil aviation, JSC Kiev Radar Plant website. Retrieved from www.radar.net.ua/catalogue.php?lang=en&root=1

Kessler, E., J. T. Lee, and K. E. Wilk. 1965. "Associations between aircraft measurements of turbulence and weather radar measurements." *Bulletin of American Meteorological Society* 46 (8): 433–47.

Kolchinsky, V. E., I. A. Mandurovsky, and M. N. Konstantinovsky. 1975. *Autonomous Doppler Devices and Systems for Navigation of Flying Vehicles* (p. 432). Moscow: Sovetskoe Radio Press. (In Russian.)

Kontur-NIIRS. 2008. Weather & Navigation Radars, Kontur-NIIRS, Ltd website. Retrieved from http://www.kontur-niirs.ru/

Korablev, A. V., F. J. Yanovski, and L. P. Ligthart. 2000. "New method for passive determination of the distance up to lightning source." *Proceedings of the 2000 International Symposium on Antennas and Propagation* (ISAP2000), Vol 2, Fukuoka, Japan, pp. 521–5.

Kropfli, R. A., R. F. Reinking, B. W. Bartram, S. Y. Matrosov, and B. E. Martner. 2002. Icing hazard avoidance system and method using dual-polarization airborne radar, Patent Application No 09/534069, USA, Priority 24 March 2000, Publication date 04/23/2002.

Kuchar, J. K. 2001. "Markov model of terrain for evaluation of ground proximity warning system thresholds." *Journal of Guidance, Control, and Dynamics* 24 (3): 428–35. DOI: 10.2514/2.4748.

Kuhn, P. M., R. L. Kurkowski, and F. Caracena. 1982. Airborne operation of an infrared low-level wind shear prediction system, AIAA-1982-153, American Institute of Aeronautics and Astronautics, Aerospace Sciences Meeting, 20th, Orlando, FL, Jan 11–14, 1982, 7 p.

Lhermitte, R. M. 1973. "Meteorological Doppler radar." *Science* 182 (4109): 258–62. DOI: 10.1126/science.182.4109.258.

Ligthart, L. P., F. J. Yanovsky, and I. G. Prokopenko. 2003. "Adaptive algorithms for radar detection of turbulent zones in clouds and precipitation." *IEEE Transactions on Aerospace and Electronic Systems* 39 (1): 357–67. DOI: 10.1109/TAES.2003.1188918.

Liu, H., and V. Chandrasekar. 2000. "Classification of hydrometeors based on polarimetric radar measurements: Development of fuzzy logic and neuro-fuzzy systems, and in situ verification." *Journal of Atmospheric and Oceanic Technology* 17: 140–64. DOI: 10.1175/1520-0426(2000)017<0140:COHBOP>2.0.CO;2.

Maracich, F. 2005. "Flying free flight: Pilot perspective and system integration requirement." 24th Digital Avionics Systems Conference, DASC 2005, Vol. 1, pp. 2.A.4–2.1–7.

Markson, R., J. W. Warwick, A. Uhlir, Jr. 1990. Interferometric lightning ranging system, United States Patent 4972195, November 20, 1990.

Melvin, W. 1987. "Terminal weather." *Flight International Magazine* 131 (4063): 44–6, 48.

Mid-air collision. 2004. *Flight Safety Australia* (July-August 2004), 27: 22–9.

Mielel, A., T. Wang, and W. W. Melvin. 1995. "Real-time onboard wind and windshear determination, part 2: Detection." *Journal of Optimization Theory and Applications* 84 (1): 39–63. DOI: 10.1007/BF02191734.

Minimum Performance Standards. 1975. RTCA DO-158, RTCA Document 158, Minimum Performance Standards - Airborne Doppler Radar Navigation Equipment, Issued 10/75 (TSO C65a, C68a).

Moore, R. K. 1990. "Ground echo." Chapter 12 In *Radar Handbook*, edited by M. I. Skolnik. New York: McGraw-Hill.

MOPS. 1993. DO-220, Minimum Operational Performance Standards (MOPS) for Airborne Weather Radar with Forward-Looking Windshear Detection Capability. Retrieved from http://www.rtca.org/downloads/DEC%202004%20-%2005-01-06.htm

Nosich, A. I., Y. M. Poplavko, D. M. Vavriv, and F. J. Yanovsky. 2002. "Microwaves in Ukraine." *IEEE Microwave Magazine* 3 (4): 82–90. DOI: 10.1109/MMW.2002.1145680.

Ostrovsky, Ya. P., F. J. Yanovsky, and H. Rohling. 2007. "Turbulence and precipitation classification based on Doppler-polarimetric radar data." *Proceedings of the European Microwave Association* 3 (1): 57–61.

PANS-ATM. 2007. Air Traffic Management. (Doc 4444), 15th edition, incorporating Amendments 1–5, p. 432. Retrieved from http://www.icao.int/ESAF/Documents/meetings/2011/workshop_fpp/docs/onda_fpl2012_presentation_nairobi_2011_%20updated%20(morocco).pdf

Pitertsev, A. A., and F. J. Yanovsky. 2006. "Detection of potential aircraft icing zones by remote sensing of meteorological objects." *Telecommunications and Radio Engineering* 65 (7): 633–40. DOI: 10.1615/TelecomRadEng.v65.i7.50.

Pokrovsky, V. I., V. V. Belkin, and F. J. Yanovsky. 2005. "Airborne weather radar with wind shear detection ability: Influence of generator instability." Proceedings International Radar Symposium IRS-2005, Berlin, Germany, pp. 505–8.

Proctor, F. H., D. A. Hinton, and R. L. Bowles. 2000. "A windshear hazard index." Ninth Conference on Aviation, Range and Aerospace Meteorology, American Meteorological Society, pp. 482–87.

RDR–4B. 2007. Honeywell Aerospace website. Retrieved from http://www.honeywell.com/sites/aero/Radar3_C867EC130-221E-7DEE-00E1-9B9088CBF060_H5CBA7513-E2B2-3320-D5A0-AF32226E4F40.htm

RDR–4000. 2007. Honeywell Aerospace website. Retrieved from https://commerce.honeywell.com/webapp/wcs/stores/servlet/eSystemDisplay?catalogId=10251&storeId=10651&categoryId=35414&langId=-7

Regulations. 2005. Regulations for Terrain Awareness Warning System and Airborne Collision Avoidance System, Commercial and Business Aviation Advisory Circular No. 0236, 2005.07.29. Retrieved from http://www.tc.gc.ca/CivilAviation/commerce/circulars/includes/circulars.asp?lang=en

Ryan, P., and and N. Spitzer. 1977. Stormscope, United States Patent 4023408, May 17, 1977.

Saunders, W. K. 1990. "CW and FM radar." Chapter 14 In *Radar Handbook*, edited by M. I. Skolnik. New York: McGraw-Hill.

Schetzen, M. 2006. "Airborne Doppler radar." In *Applications, Theory and Philosophy* (Progress in Astronautics and Aeronautics) (Volume 215), editor-in-chief F. K. Lu. Reston, VA: American Institute of Aeronautics and Austronautics.

Schwiesow, R. L., and P. A. Lightsey. 1986. "The NCAR airborne infrared lidar system (NAILS)." Langley Research Center 13th International Laser Radar Conference, NASA, National Center for Atmospheric Research, Boulder, CO, p. 3 (N87-10263 01-35).

Seymour, T. J., and R. K. Baum. 1978. Evaluation of the Ryan Stormscope as a severe weather avoidance system for aircraft, FAA-Florida Institute of Technology. Workshop on Grounding and Lightning Technol. pp. 29–35.

Shirman, Ya. D., V. N. Golikov, I. N. Busygin, G. A. Kostin, and V. N. Manshos. 1987. *Theoretical Bases of Radar* (selected pages). Translated into English from *Teoreticheskiye Osnovy Radiolokatsii*, (Moscow, USSR), 1970 pp. 1–420, 483–541, 550–560, Publication Date: 06/1987.

Shupiatsky, A. B., and F. J. Yanovsky. 1990a. Method of determining the location of dangerous hail zones by airborne radar, Patent application No 4793352, USSR, Priority 23 Jun 1990, Positive decision 1991.

Shupiatsky, A. B., and F. J. Yanovsky. 1990b. Radar method for localization of aircraft icing zones, Patent Application No 4898827, USSR, Priority 14 May 1990, positive decision 1991.

Shupiatsky, A. B., and F. J. Yanovsky. 1994. "Some results of sounding thunderstorm and hail clouds by dual-polarization airborne radar." URSI Radio Science Meeting, June, 1994. Seattle, WA, p. 37.

SoutheastAerospace. 2004. KGP-860. Retrieved from http://www.seaerospace.com/king/kgp860.htm

Spitzer, C. R. 1997. "Avionic interface." In *Avionics Navigation Systems* (2nd edition, pp. 691–704), edited by M. Kayton and W. R. Fried. New York: John Wiley & Sons, Inc.

Spitzer, C. R. 2007. *Digital Avionics Handbook* (2nd edition). CRC / Taylor & Francis. DOI: 10.1002/9780470172704.ch15.

Stead, R. P. G. P. Gambarani, and D. H. Tillotson. 1995. "Traffic alert and collision avoidance system (TCAS) transition program (TTP): A status update." *14th Digital Avionics Systems Conference*, DASC, 5–9 Nov 1995, pp. 135–9.

Stevens, M. C. 1988. *Secondary Surveillance Radar* (p. 316). Norwoord, MA: Artech House Publishers.

Stormscope®. 2007. Retrieved from http://www.stormscope.net/

Targ, R., B. C. Steakley, J. G. Hawley, L. L. Ames, P. Forney, D. Swanson, R. Stone et al. 1996. "Coherent lidar airborne wind sensor II: Flight-test results at 2 and 10 vm." *Applied Optics* 35 (36): 7117–27. DOI: 10.1364/AO.35.007117.

TCAS II version 7.1. 2008. Decision criteria for regulatory measures on TCAS II version 7.1, SIRE+/WP7/69/D, 25-07-2008, Retrieved from http://www.eurocontrol.int/msa/public/standard_page/ACAS_Upcoming_Changes.html

Traffic Collision Alerting System. 2008. Honeywell website. Retrieved from www.honeywell.com/

Traffic Surveillance Systems. 2008. Rockwell Collins website. Retrieved from www.rockwellcollins.com/

TSO.C65A. 1983. Airborne Doppler Radar Ground Speed and/or Drift Angle Measuring Equipment, Technical Standard Orders TSO.C65A C65a (For Air Carrier Aircraft). Retrieved from http://www.aero.polimi.it/~rolando/bacheca/imprimatur/TSO_Elenco.pdf

TSO C151a. 1990. Technical Standard Orders (TSO) Terrain Awareness and Warning System (TAWS).

Unal, C. M. H., D. N. Moisseev, F. J. Yanovsky, and H. W. J. Russchenberg. 2001. "Radar Doppler polarimetry applied to precipitation measurements: Introduction of the spectral differential reflectivity." *30th International Conf. on Radar Meteorology*, AMS, Munich, Germany, pp. 316–18.

US Congress. 1989. Office of Technology Assessment, Safer Skies With TCAS: Traffic Alert and Collision Avoidance System–A Special Report, OTA-SET-431. Washington, DC: U.S. Government Printing Office, February 1989.

Wallace, L. E. 1994. Airborne Trailblazer. Retrieved from http://www.univelt.com/univeltdist/nasahiss.htm

Weather Detection & Avoidance. 2008. Honeywell Aerospace website. Retrieved from http://www.honeywell.com/sites/aero/Weather_Detection.htm?c=21

Weather radars. 2008a. Honeywell website. Retrieved from www.honeywell.com/

Weather radars. 2008b. Rockwell Collins website. Retrieved from www.rockwellcollins.com/

Weissmann, M., R. Busen, A. Dörnbrack, S. Rahm, and O. Reitebuch. 2005. "Targeted observations with an airborne wind lidar." *Journal of Atmospheric and Oceanic Technology* 22: 1706–19. DOI: 10.1175/JTECH1801.1.

Weitkamp, C. 2005. *Lidar: Range-Resolved Optical Remote Sensing of the Atmosphere*, Springer Series in Optical Sciences, Volume 102, Springer Science + Business Media Inc.

Wikipedia. 2008. Ground proximity warning system. Retrieved from http://en.wikipedia.org/wiki/Ground_proximity_warning_system

Williamson, T., and N. A. Spencer. 1989. "Development and operation of the Traffic Alert and Collision Avoidance System (TCAS)." *Proceedings of the IEEE* 77 (11): 1735–44. DOI: 10.1109/5.47735.

Wolfson, M. M., R. L. Delanoy, B. E. Forman, R. G. Hallowell, M. L. Pawlak, and P. D. Smith. 1994. "Automated microburst wind-shear prediction." *The Lincoln Laboratory Journal* 7 (2): 399–426.

Yanovsky, F. J. 1974. "The use of airborne radar to estimate the parameters of cloud turbulence." *Radio Engineering and Electronics* 19 (8): 1963–5. (In Russian.) Translated into English in *Radio Engineering and Electronic Physics* 9: 132–4.

Yanovsky, F. J. 1991. *Localization of Hazardous Meteorological Phenomena Onboard the Aircraft* (pp. 1–28). Kiev: Znanie of Ukraine. (in Russian).

Yanovsky, F. 1997. "Methods and means of remote definition of clouds' electrical structure." *Physics and Chemistry of the Earth* 22 (3–4): 241–5. DOI: 10.1016/S0079-1946(97)00139-0.

Yanovsky, F. J. 2003. *Meteorological and Navigation Radar Systems of Air Vehicles* (p. 302). Kiev, Ukraine: NAU Publishing House. (In Ukrainian.)

Yanovsky, F. J. 2004. "Doppler-polarimetric approach for supercooled water detection in clouds and precipitation by airborne weather radar." *Proceedings International Radar Symposium* IRS-2004, pp. 93–100, Warsaw, Poland.

Yanovsky, F. J. 2006. "Airborne weather radar as instrument for remote sensing of the atmosphere." 3rd European Radar Conference. EuRAD-2006, Manchester, Sept. 2006, pp. 162–5.

Yanovsky, F. J. 2008. Automated Dependent Surveillance: Aircraft Position and Weather Data, submitted to 2008 Tyrrhenian International Workshop on Digital Communications - Enhanced Surveillance of Aircraft and Vehicles, Capri, Italy, September, 2008, pp. 132–7.

Yanovsky, F. J., and V. V. Belkin. 1977. "Characteristics of turbulent zones detection by measuring reflectivity factor of clouds." *Theory and Engineering of Radar, Radio-navigation and Telecommunications* (2): 41–4. (In Russian).

Yanovsky, F. J., and B. E. Fishman. 1986. "Tendencies of the development of airborne radio engineering means of flight safety." In *Theory and Practice of Functional Usage and Maintenance of Radio and Electronic Aviation Systems* (pp. 80–4). Moscow: MIIGA. (In Russian.)

Yanovsky, F. J., V. Y. Golubchik, and B. E. Fishman. 1985. "Basic operational requirements to civil aviation airborne systems for weather information display." In *Problems of Optimal Maintenance and Repair of Aircraft Equipment* (pp. 96–100). Kiev, Ukraine: KIIGA. (In Russian.)

Yanovsky, F. J., and A. V. Korablev. 2000. "Airborne sensor for passive determination of the distance up to lightning source." *IEEE 2000 International Geoscience and Remote Sensing Symposium*, Honolulu, Hawaii, 24–28 July 2000, Volume VII, pp. 3172–4.

Yanovsky, F. I., and V. A. Panits. 1996. "The use of antenna with controlled polarization to detect areas of hail and icing. *Radio Electronics and Communications Systems* 39 (10): 20–5.

Yanovsky, F. J., R. B. Sinitsyn, and I. M. Braun. 2002. "Recognition of hail areas with polarimetric radar by the method of potential functions." *Proceedings of the IEEE International Geoscience and Remote Sensing Symposium*, IGARSS '02, Toronto, Canada, Volume 5, pp. 2835–7.

Yanovsky, F. J., A. B. Shupiatsky, and I. P. Kapitalchuk. 1995. "Radar recognition of hail areas." *IEEE Antennas and Propagation Society International Symposium Digest* (Volume I, pp. 290–3), Newport Beach, CA. DOI: 10.1109/APS.1995.530017.

Yanovsky, F. J., C. M. H. Unal, and H. W. J. Russchenberg. 2005. "Retrieval of information about turbulence in rain by using Doppler-polarimetric radar." *IEEE Transactions on Microwave Theory and Techniques* 53 (2): 444–51. DOI: 10.1109/TMTT.2004.840772.

CHAPTER 5

# Devices and Sensors for Linear Acceleration Measurement

S. F. Konovalov
*Bauman Moscow State Technical University*
*Moscow, Russia*

## 5.1 INTRODUCTION

Accelerometers are devices used to measure acceleration, and can be divided into two groups according to their ability to measure constant acceleration: The first group consists of devices measuring vibration acceleration within the frequency band from a fraction of a hertz up to kHz. These accelerometers are unable to measure constant accelerations and, depending on the design and application, have some special names: vibrosensors, piezoaccelerometers, geophonic sensors, and so forth. Accelerometers in the second group are devices capable of measuring both constant and variable accelerations. The accuracy requirements for accelerometers in this group, and also their costs, are given in Figure 5.1 according to the field of application (Barbour et al. 1996; Lawrence 1993). Note that accuracy limits for various types of accelerometers given in Figure 5.1 have a tendency to improve with time.

## 5.2 TYPES OF ACCELEROMETERS

### 5.2.1 *LINEAR AND PENDULOUS ACCELEROMETERS*

These accelerometers differ in the moving element of the electromechanical component. Linear accelerometers have an advantage over pendulous accelerometers in having lower sensitivities to any transverse accelerations oriented normally to the acceleration being measured. However, it is much more difficult to minimize the effects of perturbations such as elastic forces, friction, and so forth (Abbaspour-Sani, Huang, and Kwok 1994). Pendulous accelerometers allow the measurement of a wider range of accelerations. Therefore, precision instruments involve only

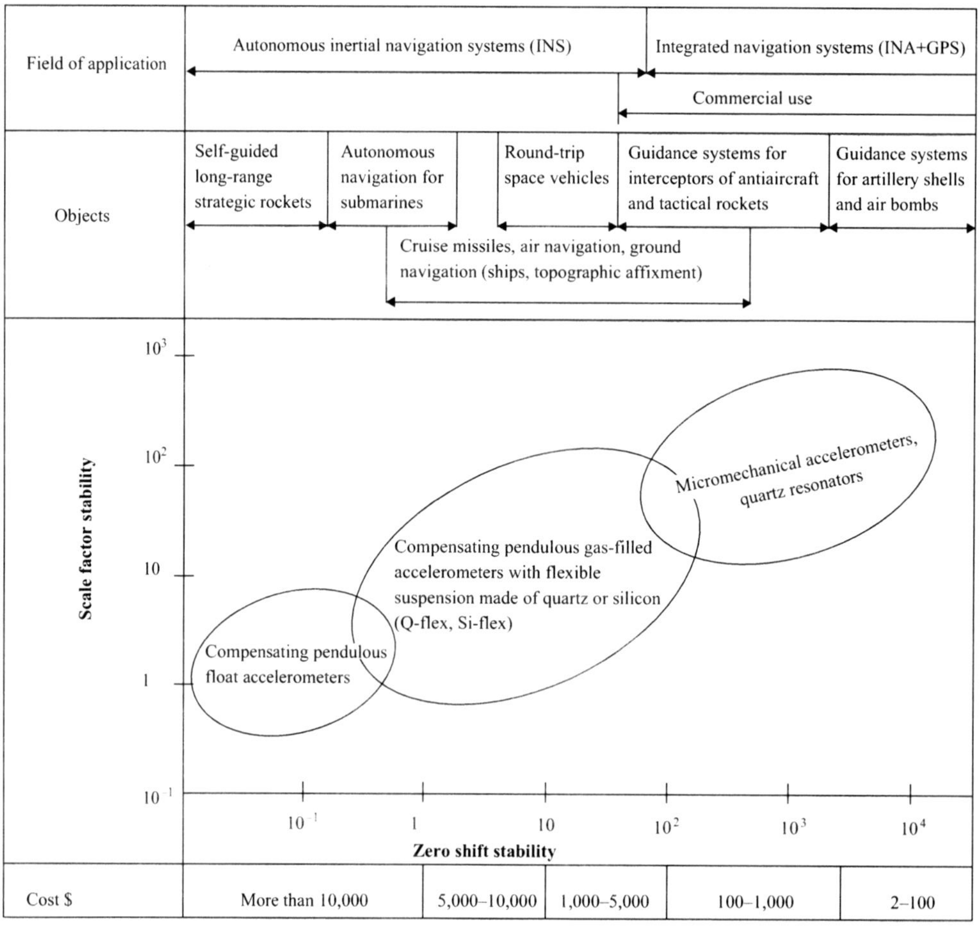


**Figure 5.1.** Accelerometer accuracy parameters depending on type, cost, and application area.

pendulous systems, whereas systems with linear moving components are used in relatively rough micromechanical devices and quartz resonators.

## 5.2.2 DIRECT CONVERSION ACCELEROMETERS AND COMPENSATING ACCELEROMETERS

Accelerometers also differ in the ways their output signals are formed. There are direct conversion devices (feed-forward devices or devices with mechanical springs) and compensating devices (feedback devices or devices with electrical springs). Both are considered below.

### 5.2.2.1 Direct Conversion Accelerometers

A micromechanical accelerometer with a silicon pendulum is shown in Figure 5.2.

The device consists basically of the pendulous unit and two insulating plates (1) and (3) (Barth et al. 1988). Evaporated onto each of these are metallic electrodes (7) forming a capacitive

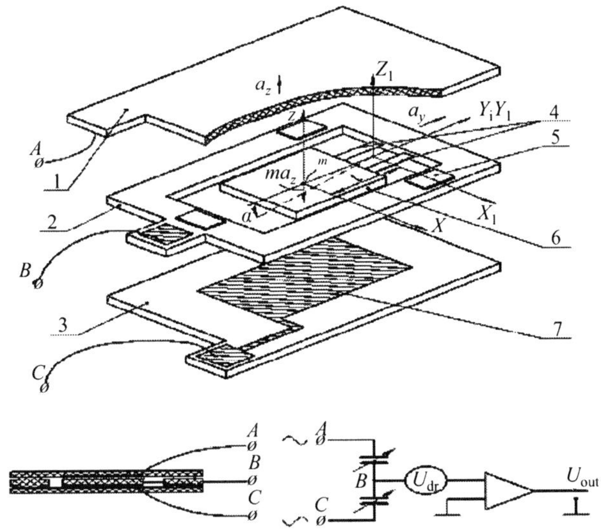


**Figure 5.2.** Direct conversion micromechanical accelerometer: 1,3—insulating plates with metallic electrodes 7; 2—silicon pendulous unit frame with base lugs (5); 4—flexors; 6—pendulum.

pick-off. In the pendulous unit is the frame (2) with base lugs (5) placed symmetrically on each side, and the flexors (4). The pendulum (6) itself is made from a silicon plate by anisotropic etching with shear plane orientation (001). This silicon is doped with phosphorus and hence has *n*-type conductivity. The contacts A, B, and C are connected to the electrodes (7) and to the body of the pendulous unit. When the accelerometer moves along the *z*-axis due to an acceleration, $a_z$, the pendulum, under the influence of the inertial force $ma_z$ acting at arm *l*, tilts to an angle $\alpha$. While tilting, the pendulum deforms the flexors which provide both the pendulum suspension and the spring that opposes the tilt (with a spring rate $K_{spr}$). When the spring moment $K_{spr} \times \alpha$ is equal to the m oment $ml\,a_z$ the pendulum will stop tilting. The viscous gas in the space between the pendulum and the electrodes (7) guarantees damping of the pendulum motion with a damping coefficient, *D*.

The electrodes on the top and the bottom plates (1 and 3) are energized by two antiphase voltages so that when the pendulum is in the central position it has zero potential. When tilt occurs, an induced signal voltage, $U_{out}$, appears on the pendulum which is proportional to the pendulum displacement, $\alpha$; and the phase of this signal corresponds to the direction of the pendulum displacement. The transfer ratio of the pick-off is $K_{po}$ and the operation of the accelerometer is characterized by the following equations:

$$\begin{aligned} J\ddot{\alpha} + D\dot{\alpha} + K_{spr}\alpha &= ml(a_z - a_y\alpha); \\ U_{out} &= K_{po} \cdot K_{amp} \cdot \alpha + K_{amp} \cdot U_{dr} \end{aligned} \tag{5.1}$$

where $J$ is the moment of inertia of the pendulum,
$K_{amp}$ is the amplifier gain coefficient, and
$U_{dr}$ is the sum of the amplifier drifts referred to the input, that is, the inherent amplifier drift and the drift arising in the pick-off due to factors such as inequality between the antiphase electrode voltages.

For the case where $a_z$ and $a_y$ are constant,

$$U_{out} = \frac{ml}{K_{spr} + ml\,a_y} K_{po} \cdot K_{amp} a_z + K_{amp} \cdot U_{dr} \tag{5.2}$$

Equation (5.2) shows that the accuracy of the direct conversion accelerometer depends on the voltage $U_{dr}$ and the stability of the coefficients $K_{po}$ and $K_{amp}$. Moreover, the cross acceleration $\boldsymbol{a}_y$ also has an influence on the device accuracy increasing with the amplitude $a$. This influence can be decreased by reducing the coefficient $K_{spr}$ but this leads to a higher error in $U_{dr}$. To reduce the $U_{dr}$ error $K_{spr}$ should be reduced, but this produces a growth in the $\boldsymbol{a}_y$ error and, in addition, the bandwidth of the device decreases. (The natural frequency of the undamped oscillations of the accelerometer pendulum is calculated in accordance with the formula $f_0 = \sqrt{\frac{K_{spr}}{J}}$.)

For these reasons it is impossible to achieve high accuracies in direct conversion accelerometers, which is why they are usually used when the permissible error exceeds 0.1–0.5% of the maximum acceleration, $a_{max}$, measured by the device.

### *5.2.2.2 Compensating Accelerometers*

An example of this type of accelerometer is the Q-flex accelerometer shown in Figure 5.3. (Cardy 1984).

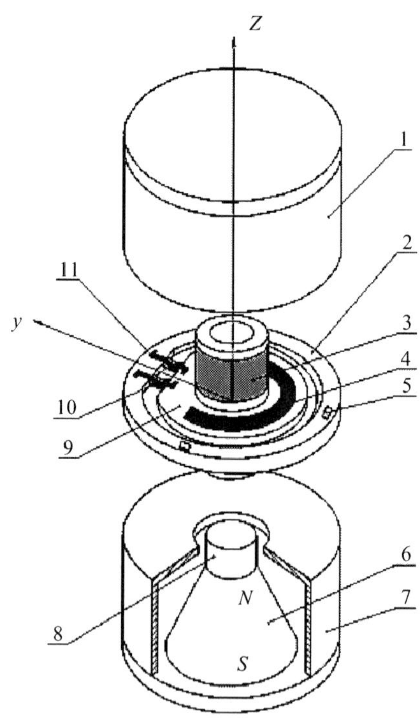


**Figure 5.3.** A Q-flex compensating accelerometer: 1,7—magnetic conductors; 2—quartz frame for pendulous unit with base lugs (5); 3—torque motor coil; 4—capacitance pick-off electrode; 6—permanent magnet; 8—pole piece; 9—movable pendulum plate; 10—flexors; 11—evaporated metal connectors.

This device incorporates two magnetic conductors (1 and 7), which clamp the pendulous unit, this being fabricated from a fused quartz block using chemical etching in hydrofluoric acid. The pendulous unit includes a circular frame (2) with base lugs (5) placed symmetrically on the both sides, the flexors (10), and the movable pendulous plate (9). On each side of this plate are evaporated metal electrodes (4) forming capacitance pick-offs, and also the coils (3) of the torque motor. The total mass of the pendulum plate together with the elements installed on it is $m$. The electrodes (4) and the coils (3) are connected to the electronic circuitry via metal connectors (11) evaporated onto the flexors. The stationary electrodes of the capacitance pick-off are the grounded magnetic conductors (1 and 7).

The pick-off capacitors are connected into a bridge circuit energized by a sine wave voltage generator. The outputs of this bridge circuit and the voltage across a reference resistor, $R_{ref}$, are compared, amplified, and applied to the coils of the compensating torque motor (3). The resistance of these coils is $R_{TM}$. The magnetic circuit of the torque motor, the magnetic conductors (1 and 7), and the permanent magnets (6), along with the poles (8), create magnetic fields in the annular gaps.

When the accelerometer moves along the $z$-axis with an acceleration $a_z$, the pendulum, under the influence of the inertial force acting at an arm $l$, tilts at an angle $\alpha$. This deforms the flexors, but in the compensating accelerometer their only function is the suspension of the pendulum because the feedback provided by the torque motor brings them back to the null position. Hence, these flexors can be very thin in order to minimize the moment $K_{spr}\,\alpha$. This is achieved as follows.

The displacement of the pendulum causes imbalance in the capacitive pick-off, the signal from which is applied to the amplifier to provide a current $\boldsymbol{i}_c$ in the torque motor coils. The resultant magnetic field produced by the torque motor, and that in the magnetic field in the annular gaps, produces a compensating force $F_c = B \cdot i_{TM} \cdot w \cdot l_c$. Here, $B$ is the magnetic flux density, $w$ is the number of turns in the coils (3) within the gap of the magnetic conductor, and $l_c$ is the length of one turn of the coil. The torque motor is designed so that the compensating force is exerted on the pendulum at its center of gravity. When the pendulum is deflected, the inertial force moment is balanced by the compensating moment $M_{TM} = F_c \cdot l = K_{TM} \cdot i_{TM}$. Pendulum oscillations are damped by the moment $D \cdot \dot{\alpha}$ where $D$ is the viscosity of the gas in the sensor.

The operation of the accelerometer is defined by the following equations:

$$\begin{cases} J\ddot{\alpha} + D\dot{\alpha} + K_{spr}\,\alpha = ml(a_z - a_y \cdot \alpha) + M_{TM} + M_{dist}; \\ M_{TM} = K_{TM} \cdot i_{TM}; \\ i_{TM} = (K_{po} \cdot K_{amp} \cdot \alpha + K_{amp} \cdot U_{dr}) / (R_{TM} + R_{ref}); \\ U_{out} = i_{TM} \cdot R_{ref}; \end{cases} \tag{5.3}$$

In these equations $M_{dist}$ is the sum of the disturbing moments affecting the pendulum, which comprise the following effects:

- The electrostatic forces affecting the pendulum electrodes when the pendulum tilts from the neutral position
- The magnetic interaction forces caused by the existence of any magnetic inclusions in the elements of the pendulum
- The hysteresis moments appearing during flexor deformation together with the evaporated conductors

- The flexor moments that appear in the frame (2) caused by inequalities in the temperature expansion factors of the quartz and the material contacting with the magnetic conductor frame (1 and 7)
- The temperature of the gas flow.

For the case where $a_z$ = const, $a_y$ = *const*, symbolizing $K = \dfrac{K_{po} \cdot K_{amp} \cdot K_{TM}}{R_{TM} + R_{ref}}$ and taking into consideration that:

$$K >> K_{spr};\ K >> ml a_y;\ U_{dr} << \frac{ml a_z}{K} \cdot K_{po};\ M_{dist} << ml a_z,$$

then

$$U^*_{out} = \frac{ml R_{ref}}{K_{TM}}(a_z - \alpha^* a_y) + \frac{U_{dr} K_{spr} R_{ref}}{K_{po} \cdot K_{TM}} + \frac{M_{dist} R_{ref}}{K_{TM}};\ \alpha^* = \frac{ml}{K} a_z + \frac{U_{dr}}{K_{po}} = \alpha_1^* + \alpha_2^* \qquad (5.4)$$

If there is no cross acceleration $a_y$ and the elastic connection between the pendulum and the accelerometer is weak ($K_{spr} \to 0$), $M_{dist}$ = 0, the output accelerometer signal is:

$$U_{out} = \frac{ml R_{ref}}{K_{TM}} a_z \qquad (5.5)$$

Thus, in contrast to the direct conversion accelerometer, the accuracy of a compensating accelerometer with a low $K_{spr}$ is practically independent of transfer coefficients and zero drifts in the amplifier and the pick-off. As a result, it is easier to maintain high accuracy in the device. That is, the smaller the angle $\alpha^*$, the less will be the effect of any $a_y$ acceleration.

In Equation 5.4 the first component $\alpha_1{}^*$ is the static error of the device feedback loop. This component produces the pendulum deflection that is necessary to maintain a current $i_{TM}$ for balancing the compensating moment and the inertial moment. To reduce the angle $\alpha_1{}^*$, the rigidity $K$ of the compensating circuit should be increased. Thus, unlike the direct conversion accelerometer, any growth in the coefficient $K$ does not lead to growth in the error caused by $U_{dr}$. In the compensating accelerometer it is possible to eliminate the effect of the angle $\alpha_1{}^*$ by including in the compensating circuit an integrating element in parallel or series with the amplifier. This inclusion will make the device loop astatic, which is useful for the purpose of reducing cross-acceleration effects and for minimizing and stabilizing the disturbing moment. The effect of the $U_{dr}$ voltage on the device accuracy (at $K_{spr} \to 0$) is negligible, but it is worth taking into account that $U_{dr}$ influences the zero position of the pendulum (the second component $\alpha_2$*) and this in turn causes a cross-acceleration effect and instability in the moment $M_{dist}$.

Compensating accelerometers are used where high accuracy and dynamic characteristics are required. The accuracy of these accelerometers is determined by the pendulum stability *ml*, the stability of the reference resistor $R_{ref}$, and the transfer coefficient of the compensating sensor $K_{TM} = B \cdot w \cdot l_t$, and this accuracy is consequently determined by the stability of the permanent magnets employed. It is also necessary to minimize the $U_{dr}$ voltage and the moment $M_{dist}$. Furthermore, in order to eliminate temperature deformation effects on the $M_{dist}$ moment stability in devices with elastic pendulum suspension (like the Q-flex device of Figure 5.3), it is useful to make a cut in the frame section between the flexors (10) and one of the closest lugs (5) (see Figure 5.4) so that the frame deformation does not influence these flexors. The other parameters are stabilized by appropriate choice of materials and technological processes.

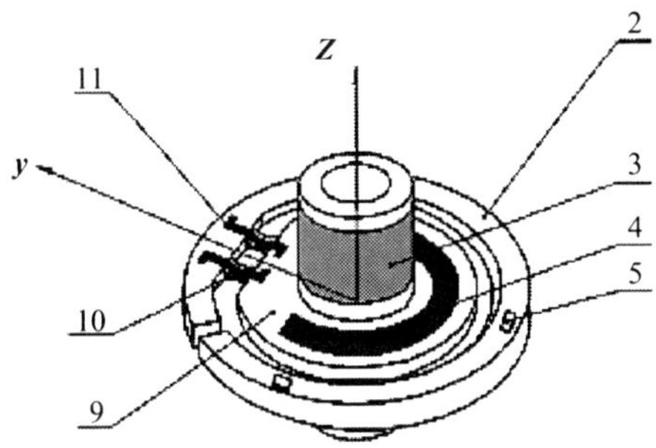


**Figure 5.4.** Q-flex type accelerometer pendulous unit with cut in frame.

## 5.3 ACCELEROMETER PARAMETERS

Accelerometers are usually characterized by the following parameters (Cardy 1984; de Coulon et al. 1993; Olvey, Knox, and Cohn 2004).

### 5.3.1 *ACCELERATION MEASUREMENT RANGE* $a_{zmax}$

$a_{z\max}$ is the maximum acceleration the device can measure.

### 5.3.2 *RESOLUTION* $a_{zmin}$

$a_{z\min}$ is the minimum acceleration increment the user is able to recognize against the noise background and some static phenomena caused by frictional moments in the bearings (if sleeve bearings are used), hysteresis effects in the magnetic supports, and the elastic suspensions of the pendulum, and so forth. The IEEE STD 530-1978 standard defines the following error components for $a_{z\min}$: the hysteresis at switching on, the threshold, the resolution, and the dead band. For each of these components an experimental calculation method exists, but some errors in $a_{z\min}$ cannot be eliminated by algorithmic compensation.

### 5.3.3 *ZERO SIGNAL (BIAS)* $a_0$

$a_0$ is the acceleration value corresponding to the output signal, $U_0$, at the output of the accelerometer in the absence of a measured acceleration $a_z$. In a compensating accelerometer, this bias depends mostly on the disturbing moment $M_{dist}$. In direct conversion devices the zero signal $a_0$ depends mostly on the $U_{dr}$ voltage. This zero signal $a_0$ can change in both reversible and irreversible ways under the influence of temperature changes, shock, and vibration impacts.

Usually, the constant component and the reversible changes are given in technical documentation in the form of a time dependence $a_0(t)$ with a power series approximation. These constant and reversible changes are subject to algorithmic compensation and for this purpose a temperature sensor is installed inside the accelerometer's case. The irreversible changes in zero-shift stability along with the scale factor (see Figure 5.1) determine the accuracy of autonomous navigation systems. These changes are carefully regulated and the equipment must be periodically tested and recalibrated. During tests for the evaluation of the acceleration $a_0$, the accelerometer output signals $U_{(+1g)}$ and $U_{(-1g)}$, corresponding to gravitational accelerations $+1g$ and $-1g$ are measured. The approximate value of $a_0$ is then calculated by the following formula:

$$a_{0c} = \frac{U_{(+1g)} + U_{(-1g)}}{U_{(+1g)} - U_{(-1g)}}$$

However, owing to the nonlinearity of the accelerometer output characteristic, this *calibrated value* can differ from the true value. Nevertheless, the calibrated value is used for device calibration because the difference between $a_0$ and $a_{0c}$ is negligible, and the accuracy of the calculation is almost independent of the error of the device when measuring vertical axis orientation (the cosine dependence between the gravity acceleration projection and the orientation error).

The zero-signal can also change because of vibration effects (Konovalov 1991) called *vibration errors*. The reasons for vibration errors involve both methodical and instrumental phenomena, that is, they can be caused by imperfections in the accelerometer elements. Among the methodical errors, special attention should be given to the single-side deflection effect caused by skew vibrations (see Figure 5.5).

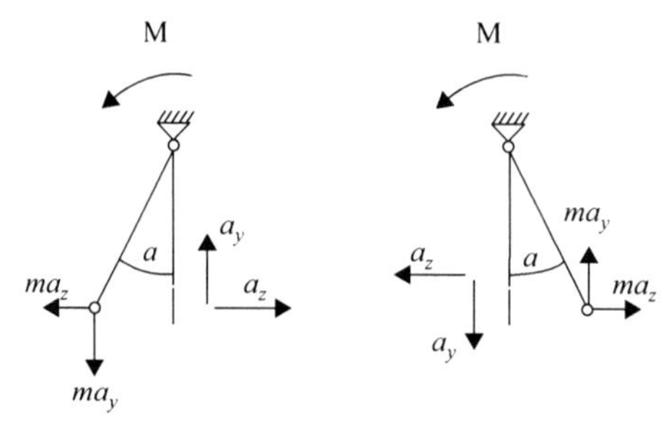


**Figure 5.5.** Single-side deflection effect caused by skew vibrations.

In Figure 5.5, extreme deflections of the pendulum caused by the acceleration $a_z = a_{zA} \cdot \sin \omega t$ during oscillations are shown (Figure 5.5 represents a low-frequency vibration situation where the pendulous oscillation is in phase with the inertial force $m \cdot a_z$). Thus, the moment $M$ created by the inertial force has the same sign in both positions and causes zero-signal changes. The main reasons for the instrumental errors are nonlinear effects in the compensating torque motor, the pendulum suspension, and the damper.

Because vibration errors cannot be compensated by algorithmic means, these errors must be minimized by a strict set-up procedure.

### 5.3.4 SCALE FACTOR $K_a$

The scale factor $K_a = \frac{U_{out}}{a_z}$ is the slope of the dependence $U_{out}(a_z)$. For compensating devices $K_a \approx \frac{ml R_{ref}}{K_{TM}}$, and for direct conversion devices it is $K_a \approx \frac{K_{po} \cdot K_{amp} \cdot ml}{K_{spr}}$.

There is no strong requirement for scale factor stability in direct conversion devices, but when dealing with compensating devices the scale factor question is very important. The scale factor of compensating accelerometers is not constant and depends on the value of the measured acceleration because $K_{TM}$ and $M_{dist}$ depend on the angle $\alpha$ and the current $i_{TM}$. All this creates nonlinearity in the output characteristic of the device.

Furthermore, $K_a$ is a function of time and temperature. These two dependences are caused by the features of the torque motor magnets and the ability of these magnets to maintain magnetization. The dependences of $K_a$ and $U_{out}(a_z)$ on time and temperature have both reversible and irreversible characteristics. Insofar as the accuracy of autonomous navigation systems depends strongly on $K_a$, the reversible dependences must be evaluated and compensated by algorithmic means. Irreversible changes (Figure 5.1: scale factor stability) must be measured and subjected to regular tests. During calibration the signals $U_{(+1g)}$ and $U_{(-1g)}$ are measured and their calculated values accepted as $K_{ac} = \frac{U_{(+1g)} + U_{(-1g)}}{2g}$. The relationship (Equation 5.5) between the accelerometer output signal $U_{out}$ and the acceleration $a_z$ is presented in Figure 5.6(a).

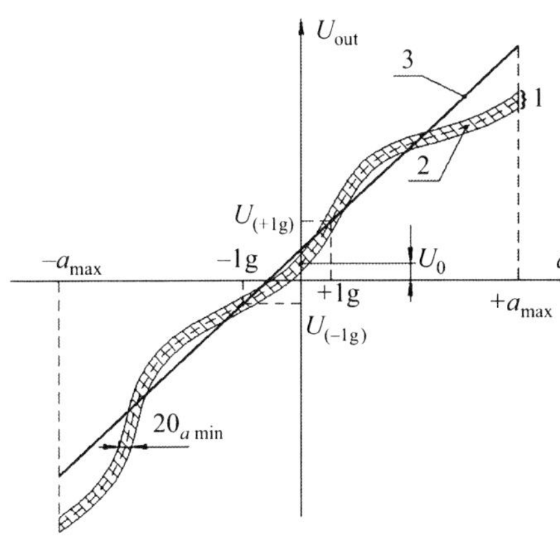


**Figure 5.6.(a)** Relationship between an accelerometer output signal $U_{out}$ (1) and the acceleration $a_z$; the true characteristic (2); and the calibrated characteristic (3).

Figure 5.6(b) shows the large scale dependence within the calibration range (±1g). The deviation of the true characteristic (2) of the device from the calculated characteristic (3) is called the *nonlinearity error* of the accelerometer.

This error is usually evaluated in comparison with the measurement range and calculated by the formula $\delta = \frac{U_{true}(a_z) - U_{calc}(a_z)}{K_{ac}} \cdot 100\%$.

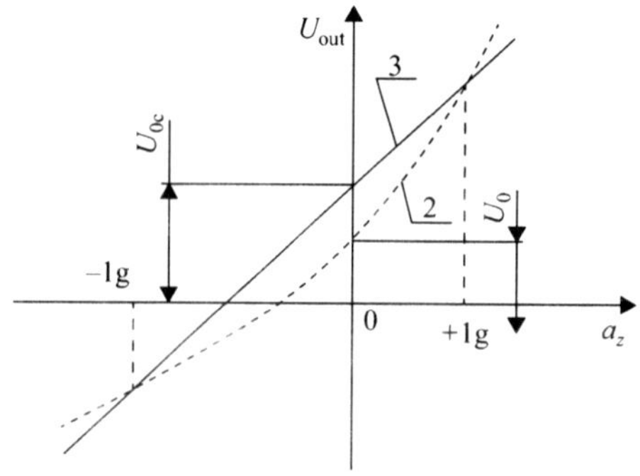


**Figure 5.6.(b)** Large scale $U_{out}$ versus $a_z$ dependence within the calibrated range ±1g.

The algorithmic compensation of the nonlinear output characteristic is approximated as a power series:

$$U_{true}(a_z) = U_{0c} + K_{ac} \cdot a_z + K_2 \cdot a_z^2 + K_3 \cdot a_z^3 \tag{5.6}$$

In the device specifications the nonlinearity error is given by the values $\frac{K_2}{K_{ac}}$ and $\frac{K_3}{K_{ac}}$ which are usually measured in $\frac{\mu g}{g^2}$ and $\frac{\mu g}{g^3}$, respectively.

Among the parameters mentioned, the technical specifications include the reversible temperature dependences $K_{ac}$, $K_2$, and $K_3$ which should be taken into account for the algorithmic compensation of the errors.

### *5.3.5 BIASING ERROR (MISALIGNMENT)*

The case of the accelerometer has several mounting faces used for installation into the measurement system. These faces usually include three lugs (see Figure 5.7) forming the base plane, or may be in the form of cylinder with a flange and a bracket with a lug or a groove (see Figure 5.8).

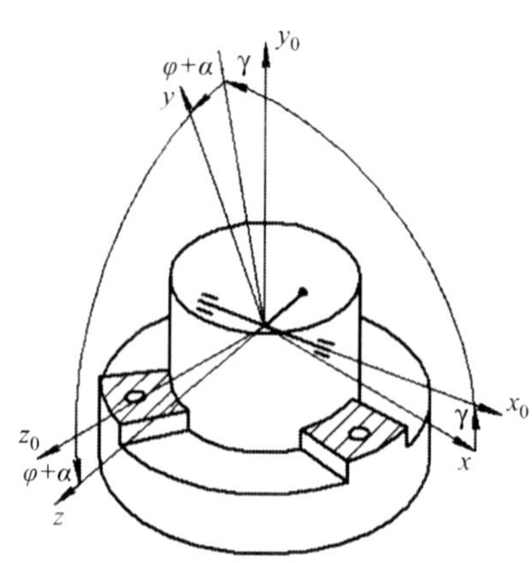


**Figure 5.7.** Mounting face with three lugs forming the base plane.

In a properly manufactured accelerometer (Figure 5.7) the zero-angle $\alpha$ of the pendulum deviation, the $x$-axis of the pendulum suspension, and the $y$-axis directed along the pendulum arm, all coincide with the lug plane. The accelerometer sensitivity axis $x$ coincides with the accelerometer measurement axis $z_0$ normal to the lug plane.

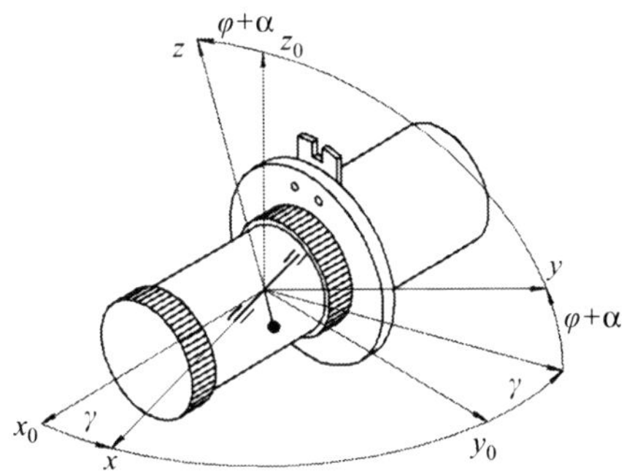


**Figure 5.8.** Mounting face in the form of a cylinder with a flange and grooved bracket.

In the properly manufactured accelerometer of Figure 5.8, at the angle $\alpha = 0$, the $x$-axis coincides with the base cylinder axis and the $z$-axis coincides with the axis $z_0$ which lies in the plane traversing the center of the base groove. Owing to technological peculiarities, when tilting the pendulum at the angle $\alpha$, the sensitivity axis $z$ deviates from the measurement axis $z_0$ by the angles $\gamma$ and $\varphi + \alpha$. The angles $\gamma$ and $\varphi$ are called the accelerometer *biasing errors*. Changes in these angles can be reversible (e.g., when caused by working temperature changes) and irreversible (when caused by permanent deformations of the device elements after shock and vibration effects; or by long-time exposure in conditions of extreme temperature; or by zero drift in the amplifier and the pick-off). The constant components of the angles $\gamma$ and $\varphi$ and their reversible temperature dependence are given in the technical certificate (data sheet) and are subject to algorithmic compensation during service. The irreversible changes are strictly defined and are given in the technical parameters of the device in the form of biasing error instability figures.

### *5.3.6 ACCELEROMETER FREQUENCY CHARACTERISTICS*

Usually, the bandwidth of the accelerometer is known, being determined in accordance with its amplitude–frequency characteristic (response). This characteristic is defined by the frequency at which the variable sinusoidal acceleration amplitude becomes equal to 0.7 of its static value $K_a$. Also, the ripple in the amplitude–frequency characteristic of the accelerometer is usually quoted. However, insofar as navigational grade accelerometers are concerned, and especially the types used in platform inertial navigation systems, the dynamic characteristics of such devices are of secondary importance and not subject to strict regulations.

### *5.3.7 SPECIAL ACCELEROMETER PARAMETERS*

The characteristics described in Sections 5.3.1 through 5.3.6 are almost always given in the device specifications, and in a number of cases the following parameters are also provided.

### 5.3.7.1 *Magnetic Leakage*

It is usually indicated that this parameter may not exceed a specified level at a given distance from the device in all directions. This parameter is of importance in cases where accelerometers are used in systems involving magnetoelectric directional sensors.

### 5.3.7.2 *Electromagnetic Noise*

An accelerometer is an electromechanical device that produces electromagnetic noise that may lead to electromagnetic pickup by other elements in the relevant systems. Usually, the level of this electromagnetic noise is small, but it must nevertheless be determined in especially crucial cases.

### 5.3.7.3 *Readiness Time*

This parameter may have two meanings, either the time for transient process completion in the relevant electronic circuits, or the time of completion of the transient process in the compensating circuit.

The heating time of an accelerometer is that period from switch-on until the time when the thermal field in the accelerometer case is fully formed and the thermal drifts of the device parameters are complete. It is worth mentioning that thermal drifts in the device parameters during heating can be algorithmically compensated according to information from temperature probes installed within the device case. However, the accuracy of such compensation during the temperature transient process is less than the accuracy of algorithmic compensation carried out at the steady-state temperature value.

### 5.3.7.4 *Noise Level in the Accelerometer Output*

The basic source of electromagnetic noise is the amplifier within the compensation feedback loop. The presence of the noise component in the signal $U_{out}$ limits the accelerometer's capability of measuring a variable acceleration component, and causes instability in the analog–digital converter that provides the connection between the accelerometer and a computer. Usually the noise component is specified in the accelerometer datasheet either as a noise distribution frequency spectrum or as a noise $\frac{g}{\sqrt{Hz}}$.

### 5.3.7.5 *Sensitivity to External Constant and Variable Magnetic Fields*

If there is insufficient magnetic shielding, both constant and variable external magnetic fields can penetrate into accelerometer working gaps and, being summed with the permanent magnet flux in the magnetic gap, can influence the Scale Factor $K_{ac}$.

A variable magnetic field induces an electromagnetic interference signal in the accelerometer circuits and so increases the noise level in the output signal. At high field intensities it can cause partial magnet demagnetization that can exert an influence on the $K_{ac}$ stability.

#### 5.3.7.6 Sensitivity to Changes in Power Supply Voltage

As follows from Equation 5.4, the influences of the amplifier parameters and those of the generator powering the compensation accelerometer pick-off are weak but cannot be completely eliminated. As a result, any instability in the source powering the accelerometer electronics can exert an influence on the accelerometer accuracy. It is usual to estimate the changes in $K_{ac}$ and $U_0$ due to ±10% voltage changes in the power supply.

#### 5.3.7.7 Sensitivity to External Pressure, Humidity, and Radiation

All these external factors can influence accelerometer operation. An external pressure change causes deformation of the case and hence can influence the value of $U_0$. Humidity changes can result in conductivity changes in external accelerometer circuits that may result in variations in $K_{ac}$ (e.g., owing to changes in the value of $R_{ref}$). Radiation influences the properties of evaporated metal electrodes and can cause parameter changes in the accelerometer module electronic components. A number of additional accelerometer parameters should be determined in the cases of special accelerometer applications which are not considered in this chapter.

## 5.4 FLOAT PENDULOUS ACCELEROMETER (FPA)

As mentioned in the introduction, float pendulous accelerometers are intended for use in high accuracy inertial navigation systems. The best types of FPAs have the following characteristics:

- zero-signal $a_0 \approx 1...10\mu g$
- stability of the zero-signal $a_0 \approx 0.1...1\,\mu g$
- scale factor stability 1 ppm
- output characteristic nonlinearity $\sim 10^{-4}$ % of the measurement range
- sensitivity threshold $a_{min} \approx 10...100\eta g$
- stability of the measurement axis angular position (the basic error stability) ~ a few angular seconds (Barbour et al. 1996).

Such characteristics make possible the use of the FPA as the sensitive element in autonomous navigation systems for rockets. These autonomous systems are independent of any external adjusting signals and are consequently absolutely jam proof. Due to high shock and vibration resistance and tolerance, the FPA can be used in conditions of the extreme dynamic loads that are typical in vehicles for military use.

However, there are several factors limiting the wide use of the FPA including high manufacturing complexity and consequently high prices (tens of thousands of dollars), and also a limited temperature span. When the temperature is lower than +5°C the heavy float liquids (density $\rho \approx 2$ g cm$^{-3}$) become too viscous and malfunctions may occur. Therefore, it is impossible to use the FPA in instantaneous readiness systems without preliminary thermostating of the device or using a heated instrument module.

The FPA consists of an *electromechanical unit* (EMU) and an *electronic compensating circuit* that can be either analog or digital. This circuit can be installed inside the EMU case as well as partly external to it. When choosing a location for the compensating circuit several facts should be taken into consideration, as follows. The closer the preamplifier is placed to the pick-off of the device, the lower the levels of noise and electrostatic pickup. That is why it seems advisable to install the preamplifier directly on or inside the accelerometer case. However, the less heat emitted inside the device, the higher is its accuracy, so it may actually be better to install the electronic compensating circuit elements emitting any appreciable heat separately from the accelerometer case.

### *5.4.1 BASIC EMU DESIGN SCHEMES*

The basic design of EMUs depends mainly on the type of torque motor used.

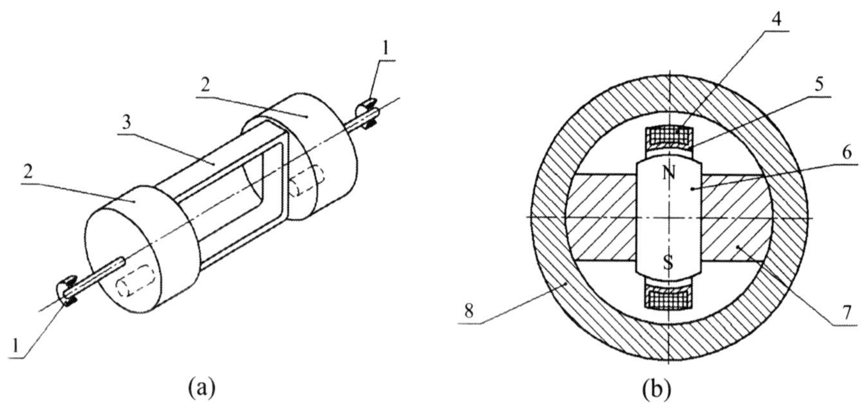


**Figure 5.9.** Design scheme for an electromechanical FPA unit (a), and cross-section of a torque motor (b). 1—sleeve bearings; 2—floats; 3—torque motor frame with coil; 4—coil; 5—frame; 6—permanent magnet; 7—nonmagnetic fixing inserts; 8—magnetic conductor ring (usually the case).

Figure 5.9 is a basic schematic for a two-float pendulous unit showing (a) an isometric view and (b) a sectional view showing the torque motor. The two-float FPA scheme is used when the device employs a magnetoelectric torque motor with a two-pole magnet and a rectangular frame. This design is more complicated than for the single-float equivalent, but the two-float design has a number of advantages relevant to the symmetry of the pendulous unit. The two-pole magnet torque motor exhibits only a weak dependence on the slope of the frame turn angle, and it has short front and rear parts of the frame and coil (the parts of the frame placed outside the air-gap of the magnetic system that do not create the moment). Consequently, such a sensor can create the necessary moment without high power consumption. This ability is evaluated by the ratio $\frac{M}{\sqrt{P}}$, where $M$ is the moment generated by the torque motor for a power consumption $P$. Because of these merits, devices with two-float pendulous units form the prevailing choice for precise accelerometers.

The single-float scheme is used for pendulous units in two situations. In the first, disassembled magnetoelectric torque motors are used (see Figure 5.10). In such sensors the frame with the coils can be taken out of the air-gap without the necessity of disassembling the magnet systems. In the second, electromagnetic torque motors are used (see Figure 5.11).

Figure 5.10a,b present isometric views of the float for both six-pole and two-pole magnets. Figure 5.10c is a cross-section of the pendulous unit through the torque motor showing the fixing of its coils to the float and the stationary magnetic system. Figure 5.10d is a cross-section of the pendulous unit through the torque motor showing the fixing of the magnet and the ring-type magnetic conductor to the float and the stationary coils.

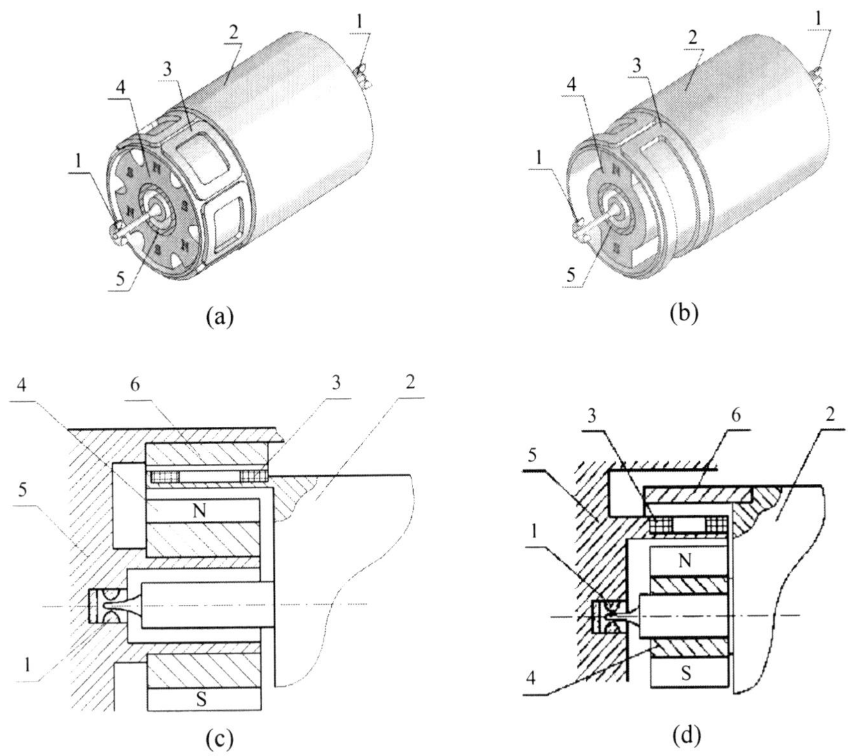


**Figure 5.10.** An FPA float with a magnetoelectric torque motor. (a) Isometric view of a torque motor with six-pole magnet, (b) isometric view of a torque motor with two-pole magnet, (c) cross section of float torque motor with movable coil, (d) cross section of a float torque motor with movable magnet and magnetic conductor. 1—bearings; 2—float; 3—torque motor coils; 4—permanent magnet; 5—case face; 6—ring-type magnetic conductor.

Figure 5.10a shows a pendulous unit with a six coil torque motor installed into a cylindrical framework complete with float. The six-pole magnet of the torque motor (4) and the ring-type magnetic conductor are fastened to the case of the device (this being the variant using a torque motor with a stationary magnetic system). Some advantages and disadvantages of this design are as follows.

### 5.4.1.1 Advantages

1. The absence of any magnets and magnetic elements within the float construction eliminates spurious moments and their interactions with external magnetic fields and magneto-conductive case elements (such detrimental moments are usually called *spurious magnetic interaction moments*).

2. Light-weight torque motor mobile elements (the coils) do not require a large float in order to maintain neutral flotation.
3. The coils have short frontal parts, and as a result the sensor has a high value of the ratio $\frac{M}{\sqrt{P}}$.

### *5.4.1.2 Disadvantages*

1. The mobile coils require current-carrying wires.
2. For a six-pole magnet the induction in the air-gap changes according to the law $B = B_0 \cos(3a)$, whilst in the case of a two-pole magnet the law $B = B_0 \cos a$ becomes valid. Because of this, the torque motor would have a more evident dependence on the slope of the pendulous unit turn.

The variant shown in Figure 5.10b differs by using a torque motor with a two-pole magnet. Because of this, the slope of the sensor is only weakly dependent on the pendulous unit turn (induction changes according to the law $B = B_0 \cos a$), but the coils have long frontal parts and therefore such sensors have a small value of the ratio $\frac{M}{\sqrt{P}}$.

Single-float pendulous units can be built with another variation in the torque motor elements wherein the coils are stationary and the float is fixed to the magnet and the magnetic conductor ring (see Figure 5.10d). In this case, flexible current-carrying wires are not required, but the pendulous unit is heavy (the magnet and the ring are heavier than the coils) because in order to maintain zero-flotation the dimensions of the float must be increased. Furthermore, in the presence of the mobile magnet, spurious magnetic moments appear, influencing the value and stability of the acceleration $a_0$.

There is also a third variant, consisting of a torque motor with a stationary magnetic conductor ring, a stationary coil and a mobile magnet. However, this variant is vigorously rejected for use in precise FPA designs because of very high and ambiguous spurious magnetic forces and moments influencing the pendulous unit. Thus, for use in an FPA, bearings with high radial rigidity (e.g., taut bends or magnet suspensions) in a mobile six-pole magnet construction with stationary coils and a ring-type magnetic conductor, the pendulous unit would have six stable positions relative to the case of the device. While moving from one position to another the device characteristics would change and consequently a high FPA precision could not be maintained.

In an FPA with an electromagnetic torque motor, the float should be built symmetrically; an isometric view of such a device is shown in Figure 5.11a (Draper 1998).

Figure 5.11b is a sectional view of the pendulous unit through the torque motor. The rotors of the magnetic suspensions are installed inside the float on both sides, and on a non-magnetoconductive collar. On the outside is the rotor of the electromagnetic torque motor on one side, and the rotor of the induction pick-off on the other. Together with the stators, these rotors form the so-called *dual coplanar microsin (ducosin),* which consists of torque motors and pick-offs of the same size and construction. Because the torque motor (or the pick-off) and the magnetic suspension systems enveloping them are placed in the same plane, they do not influence each other and can be considered as independent elements. This constitutes a reasonable use of the space inside the float cell. The jewel sleeve bearing is placed in the same plane

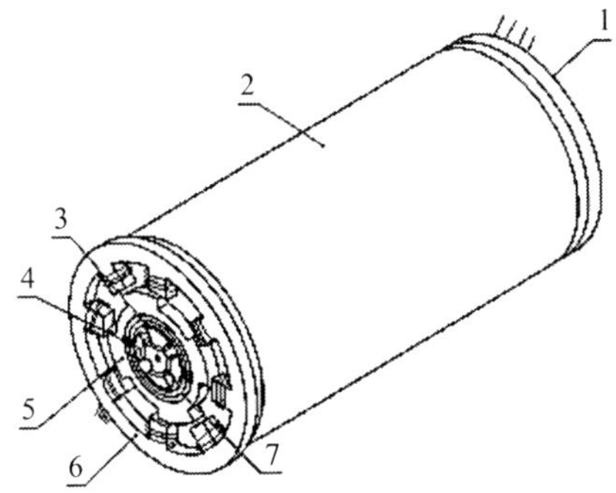


**Figure 5.11 (a).** An FPA with magnetic suspensor, induction pick-off, and electromagnetic torque motor. 1— "Ducosin"-type pick-off; 2—float; 3, 4—rotor and stator of the magnetic suspensor; 5, 6—rotor and stator of the torque motor; 7—float non-magnetoconductive collar; 8—jewel sleeve bearing with large gap; 9—FPA cover; 10—FPA case.

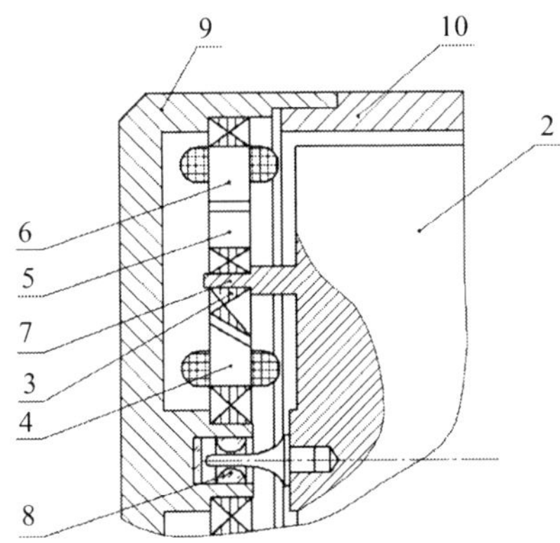


**Figure 5.11 (b).** Cross-section of a magnetic suspensor FPA. 1— "Ducosin" pick-off; 2—float; 3, 4—rotor and stator with magnetic suspension coils; 5, 6—rotor and stator with electromagnetic torque motor coils; 7—float non-magnetoconductive collar for fastening the rotors of the magnetic suspension and the torque motor; 8—jewel sleeve bearing with large gap between the journal and the bushing (float movement limiter); 9—cover with special sites for magnetic suspension stators; 10—case.

as the stators of the magnetic suspension. The gap between the journal and the bushing is about 20–25 μm, that is 4 to 5 times more than the usual gap in bearings. When the magnetic suspension is operating, the journal has no contact with the bushing jewel. The purpose of this jewel bearing is actually to limit the maximum displacement of the float when the magnetic suspension is off, and to reduce the readiness time of the device by reducing the float centering time of the magnetic suspension.

This construction has the following advantages:

1. There is no need for current-carrying wires for the FPA unit.
2. The device is easy to assemble.
3. The construction is symmetrical.

Unfortunately, it must be noted that some ferromagnetic elements in the suspension can cause spurious magnetic moments; and even more important, there are some hysteresis phenomena in the material of the electromagnetic torque motor rotor resulting in the appearance of unstable residual moments caused by previous control moments created by the torque motor. This hysteresis moment influences the stability of $a_0$ and to reduce this effect bucking windings are used, but are not very effective. Therefore there are a few rules for rotor manufacture. First, materials with small magnetic hysteresis loop areas and high magnetic permeability should be used (supermalloy-type materials). Secondly, all the elements should be thoroughly heat-treated. Finally, the rotor should be assembled carefully and without any mechanical stress that could influence the magnetic characteristics of the supermalloy-type materials.

Practice shows that if an accelerometer using any of the above schemes is designed and manufactured in a workmanlike manner, an accuracy corresponding to area 1 in Figure 5.1 can be achieved.

### *5.4.2 HYDROSTATIC ACCELEROMETER SUSPENSIONS*

Figure 5.12 shows an FPA cross-section consisting of a pendulous float unit (1) centered relative to the case by means of precision supports (2); a case (4) with hermetically sealed lead-out wires and an elastic compensator volume having a liquid filling (6) between the case and the float. (The torque motor and pick-off are not shown for clarity.)

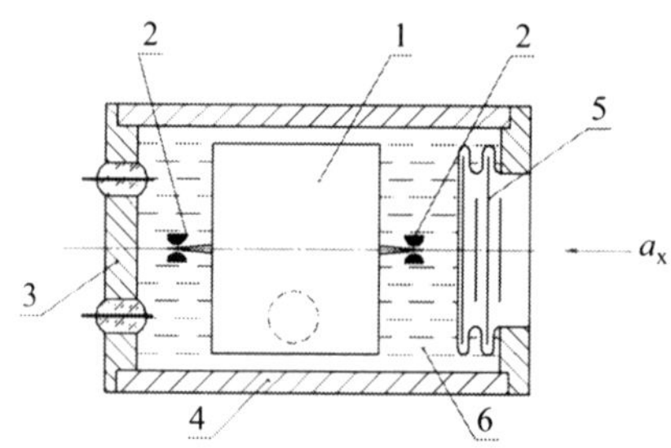


**Figure 5.12.** An FPA with hydrostatic suspension. 1—pendulous mobile unit (the float); 2—precision supports; 3—cover with the hermetic lead-out wires; 4—case; 5—elastic compensator (bellows); 6—liquid (Torque motor and pick-off are not shown).

The float cell liquid maintains neutral floatability, and provides both float oscillation damping and heat transfer to the case from the torque motor, pick-off, and magnetic suspension coils. If sleeve bearings are used the liquid also operates as a lubricant. In order to maintain neutral floatability with minimal float dimensions (and hence the dimensions of the entire device) heavy liquids such as fluorinated hydrocarbons with densities of about 2 g $cm^{-3}$ are used. Such liquids should be Newtonian, that is, there should not be any nonlinear viscous characteristics; and for correct functioning of the FPA the liquid must be homogeneous. There should not be exfoliation under any operating conditions and during extended storage (tens of years in storage temperatures ranging up to +60°C). The liquid must keep its physical characteristics during the whole service life period and should be chemically neutral to all of the constructional materials with which it is in contact, including varnishes, wire insulation, glue, and so forth. The liquid should not contain any fractions which are able to exude and polymerize on the surfaces of the float, the bearings, and the case of the device. It must also be filtered in order to eliminate any small inclusions that will be able to settle on the float surface, causing imbalance, and to get into bearing gaps, or causing an increase in friction. Before filling with the liquid, the float cell should be evacuated in order to exclude any appearance of gas bubbles that will be able to influence the float surface and cause disturbing moments.

Under all operational conditions the liquid must be under pressure. This pressure is created by the elastic compensator (an accordion boot or diaphragm) before filling the device. This pressure dissolves any gas bubbles left in the liquid after the evacuating process or exuded from the device elements (e.g., from gaps in the windings of the torque motor or pick-off). The pressurizing of the liquid hampers the appearance of air bubbles when the device is moving with an acceleration $a_x$. Liquid detachment can be caused by the influence of the inertial force $(m_f + m_l) \cdot a_x$ where $m_f$ is the float mass, and $m_l$ is the mass of the liquid displaced by the float.

The surfaces of the float and the float chamber should be absolutely clean. There should be no traces of an assembler's fingerprints (fats, salts) or any other dirt that may dissolve into the liquid and then settle on the surfaces (e.g., on the surface of the sleeve bearings, causing an increase in $a_{min}$ or an $a_0$ instability). Furthermore, the presence of any spots of fat on the float surface, and their partial dissolution in the liquid, can cause disturbing moments due to surface tension forces resulting in $a_0$ changes.

Finally, it is worth noting that the accuracy of the device depends by about 80–90 percent on the cleanliness of the device elements and the float liquid as well as on the quality of the evacuation process. It happens quite often that the device first shows low accuracy characteristics after assembly, but then exhibits high accuracy after several ablutions with solvent (which must also be of high purity), and further float liquid fills. Such high requirements for cleanness and quality of assembly as well as greater complexity in the design in comparison with gas-filled devices account for the high cost of this form of FPA.

### 5.4.3 FPA FLOAT BALANCING

Consider the forces and the moments influencing an accelerometer float in the case of constant acceleration with projections $a_x$, $a_y$, $a_z$ (see Figure 5.13).

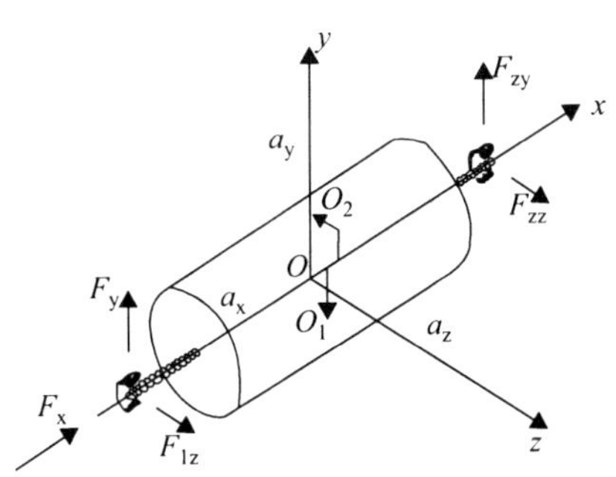


**Figure 5.13.** Accelerometer float under the action of acceleration projections $a_x$, $a_y$, $a_z$.

The point of origin $O$ is in the center of the float. The coordinates of the center of pressure $O_2$ are $x_2, y_2, z_2$ and the coordinates of the center of gravity $O_1$ are $x_1, y_1 = 0, z_1 = -l$. The conditions for float balancing are the balancing of inertial and Archimedean forces by reaction forces in the bearings and the balancing moments created by inertial and Archimedean forces, by the moment reactions in bearings, the moment of the torque motor $K_{TM} \cdot i_{TM}$ and also the moment of friction $M_{fr}$ in the bearings. That is:

$$\begin{cases} (-m+m_l)\cdot a_x = F_x \\ (-m+m_l)\cdot a_y = F_{1y} + F_{2y} \\ (-m+m_l)\cdot a_z = F_{1z} + F_{2z} \\ (ml+m_l y_2)a_z - m_l z_2 a_y = K_{TM} i_{TM} + M_{fr} \\ (mx_1 - m_l x_2)\cdot a_z + m_l z_2 \cdot a_x = (F_{1z} - F_{2z})\dfrac{L}{2} \\ (-ml - m_l y_2)\cdot a_x + (-m\cdot x_1 + m_l \cdot x_2)\cdot a_y = (F_{2y} - F_{1y})\cdot\dfrac{L}{2} \end{cases} \tag{5.7}$$

Here $m$, $m_l$ are mass of the float and the liquid displaced by the float.

In the ideal accelerometer, neutral floatation is maintained ($m = m_l$); the coordinates of the center of pressure $O_2$ are $x$, 0, 0; and the coordinates of the center of gravity $O_1$ are $x_1$, $-l$, 0. When $a_x = 0$ there are no reactions in the bearings (and consequently the friction moment is close to zero if sleeve bearings are used), the current in the torque motor $i_{TM} = \dfrac{ml}{K_{TM}} \cdot a_y$, and the output signal of the accelerometer from the reference resistance $R_{ref}$ depends on the acceleration $a_y$.

The presence of an acceleration $a_x$ results in the appearance of bearing reactions $F_{2y} = -F_{1y} = m \cdot a_x \cdot \dfrac{l}{L}$, (for the case when $x_1 = 0$). These reactions are unavoidable, and the situation $z_1 = 0$ is not of interest as in this case the accelerometer is unable to measure the acceleration $a_y$. The load-carrying ability and rigidity of the float bearings should be chosen taking these reactions into account. If sleeve bearings are used in the FPA, these reactions determine the friction moment in these bearings. It follows that the FPA should be positioned in such a way that the value of the acceleration $a_x$ would be minimal.

The mutual displacement of the center of pressure and the center of gravity ($x_1 - x_2 \neq 0$) along the axis $x$ is called the *trim* of the float and it is this that results in bearing reactions during accelerations $a_y, a_z$. This trim should be eliminated during manufacture.

Displacement of the center of gravity and the center of pressure ($z_1 \neq 0$; $z_2 \neq 0$) causes a base error in the accelerometer and creates a sensitivity in the FPA to the cross acceleration $a_y$. These displacements should also be eliminated during manufacture.

The mass of liquid forced out by the float is:

$$m_l = m_{l0}(1 - K_l \Delta t^\circ) \tag{5.8}$$

where $m_{l0}$ is the mass of the liquid at the temperature of neutral floatation,

$K_l$ is the temperature expansion factor of the float liquid (the value of the expansion factor is approximately the same for all float liquids at about $10^{-3}$ per °C)

$\Delta t^\circ$ is the change in temperature.

Therefore, neutral floatation and consequent zero reactions in the bearings of the FPA can be maintained at only one value of the working temperature. Deviation of the temperature by 10°C during acceleration causes reactions in the bearings. These will be equal to 1% of the corresponding reactions in the device without any hydrostatic load on the bearings. Consequently, the accuracy of a float device, even at serious temperature deviations from the neutral floatation temperature, is much higher than in gas-filled devices of the same construction.

The dependence Equation (5.8) in the presence of a pressure center displacement $y_2$ leads to the appearance of a temperature dependence in the torque motor current. When $m = m_{l0}$ Equation (5.7) gives:

$$i_{\text{TM}} = \frac{m\left[l+(1-K_l\Delta t^\circ)y_2\right]}{K_{\text{TM}}}a_z = \frac{m}{K_{\text{TM}}}(l+y_2)\left(1-\frac{y_2K_l}{l+y_2}\Delta t^\circ\right)a_z \qquad (5.9)$$

The component $\frac{y_2 \cdot K_l}{l+y_2}\cdot\Delta t^\circ$ creates the temperature error in the device's scale factor, therefore the displacement $y_2$ is undesirable. However, this displacement is sometimes created on purpose for compensating the temperature dependence of the slope of the torque motor $K_{\text{TM}} = K_{\text{TM0}}(1-K_{\text{M}}\cdot\Delta t^\circ)$, where $K_{\text{M}}$ is the temperature coefficient of the magnetization change. For magnets made of Alnico (an alloy based on Fe, Al, Ni, and Co with some other elements such as Cu and Ti), this coefficient is $K_{\text{M}} \approx 2\cdot 10^{-4}\frac{1}{°C}$. For magnets made of Recoma (an alloy based on Sm and Co) the coefficient is $K_{\text{M}} \approx 2\cdot 10^{-2}\frac{1}{°C}$. For magnets made of Recoma "stab 0" (materials usually customized to individual requirements) $K_{\text{M}} \approx \pm 5\cdot 10^{-5}\frac{1}{°C}$.

When $y_2 = y_{\text{comp}}$

$$K_{\text{M}} = \frac{y_{\text{comp}}\cdot K_l}{l+y_{\text{comp}}} \qquad (5.10)$$

and the output signal of the accelerometer loses its dependence on temperature.

While setting up the device, it is difficult to ensure the exact equality $y_2 = y_{\text{comp}}$. Therefore $y_2$ is made greater than is required, that is, the FPA is overcompensated. After all that, $\frac{y_2\cdot K_l}{l+y_2} = K_{\text{m}} + K_l$ and the current in the torque motor becomes:

$$i_{\text{TM}} = \frac{m(l+y_2)}{K_{\text{M0}}}\cdot a_z\left(1-\frac{K_l\Delta t^\circ}{1-K_{\text{M}}\Delta t^\circ}\right)$$

or, if $K_{\text{M}}\Delta t \ll 1$ $i_{\text{TM}} = \frac{m(l+y_2)}{K_{\text{M0}}}\cdot a_z(1-K_l\Delta t^\circ)$

To eliminate the temperature dependence of $i_{\text{TM}}$ in an overcompensated FPA, a shunt resistor $R_{\text{sh}}$ should be included in the device feedback circuit parallel to the torque motor winding as in Figure 5.14.

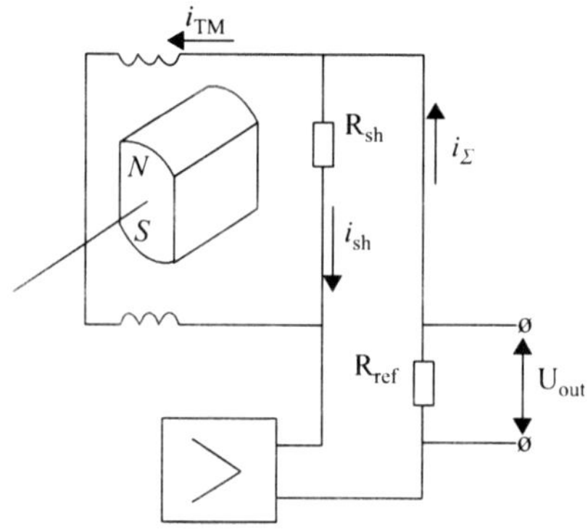


**Figure 5.14.** An accelerometer electrical schematic with a shunt resistor $R_{sh}$ used for eliminating $i_{TM}$ temperature dependence.

The torque motor winding has a resistance $R_{TM} = R_{TM0}(1 + K_{cu}\Delta t°)$, where $K_{cu} = 0.004\frac{1}{°C}$ —the resistance temperature coefficient of copper. For the circuit shown,

$$U_{out} = \frac{m(l + y_2)}{K_{M0}} R_{ref} \cdot a_z \cdot (1 + R_{TM0}/R_{sh})(1 - K_l\Delta t°)\left(1 + \frac{R_{TM0}/R_{sh}}{1 + R_{TM0}/R_{sh}} \cdot K_{cu}\Delta t°\right) \tag{5.11}$$

$R_{sh}$ is selected from the equality $K_l = \frac{R_{TM0} / R_{sh}}{1 + R_{TM0} / R_{sh}} \cdot K_{cu}$. Thus, changes of $U_{out}$ are determined by the multiplier $(1 - K_l^2\Delta t°)$. When $K_l\Delta t° << 1$, $U_{out}$ is practically constant over a wide range of temperatures. In addition $R_{sh} >> R_{TM}$, and as a result of this, the presence of $R_{sh}$ does not imply any additional requirements for the FPA amplifier compensating circuit.

The process of FPA float regulation at which the equalities $x_1 = x_2 = 0$; $y_2 = 0$ (or $y_2 = y_{comp}$); $z_1 = z_2 = 0$ are maintained, is called the *volume balancing*. To carry out this balancing it is necessary to regulate the float parameters using two values for the float liquid density (this regulation is usually carried out at two values of float liquid temperature).

## 5.4.4 HYDRODYNAMIC FORCES AND MOMENTS IN THE FPA

While the cylindrical float with radius $r_f$ and length $l_f$ is executing translational radial motion relative to the walls of the float chamber, it is under the influence of the following factors (in the case where the described cylindrical $h_c$ and end $h_e$ clearances are very small and the proportions $\frac{h_c}{r_f}, \frac{h_e}{r_f}$ are much less than 1) (Konovalov 1991).

The damping force is:

$$F_{damp} = \frac{12\pi\mu \cdot l_f \cdot r_f^3}{h_c^3} \cdot Q_1 \cdot \dot{e} \tag{5.12}$$

Here, $\mu$ is the dynamic viscosity of the liquid and $\dot{e}$ is the speed of the radial motion of the float relative to the case.

The coefficients $Q_1 \ldots Q_4$ take into account the influence of the axial flow of the liquid in the cylindrical clearance and through the end clearances. These coefficients depend on the proportions $\frac{l_f}{r_f}, \frac{h_e}{h_c}$. When the clearance $h_e \to 0$, then only the ring flow of the liquid occurs in the cylindrical clearance of the float, and the coefficients tend to 1 ($Q_1 \ldots Q_4 \to 1$).

For real FPA dimensions, the values of the coefficients $Q_1 \ldots Q_4$ lie within the limits 0.5–0.7.

The inertial force is:

$$F_{\text{in l}} = m_l^{\text{att}} \cdot \ddot{e} \tag{5.13}$$

Here $m_l^{\text{att}}$ is the mass of the liquid attached to the float

$$m_l^{\text{att}} = m_l \cdot \frac{r_f}{h_c} \cdot Q_2 \tag{5.14}$$

It is evident that if the solid object is moving in the liquid, then concurrently there is another object moving in the opposite direction with the same speed and acceleration. This second object is the volume of liquid actually displaced during the movement of the solid object. Therefore to impart an acceleration $a$ to the solid object of mass $m$ in the liquid, it is necessary to exert a force $(m + K_{sh} \cdot m_l) \cdot a$ on this object, where $m_l$ is the mass of the liquid displaced by the object, and $K_{sh}$ is a coefficient depending on the shape of the object. For objects moving in free liquid the coefficient $K_{sh}$ is usually slightly more than 1. The component $K_{sh} \cdot m_l$ is called the *attached liquid mass*, $m_l^{\text{att}}$. In the FPA this mass $m_l^{\text{att}}$ is very large—for example, if the float radius is $r_f = 25$ mm and the clearance is $h_c = 0.1$mm, the attached mass is 250 times more than mass of the liquid displaced by the float. This phenomenon can be explained by the fact that for small radial motions of the float, the particles of the liquid in the clearance must travel a long distance along the cylindrical surface. These particles therefore have considerable accelerations and consequently exert a very large force on the float. Thus, in hydrodynamic suspension, the acceleration reduction physically takes place via the motion of the liquid displaced by the float. However, whilst describing the motion of the float and the displaced liquid, it is convenient to consider only one acceleration, this reduction being conditionally related to the mass of the attached liquid.

While the float is executing axial motion relative to the case (the motion along the axis $x$ with speed $\dot{x}$ and acceleration $\ddot{x}$) the following influences occur.

The damping force:

$$F_{\text{damp2}} = \left( \frac{4\pi\mu \cdot r_f^4}{h_e^3} + \frac{6\pi\mu \cdot r_f^3 \cdot l_f}{h_c^3} \right) \cdot \dot{x} \tag{5.15}$$

The inertial force:

$$F_{\text{in2}} = \left( \frac{\pi\rho \cdot r_f^4}{3h_e} + \frac{\pi\rho \cdot r_f^3 \cdot l_f}{h_c} \right) \cdot \ddot{x} \tag{5.16}$$

where $\rho$ is density of the liquid.

When the float executes angular movements around the axes $z$, $y$ at a velocity $\dot{\beta}$ and acceleration $\ddot{\beta}$, the float is influenced by the following.

The damping moment:

$$M_{\text{damp1}} = \frac{\pi\mu \cdot l_{\text{f}} \cdot r_{\text{f}}^3}{h_{\text{c}}^3} \cdot \dot{\beta} \cdot Q_3 \tag{5.17}$$

The inertial moment:

$$M_{\text{in1}} = J_{\text{l}}^{\text{att}} \cdot \ddot{\beta} \tag{5.18}$$

where $J_{\text{l}}^{\text{att}}$ is an equatorial moment of inertia of the liquid attached to the float.

$$J_{\text{l}}^{\text{att}} = \rho \frac{\pi l_{\text{f}}^3 r_{\text{f}}^3}{12h_{\text{c}}} Q_4 \ddot{\beta} \tag{5.19}$$

As a first approximation $J_{\text{le}}^{\text{att}} \approx J_{\text{l}} \cdot \frac{r_{\text{f}}}{h_{\text{c}}} \Big/ \left(1 + \frac{h_{\text{e}}^3}{h_{\text{c}}^3} th \frac{l_{\text{f}}}{2r_{\text{f}}}\right)$, where $J_{\text{le}} = m_{\text{l}} \left(\frac{l_{\text{f}}^2}{12} + \frac{r_{\text{f}}^2}{4}\right)$ and is an equatorial moment of inertia of the object having the shape of the float and the density of the liquid (the equatorial moment of inertia of the liquid forced out by the float).

While moving around the axis $x$ at a velocity $\dot{\alpha}$ the float is under the influence of a damping moment:

$$M_{\text{damp2}} = \frac{2\pi \cdot \mu \cdot l_{\text{f}} \cdot r_{\text{f}}}{h_{\text{c}}} Q_5 \cdot \dot{\alpha} \tag{5.20}$$

In real devices the coefficient $Q_5$ is about 1.2–1.3 and is relevant to the influence of damping created by the liquid in the end clearances.

### 5.4.5 MOVEMENT OF FPA FLOAT UNDER VIBRATION

Analyzing the motion of the balanced FPA float under the vibration impact, $a_y = a_{\text{A}} \cdot \sin\omega t$ where reactions in the bearings are assumed to be zero. In this case the equation of the float motion is:

$$\left(m + m_l \frac{r_{\text{f}}}{h_{\text{c}}} \cdot Q_2\right)\ddot{e} + \frac{12\pi\mu \cdot l_{\text{f}} \cdot r_{\text{f}}^3}{h_{\text{c}}^3} \cdot Q_1 \cdot \dot{e} = (m - m_l) \cdot a_{\text{A}} \cdot \sin\omega t \tag{5.21}$$

so the amplitude of the float oscillations is:

$$e_{\text{A}} = \frac{(m - m_l) a_{\text{A}} \cdot h_{\text{c}}^3}{12\pi\mu \cdot l_{\text{f}} \cdot r_{\text{f}}^3 \cdot Q_1 \cdot \omega\sqrt{(T^2\omega^2 + 1)}} \tag{5.22}$$

where $T = \dfrac{\left(m + m_l \dfrac{r_{\text{f}}}{h_{\text{c}}} Q_2\right) h_{\text{c}}^3}{12\pi\mu \cdot l_{\text{f}} \cdot r_{\text{f}}^3 \cdot Q_1}$

The amplitude $e_A$ for the case where $m - m_l = 10^{-4}$ kg (the residual flotation is 1% of the float mass $10^{-2}$ kg and corresponds to the working temperature deviation from the neutral flotation temperature for 10°C); $a_A = 10$ g; $r_f = 25$ mm; $l_f = 30$ mm; $h_c = 0.1$ mm; $Q_1 = Q_2 = 0.7$; $\mu = 10^{-2}\,\text{Pa}\cdot\text{s}$; $\omega = 5$ Hz.

In this case $e = 2.5 \times 10^{-3}$ $\mu$m, so even at such a large vibration impact the float oscillation amplitude does not exceed a thousandth of the usual clearance in a sleeve bearing, that is, about 4–5 $\mu$m. If the vibration is high frequency, the amplitude of the float motion decreases rapidly.

Thus, the liquid in a float device effectively prevents the float from displacing under the influence of dynamic effects such as vibrations and shocks. Also, the bearings of the float under static influences are lightly loaded because the float weight is almost completely balanced by the Archimedean force. Thus, in the FPA it is possible to specify static (or vibrated) jewel bearings with very small journal diameters, or thin taut bends, or magnetic suspensions. And though they have limited load ability, they are able to maintain very low levels of disturbing moment.

## 5.5 MICROMECHANICAL ACCELEROMETERS (MMAs)

### 5.5.1 THE SINGLE-AXIS MMA

MMA operating principles are similar to those of the gas-filled compensating and direct-conversion devices described above. The difference lies in the manufacturing technology and it is this that determines the design peculiarities of the device. MMAs are made of single-crystal or polycrystalline silicon using manufacturing processes borrowed from integrated circuit technology such as photolithography, anisotropic and isotropic liquid or plasma etching, metal film evaporating, epitaxial growth of silicon crystals, silicon surface oxidation, sacrificial layer coating and etching, and so forth. Hence, the technological processes used in MMA manufacture require unusual approaches to device design. Following from this, there are two approaches to accelerometer manufacture (Boser 1997; Chau et al. 1995; Danel, Michel, and Delapierre 1990; van Drieenhuizen, Maluf, Opris, and Kovacs 1997; Geitner 2003; Josselin Touboul, and Kielbasa 1999; Peeters, Vergote, Puers, and Sansen 1992; Ristic et al. 1993; Rudolf, Jornod, Berqovist, and Leuthold 1990; Selvakumar, Yazdi, and Najadi 1996; Song 1997; Beeby, Ensell, Kraft, and White 2005; Xie, Fedder, and Frey 2003; Yazdi and Najafi 1997; Jiang, Du, Luo, and Li 2004), as follows:

1. The "sandwich," in which the accelerometer is made in several levels as shown in Figure 5.2. In this construction the pendulous unit usually has a thickness equal to the thickness of the wafer (about 400 $\mu$m).
2. The solid-state method, in which the thickness of the pendulous unit depends on the thickness of additional silicon layers grown on the wafer, usually amounting to several $\mu$m to 20 $\mu$m.

As a pendulum displacement detector, a capacitance pick-off or tensoresistive bridge is used in MMAs; and in the servoloop, a capacitive force generator serves as the torque motor. Almost all MMAs use gas damping and the "sandwich" devices are usually bigger than the solid-state ones, having dimensions in the horizontal projection that vary from 15 mm by 15 mm to 5 mm by 5 mm at thicknesses of about 1.5 mm. For solid-state devices the dimensions range from

2 mm by 2 mm to 0.5 mm by 0.5 mm at thicknesses of about 0.5 mm. Accordingly, all the clearances in the capacitance pick-off in "sandwich" devices are within limits from 5–20 $\mu$m, whilst in solid-state devices the clearances do not exceed one or two microns. Both types of MMA use standard semiconductor device packages. The "sandwich" type devices are much easier to manufacture by small enterprises since they require minimal production tooling. Conversely, the manufacture of solid-state MMAs needs more production tools and a wide variety of technological processes. Therefore, the manufacture of solid-state MMAs is reasonable only for mass production.

### *5.5.2 THE THREE-AXIS MMA*

There are special requirements for the design of devices forming the triads in the small-sized INS designed for artillery shell guidance systems (Lemkin, Boser, and Smith 1997). In such systems all three devices must be placed in one plane and combined on a single silicon wafer. This achieves the following results:

1. Downsizing and simplifying the triad design in comparison with installing three separate single-axis accelerometers
2. Maintaining the necessary mutual perpendicularity of the sensitive element axes and
3. Simplifying the technological manufacturing processes—all three accelerometers of a triad are manufactured simultaneously.

The three elements on the plate are a pendulous accelerometer for measuring the acceleration normal to the plane of the plate, and two linear accelerometers measuring the accelerations acting in the plane of the plate. There are several methods for manufacturing of such MMAs.

The operation of the pendulous accelerometer has been analyzed above, and the linear accelerometer with a comb-type capacitance sensor for the inertial mass position is depicted in Figure 5.15.

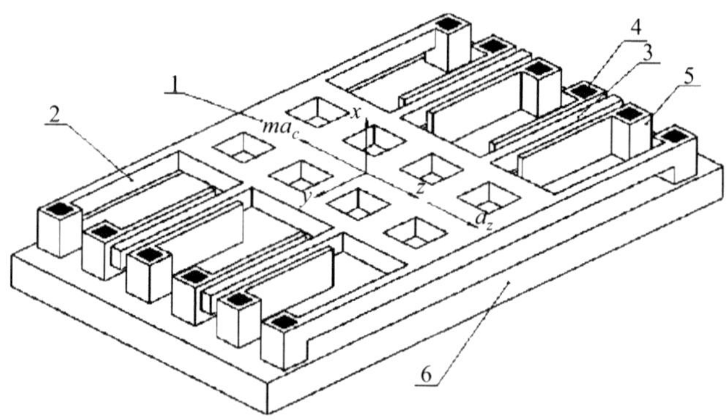


**Figure 5.15.** A micromechanical accelerometer. 1—movable silicon inertial mass; 2—flexors; 3—movable plate; 4,5—stationary electrodes; 6 —nonconductive substrate; 7—anchors.

Here the inertial mass (1) is suspended on four flexors (2) and can move along the axis $z$ under the influence of the inertial force, $ma_z$. Displacements of the inertial mass along the axes

$x$ and $y$ are negligible, as the flexors have high rigidities in these directions. When the inertial mass is displaced, the flanges of the movable plate (3) act as the movable electrodes of the capacitive pick-off (doped silicon has a conductivity sufficient for operation of the capacitive pick-off). These electrodes move relative to the stationary electrodes (4) and (5) mounted on the nonconductive substrate (6). The substrate also carries the anchors (7).

Capacitances C1 and C2 formed by the electrodes (3, 4, 5) are connected to a bridge circuit via two resistors and are fed from a sine-wave generator. In devices without feedback the signal, which is taken from the bridge diagonal and is proportional to the movable plate displacement, corresponds to the measured acceleration $a_z$.

### 5.5.3 THE COMPENSATING TYPE MMA

In devices with feedback, the signal taken from the bridge diagonal is used for forming controlling voltages on the stationary electrodes (4, 5) in Figure 5.15 (Chau 1995; Boser 1997; Smith et al. 1994). These controlling voltages create electrostatic forces that prevent plate (1) from further displacement and can be determined with reference to Figure 5.16 as follows.

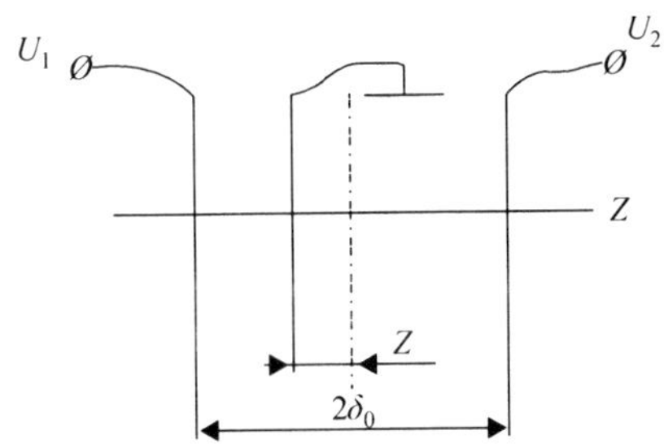


**Figure 5.16.** The capacitive pick-off.

For the case when direct voltages $U_1$ and $U_2$ are formed on the stationary electrodes (4, 5), the energy balance equation for the first capacitor is:

$$dA_m + dA_e = dP_e \tag{5.23}$$

Here, $dA_e = U_1^2 dC_1$ is the energy needed for maintaining the direct voltage $U_1$ during a change in $C_1$;

$dA_m = F_1 dz$ is the mechanical work created by the electrostatic force on an element $dz$;

$dP_e = \dfrac{U_1^2 \cdot C_1}{2}$ is the potential energy of the capacitor charge.

Inserting these into Equation 5.1 gives $F_1 = -\dfrac{1}{2}U^2 \dfrac{dC}{dz}$.

Because the capacitance of a planar capacitor is $C = \varepsilon \dfrac{S}{\delta}$ where S is the area, $\delta$ is the air gap, and $\varepsilon$ is dielectric constant, $F_1 = -\dfrac{1}{2}\varepsilon \dfrac{S}{(\delta_0 - z)^2} \cdot U_1^2$.

Similarly, for the second capacitor $F_2 = -\frac{1}{2}\varepsilon \frac{S}{(\delta_0 + z)^2} \cdot U_2^2$

and the total electromagnetic force influencing the movable electrode is:

$$F_\Sigma = F_2 - F_1 = 0.5\varepsilon S \cdot \frac{(\delta_0^2 + z^2) \cdot (U_2^2 - U_1^2) + 2\delta_0 z \cdot (U_1^2 + U_2^2)}{\delta_0^2 - z^2} \quad (5.24)$$

As seen from the above equations, the force depends nonlinearly on the controlling voltages $U_1$ and $U_2$ and on the plate displacement $z$. Both these factors must be eliminated because otherwise they make it impossible to obtain the necessary accelerometer accuracy. To decrease the dependence of the force $F_\Sigma$ on the displacement $z$, the amplification factor of the feedback circuit can be increased, or an integrating component can be included into the compensating circuit. This ensures that the displacement tends to zero, $z \to 0$.

To eliminate nonlinearities in the dependence $F_\Sigma(U_1 - U_2)$ the voltages $U_1$ and $U_2$ are formed in the following way:

$$U_1 = U_0 - \Delta U \text{ and } U_2 = U_0 + \Delta U$$

where $\Delta U$ is the controlling voltage and $U_0 = const$, $U_0 > \Delta U$. Thus,

$$F_\Sigma = 2\frac{\varepsilon \cdot S}{\delta_0^2} U_0 \cdot \Delta U \quad (5.25)$$

### 5.5.4 *SOLID-STATE MMA MANUFACTURING TECHNIQUES*

Many technologies for micromechanical device manufacture have been developed. Here, the "silicon-on-glass" method, developed in the Draper Laboratories (Barbour et al. 1996) is presented in Figure 5.17 as applied to MMA fabrication. The sequence of operations is as follows.

At the first stage the silicon wafer is doped with phosphorus. Then the area for the anchors and stationary electrodes is exposed by photolithography. An initial anisotropic etching is then carried out to a depth equal to the air-gap between the stationary plate (1) and the base (6). During the second stage, boron diffusion into the body of the wafer to a depth equal to the thickness of plate (1) is carried out. This results in a silicon layer with *p*-type conductivity. At the third stage, lithography corresponding to the forms of the movable plate, the anchors, flexors, and stationary electrodes is carried out. The fourth stage consists of ion–plasma etching to form the slits with vertical walls on the uncovered areas of the wafer. The etching depth may not exceed thickness of the layer of boron-doped silicon. During the next process, the wafer is turned over and placed on the glass plate (6) that carries the metalized areas. Welding along all the metalized areas is then carried out, and in the last stage the elements not doped with boron are removed by means of anisotropic etching. The final structure is shown as item 7 in Figure 5.17.

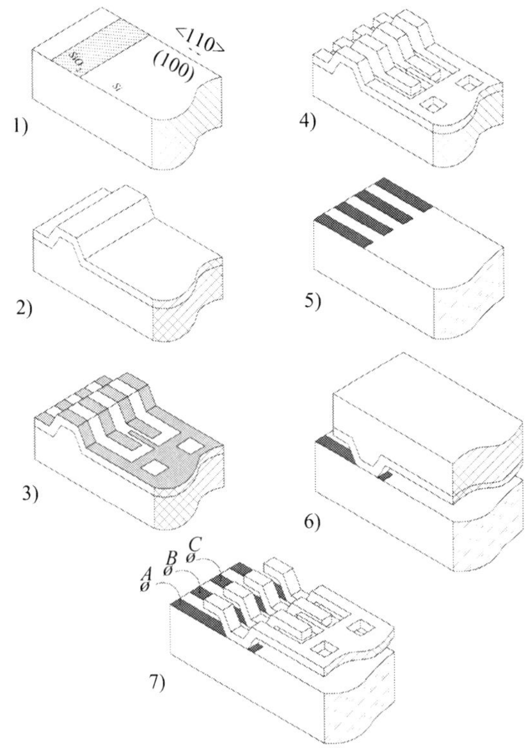


**Figure 5.17.** Manufacturing stages of a solid-state MMA. 1—the wafer after the 1st photolithographic process; 2—the wafer after the 1st anisotropic etch; 3—the wafer after the 2nd photolithographic process; 4—the wafer after an ion-plasma etch; 5—the substrate after evaporating the metal tracks; 6—the wafer turnover and bonding to the substrate; 7—the MMA overview.

## REFERENCES

Abbaspour-Sani, E., R. S. Huang, and C. Y. Kwok. 1994. "A linear electromagnetic accelerometer." *Sensors Actuators A* 44: 103–9. DOI: 10.1016/0924-4247(94)00792-6.

Barbour, N., J. Connely, J. Gilmore, P. Greiff, A. Kourepenis, M. Weinberg. 1996. Micromechanical silicon instruments and systems development at Draper Laboratory (CSDL-P-3507) presented at *AIAA Guidance, Navigation, and Control Conference*, San Diego, California.

Barth, P. W., F. Pourahmadi, R. Mayer, J. Poydock, and K. Petersen. 1988. "A monolithic silicon accelerometer with integral air damping and overrange protection." in *Tech. Dig. Solid-State Sensors and Actuators Workshop,* Hilton Head Island, SC, pp. 35–8.I. DOI: 10.1109/SOLSEN.1988.26427.

Beeby, Stepen, Graham Ensell, Michael Kraft, and Neil White. 2004. *MEMS Mechanical sensors*. Boston and London: Artech House, Inc.

Boser, B. E. 1997. "Electronics for micromachined inertial sensors," in *Tech. Dig. 9th Int. Conf. Solid-State Sensors and Actuators (Transducers'97),* Chicago, IL, USA, pp. 1169–72.

Cardy, W. 1984. "Q-Flex accelerometer, construction and principle of operation." Technical Note TN-103, Sundstrand Data Control Inc.

Chau, K., S. R. Lewis, Y. Zhao, R. T. Howe, S. F. Bart, and R. G. Marcheselli. 1995. "An integrated force-balanced capacitive accelerometer for low-g applications," in *Tech. Dig. 8th Int. Conf. on Solid-State Sensors and Actuators (Transducers'95).* Stockholm, Sweden, pp. 593–6.

de Coulon, Y., T. Smith, J. Hermann, M. Chevroulet, and F. Rudolf. 1993. "Design and test of a precision servoaccelerometer with digital output." in *Tech. Dig. 7th Int. Conf. Solid-State Sensors & Actuators (Transducers '93)*, Yokohama, Japan, pp. 832–5.

Danel, S., F. Michel, and G. Delapierre. 1990. "Micromachining of quartz and its application to an acceleration sensor," *Sensors Actuators* A21/A23: 971–7. DOI: 10.1016/0924-4247(90)87071-P.

Draper Ch. S., W. Wrighly, D. G. Hoag, R. H. Battin, J. E. Miller. 1998. *Space Navigation, Guidance and Control* (p. 326). Cambridge, MA: Massachusetts Institute of Technology Cambridge.

Geitner, H. 2003. "Considerations for soldering accelerometers in LCC-8 packages onto printed circuit boards." *Application Note AN-652*, Analog Devices.

Jiang, Yuqi, Maohua Du, Le Luo, and Xinxin Li. 2004. "Simulation of the potting effect on the high-G MEMS accelerometer." *Journal of Electronic Materials* 33 (8): 893–9. DOI: 10.1007/s11664-004-0217-4.

Josselin V., P. Touboul, and R. Kielbasa. 1999. "Capacitive detection scheme for space accelerometer application." *Sensors and Actuators* A 78: 92–8. DOI: 10.1016/S0924-4247(99)00227-7.

Konovalov, S. F. 1991. *Theory of Vibration Resistance of Accelerometers* (p. 272). Moscow: Mashinostroenie. ISBN 5-217-01273-0. (In Russian.)

Lawrence, A. (1993). Modern Inertial Technology: Navigation, Guidance, and Control. New York: Springer-Verlag.

Lemkin, M., B. Boser, and J. Smith. 1997. "A 3-axis surface micro-machined EA accelerometer." in *Tech. Digest Int. Solid State Circuits Conf. (TSSCC'97)*, San Francisco, CA, pp. 202–3. DOI: 10.1109/ISSCC.1997.585333.

Olvey, Stephen E., Ted Knox, Kelly A. Cohn. (2004). "The development of a method to measure head acceleration and motion in high-impact crashes." *Neurosurgery* 54 (3): 672–7. DOI: 10.1227/01.NEU.0000108782.68099.29.

Peeters, E., S. Vergote, B. Puers, and W. Sansen. 1992. "A highly symmetrical capacitive micro-accelerometer with single degree-of-freedom response." *Journal of Micromechanics and Microengineering* 2: 104–12. DOI: 10.1088/0960-1317/2/2/006.

Ristic, L., R. Gutteridge, J. Kung, D. Koury, B. Dunn, and H. Zunino. 1993. "A capacitive type accelerometer with self-test feature based on a double-pinned polysilicon structure," in *Tech. Dig. 7th Int. Conf. Solid-State Sensors and Actuators (Transducers '93)*. Yokohama, Japan, pp. 810–12.

Rudolf, F., A. Jornod, J. Berqovist, and H. Leuthold. 1990. "Precision accelerometers with fig resolution." *Sensors Actuators* A21/A23: 297–302. DOI: 10.1016/0924-4247(90)85059-D.

Selvakumar, A., N. Yazdi, and K. Najafi. 1996. "A low power, wide range threshold acceleration sensing system," in *Proceedings of the IEEE Micro Electro Mechanical Systems Workshop (MEMS '96)*, San Diego, CA, pp. 186–91. DOI: 10.1109/MEMSYS.1996.493851.

Smith, T., O. Nys, M. Chevroulet, Y. DeCoulon, and M. Degrauwe. 1994. "A 15b electromechanical sigma-delta converter for acceleration measurements," in *Tech. Dig. IEEE Int. Solid-State Circuits Conf. (ISSCC94)*, San Francisco, CA, pp. 160–1.

Song, C. 1997. "Commercial vision of silicon based inertial sensors," in *Tech. Dig. 9th Int. Conf. Solid-State Sensors and Actuators (Transducers '97)*, Chicago, IL, pp. 839–42.

van Drieenhuizen, B. P., N. Maluf, I. E. Opris, and G. Kovacs. 1997. "Force-balanced accelerometer with mG resolution fabricated using silicon fusion bonding and deep reactive ion etching," in *Tech. Dig. 9th Int. Conf. Solid-State Sensors and Actuators (Transducers '97)*. Chicago, IL, pp. 1229–30.

van Kampen, R. P., M. J. Velekoop, P. M. Sarro, and R. F. Wolffenbuttel. 2006. "Application of electrostatic feedback to critical damping of an integrated silicon capacitive accelerometer." *Semiconductor Electronics*. ICSE '06. IEEE International Conference.

Xie H., G. Fedder, Z. Pan, and W. Frey. 2003. "Method of fabricating micro structures and devices made therefrom", US Patent pending (10/374197), Filed on Feb. 26, 2003.

Yazdi, N., and K. Najafi. 1997. "An all-silicon single-wafer fabrication technology for precision micro-accelerometers," in *Tech. Dig. 9th Int. Conf. Solid-State Sensors and Actuators (Transducers '97)*, Chicago, IL, pp. 1181–4.

CHAPTER 6

# Gyroscopic Devices and Sensors

Vladimir N. Branets (assisted by Y.A. Bazhanov)
*Moscow Institute of Physics and Technology, Russia*

Boris E. Landau
*Concern CSRI "Elektropribor," JSC, Russia*

Yuri N. Korkishko
*Optolink, Russia*

David Lynch and
*Tula State University, Russia*

Vladimir Y. Raspopov
*Northrop Grumman Corporation, USA*

## 6.1 INTRODUCTION

### *6.1.1 PRELIMINARY REMARKS*

The term *Gyroscope,* or "Gyro," was suggested by the French scientist Leon Foucault, who in 1852 had experimentally shown the ability of a spinning metal rotor, isolated from geocyclic turning, to maintain the direction of its axis of rotation constant in absolute space.

Depending on the physical principles of their construction, gyros can be divided into two groups: mechanical gyros and quantum (wave) gyros.

With regard to both the state of the art and prospects for future developments in gyroscope technology, a gyro in the generic sense may be regarded as a device with an element that performs fast periodic movements, so permitting the detection and measurement, relative to inertial space, of the angular displacement of the platform where it is installed. These fast periodic movements may be rotational, reciprocal, or oscillatory, amongst others. In mechanical gyros, such fast periodic movements may be performed by a rigid body, a liquid, or a gas. In quantum gyros this role is taken by atomic nuclei, protons, or electrons, which possess orbital or spin magnetic and mechanical moments, and also by coherent flows of photons, phonons, and any other particles without magnetic moments.

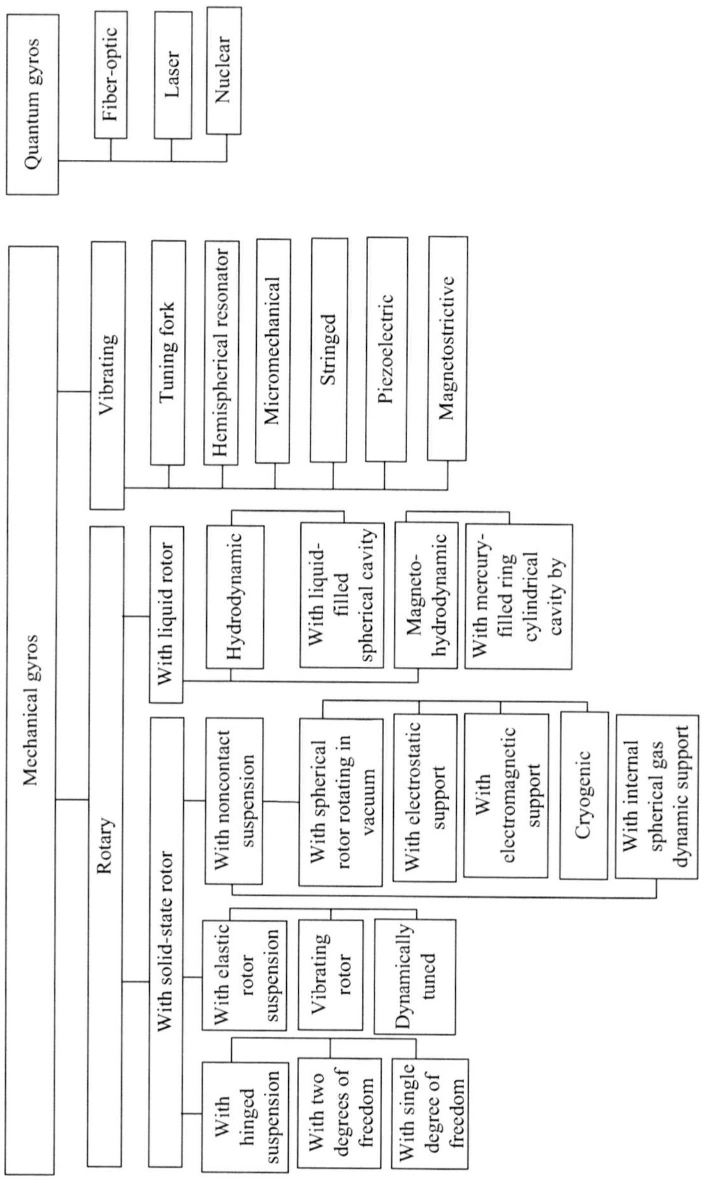


**Figure 6.1.** Classification of gyros by main distinguishing features.

### *6.1.2 CLASSIFICATION OF GYROS*

A classification of gyros by their main distinguishing features is shown in Figure 6.1. Gyro subtypes not having wide distribution or major distinction are not displayed.

The most important characteristic of any practical gyro is its number of degrees of freedom, that is, the *degree of freedom number*. Internationally, two approaches to the definition of this number are employed. In English-speaking countries, it usually refers to the degrees of freedom leading to measurements. In Russia and some other countries, however, it refers to the number of axes around which the kinetic energy carrier (usually termed the "rotor") can perform rotary motion. This latter approach is derived from classic theoretical mechanics that considers the gyro itself as a rotor, but because any three-ring gimbal suspension provides three degrees of freedom, the number of measuring degrees of freedom has to be decremented by one. This is why a gimbal suspension gyro may be described as a Two Degree of Freedom (TDF) gyro by the number of measuring axes, but a Three Degree of Freedom gyro by the number of rotation axes.

In this book, as in the majority of other English-language publications, the free gyro will be termed a *TDF gyro* taking into account the number of measuring axes. Accordingly, a classic gyro with two rotation axes will be called as Single Degree of Freedom (SDF) gyro. The term *Zero Degree of Freedom* (ZDF) gyro will refer to a rotor that has a single axis of rotation rigidly connected to its base (Magnus 1971). Gyros with liquid rotors, vibratory gyros, and quantum gyros are SDF gyros. Only gyros with rigid body rotors can be TDF gyros.

TDF gyros may be divided into *astatic* (*balanced*) and *unbalanced* types. An astatic gyro with no frictional-force moment, error torque due to gravity, or any other error torques related to the suspension axes, is called a *free gyro*. In the precessional approach frame, a free gyro keeps its initially defined orientation constant in inertial space. This property of the free gyro therefore permits it to detect a change in an aerospace vehicle's altitude referring to two axes in the inertial coordinate frame. SDF gyros permit an angular velocity measurement or its integral in only one axis in the coordinate frame.

For avoiding mechanical bearings in different SDF and TDF gyros, many kinds of special frictionless suspensions are used. Among them are liquid (floated), gas, electromagnetic, magnetoelectric, and other forms of support, all of which are of considerable importance in improving gyro performance. Those exhibiting the smallest adverse moments involve gyros using contactless supports, for example, a spherical rotor centered in suspension with the help of magnetic or electrostatic forces.

### *6.1.3 GYROSCOPIC INSTRUMENTS*

Gyros represented in Figure 6.1 should be considered as the main components of different gyroscopic equipment and vehicle-borne gyroscopic instruments (GIs) designed especially for the determination of aerospace vehicle motion or position parameters, and also for vehicle stabilization (Savet 1961). GIs may be subdivided according to their applications as follows:

1. *GIs for determining vehicle angular displacement*: Here it is necessary to consider different free and positional gyros, especially *directional gyros* for determining vehicle azimuthal deviation (yaw angle); *gyro verticals* for determining vehicle deviations with reference to the vertical/horizontal plane (pitch and roll angles); and *orbit gyros* for determining the yaw angle of Earth satellites moving in the orbital coordinate frame.

2. *GIs for determining vehicle angular velocity and angular acceleration*: These include SDF rotor, vibrational, and quantum devices for angular rate measurement, and also SDF and TDF rotor gyros for the simultaneous measurement of angular rate and acceleration.
3. *GIs for determining the integrals of input values*: Integrating gyros constructed on the basis of SDF rotor gyros with swing joint rotor suspension, and also hemispherical resonator gyros,* can be used for determining angular rate integrals. Gyroscopic linear acceleration integrators can be designed on the basis of the unbalanced TDF gyro and are intended for the determination of vehicle apparent linear velocity.
4. *GIs for the stabilization of a vehicle or devices within it*: These are also for the estimation of vehicle angular turn. These are called *gyrostabilizers*.
5. Gyroscopic instruments for solving aero navigational tasks.
6. Inertial Navigation Systems (INS) on the basis of gyro-stabilized platforms (gimbaled INS) and strapdown INS that permit determination of vehicle trajectory and angular motion without information from external sources.

### *6.1.4 POSITIONAL GYROS*

Positional gyros are for maintaining direction, and these use the property of the free gyro to keep the direction of the main axis invariant in space. A variety of positional gyros classified by functional features is listed in Table 6.1.

**Table 6.1.** Positional gyros applied in aerospace technology

| **Positional gyros** | |
|---|---|
| **Without correction** | **With correction** |
| • Free gyros<br>• Angle measurers<br>• Gyros for straight-line or computer-controlled direction of flight<br>• Free heading gyros | • Heading (azimuth) gyros<br>• Compass heading gyros<br>• Vertical (horizon) gyros<br>• Orbit gyros |

Gyroscopic equipment without correction may be used during short periods only, because disturbance torques inevitably appear during departure of the main axis from its initial position over time. To avoid this drift, control and correction of the gyro axis direction are employed.

Gyro correction is mostly achieved by the fairly rigid coupling of a directional indicator (compass, pendulum, vertical gauge, etc.) to a positional gyro and direction keeper, so forming a complete system. Via the resultant coupling adjustment, it is possible to essentially eliminate or at least provide major inaccuracy decrements in both direction indicator output fluctuations and the slow drift of the direction keeper.

---

* The hemispherical resonator gyro can be used to measure angular rate, or as an integrating gyro with unlimited allowable turn angle. Here, a stationary wave is generated in the hemispherical resonator and the measurement axis defined depending on whether the excitation is positional (for angular rate) or parametric (for angular rate integral). Further details are given in Section 6.9.

### 6.1.5 THE VERTICAL (OR HORIZONTAL) GYRO

This gyro is a device for the determination of the true vertical or horizontal plane (which are effectively the same) and also any vehicle angular turn with reference to this plane, that is, the roll and pitch angles.

A vertical gyro with a pendulum correction loop includes a pendulum for sensing the gyro axis angular drift from the local vertical, plus torque motors for applying appropriate compensatory moments to return the gyro axis to that local vertical. Here, a "local vertical" is a line drawn from an orbiting body to the center of the body being orbited. The local vertical/local horizontal (LVLH) is a frame of reference relative to that line.

The TDF astatic gyro can serve only as a short-period directional indicator (particularly with respect to the vertical) because over time its main axis will change direction with respect to the rotating Earth. Nevertheless, such vertical gyros are applied in ballistic missiles for measuring inclination angles in the vertical and horizontal planes (pitch, yaw, and roll angles) during the relatively short period occupied by the active part of the trajectory.

However, gyro-inertial vertical gyroscopic instruments have been perfected that permit estimation of the vertical with high accuracy irrespective of vehicle motion and acceleration. In these instruments the physical simulation of the horizon is implemented with correcting systems involving accelerometers installed in the internal frames of gimbal mountings—in the majority of real instruments actually on the gyro-stabilized platforms themselves.

### 6.1.6 ORBIT GYRO

A single-rotor orbit gyro is a gyroscopic instrument for measuring the yaw angle (inclination from the orbit plane) of Earth satellites. The orbit gyro is a corrected position gyro installed aboard a satellite and continually corrected by a local vertical. When the yaw angle is zero, the axis of the external frame of the gimbal mount rotation makes a tangent to the vehicle orbit, and the main gyro axis is superposed on the orbital angular velocity vector.

If a yaw angle is not zero, the plane of the outer gimbal ring goes out of superposition with the orbit plane and an angular gauge at the axis of the inner gimbal ring originates a correcting signal. Correction systems having angle sensors at the axis of the outer gimbal ring and torque motors at the axes of both rings perform the process of *gyrocompassing*. Note that if correcting signals are switched off, an orbit gyro becomes a free gyro (Raushenbah and Tokar 1974).

### 6.1.7 SINGLE DEGREE OF FREEDOM (SDF) GYROS

These represent the numerous groups of gyros with different physical principles of construction, design, and ranges of practical application. For a long time SDF gyros were used chiefly as angular rate sensors with comparatively low accuracy for the angular stabilization systems of various vehicles.

However, since the invention of float suspension, the accuracy and performance of gyros have greatly improved. Precise integrated gyros appeared and became broadly applied as sensing elements in gyro-stabilized indicators and gyro-stabilized platforms for INS.

### 6.1.8 GYRO STABILIZERS

These are divided by operating principles into *direct*, *power*, and *indicating* versions.

- *Direct gyro stabilizers* use the stabilizing property of the TDF gyro directly and are used for stabilizing the sensing elements of onboard control systems, for example aerials, infrared radiation detectors, and so forth. Such sensing elements must be installed on the internal frames of gyro gimbal mounts so that both the direction of the sensitivity axis of the element and the direction of the gyro main axis coincide.
- *Power gyro stabilizers* (*gyroframes*) are electromechanical devices that include not only gyros, but also special actuators for the compensation of external disturbance torques applied to the stabilized object. In such systems, the gyros participate in the stabilization task both as angular error sensors and as the power units that produce the gyroscopic torques applied to the stabilized object. They are used in aerospace vehicles for the stabilization of separate instruments and platforms. Furthermore, the principle of the power gyro stabilizer may be applied to some kinds of directional gyros, vertical gyros, and combined attitude-and-heading reference devices.

  Depending on the number of gyros in the frame, power gyro stabilizers may be *double-gyroscopic* or *single-gyroscopic*. Depending on the number of stabilization axes, they may be *uni-axial*, *bi-axial*, or *tri-axial.*

  A combination of two uni-axial gyro stabilizers permits the stabilization of an object in a plane, for example the horizontal plane. A combination of three uni-axial gyro stabilizers permits the realization of a three-axis gyro stabilizer, that is, a device consisting of a directional gyro (a heading gyro) and a vertical gyro (or horizontal gyro). That is, it acts to measure the three angles that determine the object attitude. The tri-axial gyro stabilizer is also used for the attitude stabilization of gyro stabilized platforms.
- *Indicating gyro stabilizers* are automatic control systems in which a GI installed on a platform (or a stabilized vehicle) performs the roles of sensing elements for platform stabilization servomechanisms. Gyros sensing angles, or more rarely angular rates of platform inclination, act as such sensing elements.

  TDF indicating gyro stabilized platforms are used in INS, and vertical gyros and heading gyros have found widespread applications in aeronavigation systems.

### 6.1.9 GYROSCOPIC INSTRUMENTS IN AERONAVIGATION

A GI aboard a vehicle forms a mechanical model of the various basic directions (vertical, geographical, magnetic pole, and others) that provide controlled flight along an *orthodrome*—that is, a Great Circle. They are all constructed on the basis of positional gyros or gyro stabilizers with corresponding correctional systems.

The construction of a local vertical is performed with the help of different kinds of vertical gyros; and the construction of azimuth direction is achieved with the help of corrections using external information methods and sensors (astro and radio navigation instruments, magnetic or inductive compasses, etc.).

In INS, built-in equations are used for the construction of the local vertical and do not take into account any gyro or accelerometer errors. That is, the local vertical is constructed purely analytically.

In contrast to positional gyros and gyro stabilizers that maintain direction by some associated means, a class of GI—*gyro compasses* and *gyro pendulums*—possess inherent directional capabilities that enable them to search for and maintain required directions relative to the Earth. Owing to the Earth's rotation, gyro compasses maintain the direction of this rotational axis. Gyro pendulums can usually automatically maintain the direction of the local vertical. These devices are constructed on the basis of unbalanced TDF gyros.

### *6.1.10 INERTIAL NAVIGATION SYSTEMS (INSs)*

These systems are designed to achieve vehicle motion simulation using only autonomously obtained information from inertial sensors that measure *apparent acceleration* and angular rate. Here, apparent acceleration is the difference between real acceleration and gravitational acceleration:

$$\overline{W} = \frac{d^2\vec{r}}{dt^2} - \overline{F}(\vec{r}) \tag{6.1}$$

where $\overline{W}$ is the acceleration vector measured by the accelerometer, $\vec{r}$ is the radius–vector of a point (center of gravity of the sensing element in the inertial coordinate frame), and $\overline{F}$ is the vector of gravitational acceleration at the considered point. Thus, the inertial method consists of using an accelerometer to determine velocity by single integration and position (displacement) by double integration of this expression.

Inertial navigation systems can be categorized by a set of features as follows:

1. By orientation of the sensitive axes of the inertial sensors by stars, by body-fixed axes, by Earth-fixed axes, and others
2. By the method of local vertical construction using an analytical or calculated vertical with the inertially formed vertical being undisturbed by horizontal accelerations
3. By associated stabilization using gyros, astronavigation or other means, or the absence of it in the case of strapdown INS.

#### *6.1.10.1 Types of INS*

1. In *Geometric INS*, a platform carrying gyros and accelerometers is stabilized in such a manner that its absolute angular rate is zero, that is, the platform orientation with reference to an inertial coordinate frame is constant.
2. In *Analytical INS*, gyros and accelerometers are rigidly mounted in a vehicle body. The vehicle angular rates (increments of angles) are measured by gyros, and translational accelerations by accelerometers. Data on vehicle motion and instant position relative to the Earth or some other celestial body become available via calculation.
3. In *Local-level INS*, a platform with accelerometers is kept in a horizontal position and in some cases even in accordance with Earth-fixed axes north-west-zenith. For some systems of this kind the platform rotation rate around the vertical is maintained at zero.

For all three kinds of INSs, a navigational computer is necessary for calculating the moving vehicle coordinates. Analytical INS requires an essentially higher-level calculation capability because the relevant computer must also solve some structural/mechanical problems.

Depending on the design and required functions, platforms are suspended in two, three, and even four frames (with three, four, or five axes) to provide full rotational freedom. For the platform suspension itself, only two frames (or three axes) are necessary. Additional frames are installed for avoiding the phenomenon of *frames folding*, which is when two frames appear in one plane so that a degree of freedom is lost.

The accuracy of the measuring devices—accelerometers and gyros—is of crucial importance for acceptable INS functioning (Broxmeyer 1964); and decreasing undesirable gyro drift, even without full drift elimination, is one of the major problems in INS practical realization.

In aviation, an onboard INS is generally composed of a two-dimensional local-level INS constructed on the basis of a gyro-stabilized platform with three installed accelerometers. The outputs of these accelerometers, whose axes of sensitivity are parallel to the horizontal axes of a platform, are used for constructing a gyro-inertial vertical with an artificially modeled period equal to the Schuler period. Because the aircraft altitude is measured by altimeter, only the latitude and longitude need be defined in the two-dimensional inertial system.

Even the best examples of aerospace INS have coordinate definition errors at levels of 1–3 km $h^{-1}$. To reduce such errors and widen the field of application, various ways of correction using satellite and astronavigational means may be applied.

### *6.1.10.2 Strapdown INS*

The practical use of strapdown INS appeared between the end of the 1960s and the beginning of 1970s. At this time the reserve strapdown INS of guidance systems used in the "Apollo" expeditionary lunar module had begun operation. In Russia, the application of strapdown INS in space technologies began in the orientation systems of the "Salute" orbital space station and then in the transport spacecraft "Soyuz-T." Both American and Russian systems have been constructed on the basis of SDF floated rate-integrated gyros with torque feedback (Branets and Shmuglevskiy 1992; Gelb and Sutherland 1967, 1968a,b; Matthews and Taylor 1969; Roantree and Kormanik 1966).

Concurrently, work was considerably intensified to perfect traditional gyros along with the creation of new gyros such as laser, fiber optic, and hemispherical resonator types, and focused on their use in strapdown INS.

Criteria for gyro (or GI) accuracy estimation are defined for each type. In particular, errors in SDF astatic gyros and gyro stabilizers can be estimated by:

1. the speed of drift in the measuring axes from the set direction (grad $h^{-1}$, angular min $min^{-1}$);
2. errors in SDF measuring instruments for angular speed by the displacement relative to a statically steady position (displacement of zero, grad $s^{-1}$, grad $h^{-1}$); and

3. by the deviation of the *scale factor* of the output signal from an established value; and errors in gyro linear acceleration integrators by deviations in that scale factor. (Here, the scale factor is indicative of the possible fluctuation of the gyro gain from its nominal value.)

The majority of instrumental errors involves the differences from the ideal characteristics of the various units, elements, materials, electronic components, and so forth within the GI.

The drift (zero displacement) has an inherent component and also a component that is dependent on linear acceleration. Both the drift and the output signal scale factor are characterized by the following components:

1. Systematic component expectations from run to run that are published in the relevant certificates
2. A random (or "casual") component from run to run (giving a ceiling on deviation from certificated value, $3\sigma$)
3. A casual component in any run (that is, the maximum deviation from the average value for one particular start).

For the platform type of strapdown INS, independent maintenance of observability is possible in runs for both drift components by the method of double gyrocompassing (see Section 6.1.6). For each run, this forms a basis on which to accept a level of INS accuracy in drift speed not dependent on linear acceleration. The carrying out of double gyrocompassing for the strapdown INS is impossible.

For the aeronautic strapdown INS it is possible to use optical (or other nonautonomous) methods for transferring azimuthal positional information for solving the problem of the observability components of gyro zero displacement in runs.

For the space-borne strapdown INS in trajectories with greater linear accelerations (especially at launch and descent from an orbit) it is preferential to use gyros tolerant to linear acceleration (fiber-optical, laser, and hemispherical resonator types) and having a minimal casual component of zero displacement from run to run. As a variant, the preliminary calibration of such a strapdown INS at the launch position before installation on a carrier rocket is possible. The carrying out of such a calibration is expedient for GI having variable values from run to run, so considerably reducing the time intervals between runs.

By not using SDF and TDF mechanical gyros in the strapdown INS of highly maneuverable space vehicles, methodical errors generated by vehicle spatial angular fluctuations are avoided. (This effect is called the *cinematic drift* of the gyro.)

The level of strapdown INS accuracy depending on the variant used can be estimated to be in a range from the value of a casual component of zero displacement up to its value from run to run.

### *6.1.11 The Scope of Gyros and Gyro Instruments of Various Types*

These are presented in Table 6.2 where accuracies in specific modes of vehicle control are defined on the basis of full models of practical GI errors.

**Table 6.2.** Gyro applications and accuracies

| **Level of accuracy, grad $h^{-1}$** | **Better than $10^{-3}$** | **$10^{-3}$–$10^{-2}$** | **$10^{-2}$–$10^{-1}$** | **$10^{-1}$–1.0** | **1.0–$10^{2}$** |
|---|---|---|---|---|---|
| Type of gyro | Electrostatic, Hemispherical resonator, Floated | Spherical aerodynamic, gas suspended, Floated, Laser, Hemispherical resonator | Dynamically tuned, Simplified floated, Laser, Hemispherical resonator, Fiber-optic | Simplified: Floated, Dynamically tuned, Laser, Fiber-optic, Vibrating | Simplified: Dynamically tuned, Pneumatic with gas suspension, Vibrating |
| Type of vehicle | Intercontinental ballistic missiles, Strategic delivery aircraft, Space telescopes and special platforms, Autonomous means of azimuth alignment. | Strategic missiles, Airliners and tactical aircraft, Space vehicles. | Long-range tactical missiles, Space vehicles, Short-haul aircraft, Helicopters. | Tactical missiles, Ballistic missile, Warheads, Mobile missile launchers | Sights, Passive bombs, Torpedoes, Cluster warheads. |
| Type of GI | Complex redundant Gimbaled INS, Advanced strapdown INS | Gimbaled and strapdown INS | Chiefly strapdown INS | Simplified strapdown INS | Autonomous gyro assemblies |
| Required uptime, hours | $3 \times 10^5 - 10^5$ | $10^5 - 5 \times 10^4$ | $5 \times 10^4 - 10^4$ | — | — |
| GI cost, USD | $3 \times 10^5 - 10^5$ | $10^5 - 5 \times 10^4$ | $5 \times 10^4 - 10^4$ | $10^4 - 3 \times 10^3$ | $5 \times 10^3 - 5 \times 10^2$ |

## 6.2 SINGLE DEGREE OF FREEDOM (SDF) GYROS

Historically, the first SDF gyros had a solid rotor suspended in a frame, but modern SDF gyros now include vibrating gyros, hemispherical resonator, and micromechanical gyros, all of which will be treated individually below.

### *6.2.1 THE SOLID ROTOR SDF GYRO*

This gyro consists of a solid rotor that rotates about a single spin axis, and which is supported by a frame that is able to rotate about another, orthogonal, single axis, as shown in Figure 6.2. This frame, or gimbal, can rotate about its *axis of sensitivity* to produce a measureable signal angle $\beta$ with respect to an outer case, or *base*, and will do so when that base is rotated. This illustrates the basic principle of the solid-rotor gyroscope. Note that when $\beta = 0$, the orthogonal spin and gimbal axes are also orthogonal with the axis around which the base is able to rotate.

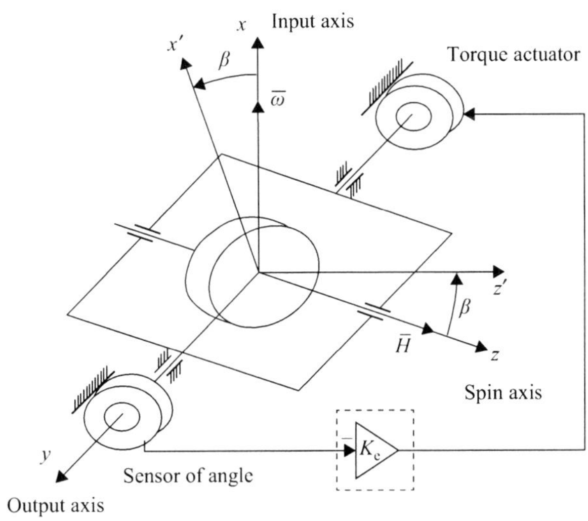


**Figure 6.2.** A Single Degree of Freedom (SDF) gyro system.

If the base rotates at an angular rate $\bar{\omega}$, the gyroscopic torque $M_G = H\omega\sin(\overline{H},\overline{\omega})$ tries to reconcile the kinetic moment vector $\overline{H}$ with the angular speed vector $\bar{\omega}$ that causes gyro rotation around the output axis $y$.

The gyro can be used either as an angular rate integrator (an *integrating gyro*) or as a measuring instrument (an *angular rate gauge*). The former are used as the sensing elements in gyro stabilizers and gyro platforms and are not normally used as independent devices. Angular rate gyros have found widespread application in systems for angular stabilization wherein information on angular movements of the base around a center of gravity is formed by separate measuring instruments. An angular rate signal is necessary in this case only for maintenance of the demanded transient quality in the system, and high angular rate gauge accuracy is not mandatory.

An integrating gyro includes a torque actuator, also shown in Figure 6.2, which carries out auxiliary functions and does not participate in a measurement. The signal from an angle sensor is amplified and applied to the torque actuator to produce negative feedback that balances the gyro moment. The current needed for this is the output signal that measures the angular rate.

In the overwhelming majority of cases, SDF gyros are of the floating type. For an integrating gyro, floating support is essential for providing the required hydronic damping; and in both cases it effectively removes the load applied by a mechanical bearing, so allowing an increase in accuracy and application exploitability of the gyro.

Angular rate sensors based on floating integrating gyros with torque feedback find major applications in strapdown INS. However, when designing these it is necessary to consider a wider spectrum of disturbing influences from which the gyros are isolated by gyro platforms, whereas in traditional angular rate instruments this is not essential.

### *6.2.2 THE INTEGRATING GYRO*

As a first approximation, the basic equation for gyro movement can be written as follows:

$$J_0\ddot{\beta} + \xi\dot{\beta} = H\dot{\psi} \qquad (6.2)$$

where $\dot{\psi} = \omega_x$ and $J_0$ is the moment of inertia of a float about an axis $y$. $\xi$ is a damping factor. This equation may be resolved as

$$\beta = \frac{H}{\xi}\psi - \frac{H}{\xi}\int_0^t \dot{\psi}(\tau)e^{-\frac{t-\tau}{T_f}}d\tau \qquad (6.3)$$

where $J_0/\xi = T_f$ is a float time constant.

The first term in this solution characterizes the integrating properties of the gyro and the second term is a dynamic integration error. To reduce this, it is necessary to reduce the moment of inertia of the float about the output axis and increase the damping factor.

$H/\xi$ is the essential transfer factor of an integrating gyro, and this is maintained constant by thermostating the device and stabilizing the frequency of the voltage powering the synchronous gyro motor.

The integrating gyro is thus accurately described by the transfer function of a delayed integrator.

### *6.2.3 RATE OF SPEED GAUGING*

As a first approximation, the equation of a gyro movement appears thus

$$J_0\ddot{\beta} + \xi\dot{\beta} + c\beta = H\omega_x \qquad (6.4)$$

where $c$ is the transfer factor of the torque feedback contour.

In the steady-state mode, the deviation of a float will be

$$\beta_{st} = \frac{H}{c}\omega_x$$

The dynamic characteristics involved in the gauging of angular rate in a feedback contour passband are defined by the transfer function of an oscillatory unit.

The inherent frequency should considerably exceed the maximum frequency of the measured angular rate $\omega_x$ change; and the attenuation factor should be within limits 0.5–0.8.

### 6.2.3.1 Feedback Contours of the Angular Rate Gauge

The contour of a measuring instrument feedback system may appear in continuous or pulse (discrete) forms. The continuous contour demands the separate conversion of the feedback current to a discrete output signal. In the majority of cases, integrating type converters are used in which the integral of the gauge current is counterbalanced by discrete amplitude and duration pulses. In a measuring instrument with a pulse contour, quantization of the feedback current is provided directly. The pulse contour can contain a two-position or a three-position relay unit, or a quasi-linear pulse-width modulator. A typical schematic of the electronic part of a discrete feedback contour is shown in Figure 6.3.

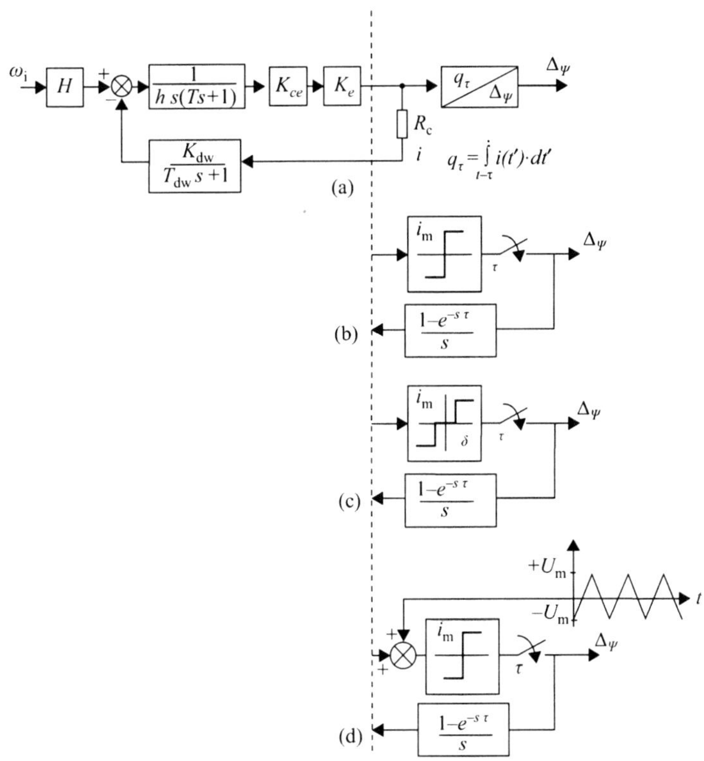

**Figure 6.3.** Feedback contour schematic. (a) a continuous (analog) contour with integrating converter; (b) a discrete (pulse) contour with on-off relay; (c) a discrete contour with three-position relay; (d) a discrete contour with linear pulse-width modulator.

The advantages of continuous and discrete contours may be combined in a discrete feedback contour using a pulse-width linear modulator. This scheme implies small angular disturbances applied to the float and provides good opportunities for dynamic correction including the application of astatic regulation by the angle of float deviation. Several advantages are conferred: the thermal emission does not depend on the measured angular speed, the fixed working points ($i_m^+$, $i_m^-$) of the torque actuator are advantageous for operation mode adjustment, and it is simple to change the scale factor if necessary. (It is important that a constant current flows through the torque actuator corresponding to the maximal measured angular speed.) For these reasons, such a feedback contour is very suitable for sensors with small measurement ranges, such as moment gauges that operate at low power levels. For example, it is often used in strapdown INS for space vehicles requiring high accuracy in angular stabilization and guidance.

#### *6.2.3.2 Design Variants*

Structurally, SDF floated gyros are designed according to an established scheme consisting of a cylindrical format along with an inverted-type synchronous-hysteresis gyro motor, and inside a tightly sealed float gyro-housing that carries out the functions of support frames. Depending on the purpose of the gyro, the rotation speed of the rotor may be selected within the limits (7.5–15) $\times$ $10^3$ rev min$^{-1}$; and the kinetic moment should be within the limits (10–200) g cm·s.

The rotor support consists of precision radial ball-bearings with fixed axial tightness or, for more accurate devices, gas-dynamic hemispherical supports. For dissipating the gyro motor heat and for maintaining the working capacity of a gas-dynamic support, the gyro-housing is filled mainly with hydrogen up to a pressure close to atmospheric.

Twisted tie-rods of various designs, ball-bearings, and steel (or composite material) pins, along with axial bearings made from a suitable stone such as agate, ruby, or corundum, are used in the gyro-housing supports. In precise devices a combination of pivot support with electromagnetic axial and radial float centering (either active or passive) is also frequently used.

The most common float turn-angle gauges are induction types and micro-selsyns, while torque gauges are electromagnetic or magnetoelectric.

The backlash space between the float and the case, and also the free space inside that case, are filled with a high density liquid to form a (specified) floated suspension. The liquid in the backlash space between the cylindrical surfaces of a float and its case provides appropriate damping. Such a suspension partially protects the gyro-housing support from the influence of impacts, vibrations, and so forth.

Neutral buoyancy in a gyro unit is attained only at a strictly defined working temperature and to maintain this, high accuracy gyros are equipped with independent thermostating systems. This working temperature is greater than the normal maximum temperatures found in compartments and is usually 60°C, or in some cases up to 100°C.

Exact maintenance of the working temperature is necessary not only for providing constant liquid density and viscosity, but also for establishing a closely controlled location for the center of gravity of the floated gyro unit. The center of pressure of the liquid at a gyro-housing rotational axis, and the geometrical sizes of the gyro units themselves, influence the casual components of drift speed (zero displacement). To be precise, for integrating gyros and measuring instruments for strapdown INS, the working temperature must be maintained with accuracies in the region of hundredths of degrees.

An important element influencing the accuracy of a gyro is found in the slip rings that conduct current via the device case to the float. For limiting angular values of gyro unit deviation within ±1°, these can be in the form of flexible conductors that create a drag torque proportional to the angle of deviation. Residual drag torque after returning a gyro unit to a starting position is one of the reasons for the appearance of corresponding casual drift speed (zero displacement) components. Instability in gyro unit balancing from run to run is a factor limiting the sphere of application of angular rate measuring instruments in strapdown INS.

Under conditions of zero linear acceleration, and at angular rates of base rotation up to ~1 grad $s^{-1}$, such measuring instruments have almost the same accuracy as the integrating gyros used in gimbaled INS, that is, about $10^{-3}$ grad $h^{-1}$.

It is possible to consider using angular rate measuring instruments on the basis of SDF floated integrating gyros in strapdown INS for space telescopes, precise platform guidance, geostationary satellites, and so forth.

## 6.3 THE TDF GYRO IN GIMBAL MOUNTINGS

There are two design versions of gimbal mountings with hinged gimbal ring supports: external and internal. Internal gimbal mounting is seldom used, and only in some special types of gyro.

### 6.3.1 PROPERTIES OF A FREE GYRO

The fundamental properties of a gyro may be listed as follows.

1. The *main axis* (the axis of gyro symmetry or rotor rotation) tries to maintain its initial direction in inertial space in spite of any rotary movements by the base upon which it is mounted.
2. An external force $\vec{F}_{ex}$ acting on the main axis (via a gimbal ring) tries to set that axis in motion (i.e., creates a torque about the center of support), but this results in the main axis of the gyro moving in a direction perpendicular to the direction of that applied force, as shown in Figure 6.4.

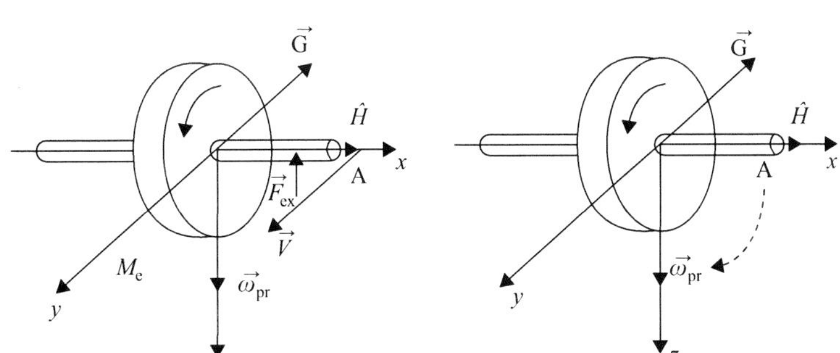


**Figure 6.4.** Action of an external force on a free gyro main axis.

According to the Rezal theorem, the end of a vector of kinetic moment $\vec{H}$ (point $A$) acquires a linear velocity $\vec{V}$ equal and parallel to the vector of the moment of the external force $M_e$, and moves with an angular rate $\vec{\omega}$ directed along the axis $OZ$. This movement is called precession and may be described by the equation:

$$\vec{\omega} \times \vec{H} = \vec{M_e} \,. \tag{6.5}$$

During precessional movement, the angular rate $\vec{\omega}$ is orthogonal to the moment of an external force $M_e$ that is possible only due to a reaction counterbalancing an external force moment. That is, the moment $\bar{G}$ is equal in value and opposite in sign to $M_e$. This moment $\bar{G}$ is a result of the Coriolis forces of inertia and is called the gyroscopic moment (Figure 6.4):

$$\vec{G} = \vec{H} \times \vec{\omega} \tag{6.6}$$

and its value is $G = H\omega \sin(\vec{H}, \vec{\omega})$.

That is, during rotation of the main axis of a gyro in space with an angular rate $\vec{\omega}$, the gyroscopic moment $\bar{G}$ is directed so as to combine the kinetic moment vector $\vec{H}$ with the angular rate vector $\vec{\omega}$ of the main axis of rotation.

3. Under the action of an impulse force (an impact), the main axis of a gyro hardly changes its initial direction but only exhibits rapidly fading fluctuations. Such fluctuations are called *nutation*.

   For the majority of practical cases, and referring to Figure 6.5, the accuracy of the gyro movement about the support center may be described by the so-called technical equations:

$$\begin{cases} J_B \ddot{\upsilon} + H\dot{\psi}\cos\upsilon = M_B \\ J_C \ddot{\psi} - H\dot{\upsilon}\cos\upsilon = M_C \end{cases}, \tag{6.7}$$

   where $J_B$ and $J_C$ are the moments of inertia of a gyro about axes BB and CC, respectively; $H$ is the kinetic moment of the gyro; $\psi$ and $\vartheta$ are angles of turn about axes BB and CC, respectively; and $M_B$ and $M_C$ are the moments of the forces operating on axes BB and CC, respectively.

   These differ from the complete equations due to the absence of higher-order terms and also the equation describing a rotor movement around its own axis $J\ddot{\phi} = M_D - M_S$, where $M_D$ and $M_S$ are the moments of rotation and resistance of rotation of the rotor, respectively.

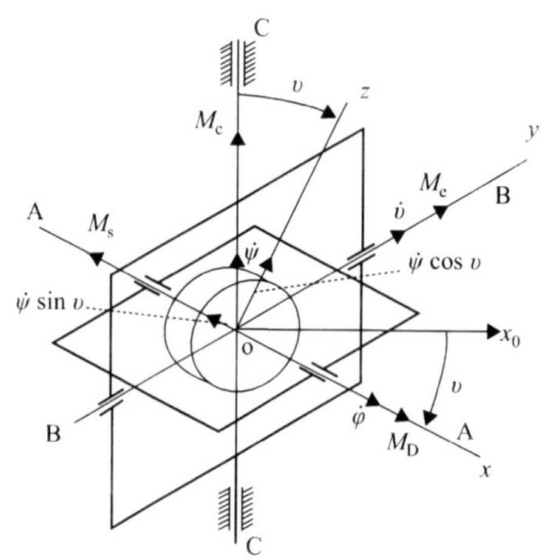


**Figure 6.5.** Action of the instantaneous moment of an impulse force on a gyro.

For an infinitesimal time of action of an impulse moment, the coordinates of any points in a system remain constant and only the speeds of their movements change. For $\upsilon(0)=0$ and $\psi(0)=0$ an impulse results in gyro angular speeds $\dot{\upsilon}(0)=\omega_B^*$ and $\dot{\psi}(0)=\omega_C^*$, after which $M_B = M_C = 0$ and the top of the gyro takes an elliptic path (at $J_B = J_C$ and at $\omega_B^* = \omega_C^*$ this becomes circular) with coordinates:

$$\upsilon^* = -\frac{J_C}{H}\cdot\omega_C\,,\ \psi^* = \frac{J_B}{H}\cdot\omega_B \tag{6.8}$$

about the motionless center, and with a frequency $\omega_H = \dfrac{H}{\sqrt{J_B J_C}}$, which is the frequency of nutation fluctuations.

As $H = J\cdot\Omega$, and $J$ has the same order as $J_B$ and $J_C$, the angular frequency of movement of the gyro apex $\omega_H$ can be hundreds or thousands of radians per second. In real designs, nutation fluctuations fade quickly because of the presence of resistive forces and the dissipation of impulse energy. Thus, a gyro maintains its main axis at a steady position in space.

4. Under the action of a constant externally applied moment where $M_B$ = *const* and $M_C$ = *const*, the main forced motion of a gyro top exhibits a constant angular rate having a vector argument equal to the argument of the external moment vector. This motion, caused by the action of an external moment on a gyro, is precession. Any high-frequency nutation fluctuations are superimposed on this main precession motion, and their character and frequency remain the same as in the absence of the external moment. However, the amplitude depends on the moment of external forces:

$$\upsilon^* = \frac{M_B J_C}{H^2} - \frac{M_C}{H}t \text{ and } \psi^* = \frac{M_C J_B}{H^2} + \frac{M_B}{H}t \tag{6.9}$$

Usually, within the precession frame approach, small high-frequency fading nutation fluctuations are not considered, in which case the precession movement has a constant angular rate $M_C / H$ and $M_B / H$, and may be considered inertialess.

An analysis of the equations of gyro movement according to the various laws of time changes in any applied external moments results in the conclusion that the moments of inertia of the gyro support rings influence only changes in nutation fluctuations and do not influence angular rates of precession.

### 6.3.2 AREAS OF APPLICATION, DESIGN FEATURES, AND ERROR SOURCES

There are two essential spheres of practical usage of free gyros, and hence of the relevant designs, as follows.

1. Free gyros may be used to measure angular deviations in systems applied to the angular stabilization of various vehicles. By setting the initial direction of the main axis, a free gyro working in such a system may provide a zero motion mechanical model in the inertial reference frame. Any object position mismatch angles relative to this reference frame are registered by angle gauges at the axes of the gimbal support rings.
2. Free gyros may also be used as sensitive elements for gyro stabilized platforms. In this case the zero motion mechanical model in the inertial reference frame is the platform itself, and the gyro is a zero-indicator in the platform stabilization system.

Installation of the main axis of a gyro in a given direction can be carried out by preliminary orientation of the object on which the gyro is installed, and also by including in the structure correcting devices the sensitive elements of which may be pendulums, levels, zero-indicators of speed, and so forth. These devices are disconnected during the operating mode.

Exact fixing of the main axis of a gyro relative to the case (object) is carried out by means of a cage that orients it in a fixed position relative to this case and quickly releases it when switching to the operating mode. Thus, in operating modes, free gyros are not corrected, and errors in object angles of turn can be divided into two basic groups:

a. methodical errors caused by the rotation and curvature of the Earth's surface (considered at a system level, if necessary); and
b. instrumental errors caused by inherent gyro drift.

Errors in group (a) arise because a free gyro keeps the direction of its main axis in space constant while the object moves relative to points located on the rotating Earth.

With the development of inertial navigating systems (particularly strapdown systems), the need to create an onboard zero motion mechanical model in the inertial reference frame has almost disappeared, and this procedure is now carried out by image computing. Nevertheless, free gyros by virtue of simplicity, relative cheapness, considerable "know-how," and various operational advantages remain adequate for applications in systems for the angular stabilization of rockets, various kinds of missiles (except for long-distance cruise types), some space vehicles, and target seekers.

Readings of angular information from a gyro are carried out by various types of angle gauge such as potentiometric, inductive, bolometric, and so forth, either from one or both gimbal support frames (rings). In the single case a working frame is an external one that offers some advantages in accuracy.

A typical example of free gyro usage as a short-term direction keeper is in the gyro horizontals (or verticals) used for the angular stabilization of ballistic missiles. Because of some forms of free gyro construction that are optimized for certain objects and certain conditions of operation, some adverse moments defining inherent gyro drift have no determinate dependence. However, it is possible to list the basic methodical adverse moments acting on a free gyro as follows.

a. Moments not dependent on linear acceleration:
    - Inherent moments arising from dry friction in a support depend not only, and not predominantly, on the absolute values of these moments, but mainly on their asymmetry with respect to movements having unlike signs
    - Spurious moments created by flexible current collectors, and also the reactions of angle and moment gauges that may have somewhat elastic characteristics
    - Moments arising from the action of external magnetic fields.
b. Moments proportional to linear acceleration:
    - Moments arising due to static unbalance
    - Moments caused by the linear expansion factors of design element materials and wear in support elements
    - Moments caused by a discrepancy in the center of gravity of the rotor and the support center.

c. Moments proportional to the square of linear acceleration:
These moments are caused by the *anisoelasticity* of a design in the directions of its main axes. *Elasticity* is the feature opposite to *rigidity*, the latter being understood as the ratio of force to the resultant displacement. Anisoelasticity in a design exhibiting sideways linear vibration results in the detection of inertial forces of unlike signs that create a constant spurious component in the moment.

It is also necessary to take free gyro methodical drifts into account as follows.

1. The drift caused by nutation fluctuations in the main axis arises around an axis of a support external frame if that main axis is not perpendicular to an axis of the external frame. The relevant drift speed increases with any increase in such deviations, though its relative value is rather small. However, it can noticeably affect the behavior of precision gyros because nutation fluctuations can be not only consequences of external moments (which fade quickly), but can also result from rotor nonideal balancing.
2. Kinematic drift (wandering of the conical movements) is generated by angular fluctuations in the base.

Free gyros intended for use as sensitive elements in gyro platforms were initially developed as alternatives to floated SDF integrating gyros. Devices having two measuring degrees of freedom within comparable overall dimensions and masses made for flexibility in the amount of redundancy in gyro platform sensitive elements.

Necessary improvement in gyro accuracy is achieved by the application of hydrostatic unloading in the support rings such as the floated support of spherical gyro housings, gas dynamical rotor support, and better constructional quality, all of which lead to better isolation of a gyro from rotary movements in the base, and to an essential reduction in a range of angular movements in the support rings.

The technical equations of a gyro using liquid damping by a floated support are

$$\begin{cases} J\ddot{\upsilon} + h\dot{\upsilon} + H\dot{\psi} = M_B \\ J\ddot{\psi} + h\dot{\psi} + H\dot{\upsilon} = M_C \end{cases}, \qquad (6.10)$$

where $J$ is the equatorial moment of inertia of a rotor in a gyrohousing; and $h$ is a liquid damping factor. The dynamic properties of a gyro are defined by a relative damping factor $h/H$, and improve with its reduction.

The remaining technical features of the gyro under consideration are similar to those of the SDF floated integrating gyro. Levels reached by the casual drift speed component (not dependent on linear acceleration) in 24-hour runs can be as low as (1 to 2)$10^{-3}$ grad $h^{-1}$.

### 6.3.3 *TWO-COMPONENT ANGULAR SPEED MEASURING INSTRUMENTS*

The free gyro designed for application as an element in gyro platforms can also be used as a TDF measuring instrument of angular rate in strapdown INS. For this purpose, and by means of external cross-link negative feedback, the gyro is switched to a mode of measurement accessing moments arising from a desired main axis rotation in space and resulting in an angular rate measurement. Figure 6.6 depicts such a mode of operation.

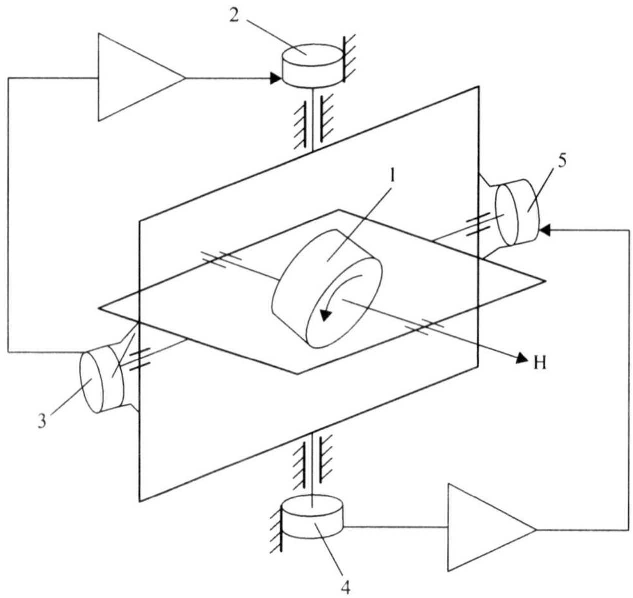


**Figure 6.6.** A free gyro used as a TDF instrument for angular rate measurement in strapdown INS.

Here, amplified signals from angle gauges (3 and 4) are applied to moment gauges (2 and 5) so that the gyro (1) main axis deviation angles around both support axes are almost zero. The angular rate of turn measurement of the main axis in space is provided by moment gauges (2 and 5). The values of these moments are given by the currents through the gauge windings, and the measured angular rate values by the turn angles of the gimbal rings.

Versions of a feedback contour and its error sources are similar to those of the strapdown INS angular rate measuring instruments designed on the basis of the SDF floated integrating gyro.

## 6.4 THE GYROSCOPIC INTEGRATOR FOR LINEAR ACCELERATION (GILA)

The GILA is intended to measure the linear speed of a vehicle's center of gravity along a given direction. For this purpose it is necessary to treat the linear acceleration as that caused by all external forces of nongravitational origin that are applied to the vehicle. This is the *apparent acceleration,* which is used by the gyroscopic integrator to determine the linear speed.

The most common applications of gyroscopic integrators are in ballistic missiles and spacecraft.

### *6.4.1 PRINCIPLES OF GILA OPERATION*

A schematic of the longitudinal acceleration gyroscopic integrator of a ballistic missile is shown in Figure 6.7. This integrator is representative of the unbalanced SDF pendulum-type gyro.

The axis of a three-ring external support frame used as the axis of sensitivity is fixed in parallel with the longitudinal axis of the vehicle.

The axis of a gyro casing does not cut across the axis of the external frame, and is in effect at a fixed distance from it.

The center of gravity of the rotor within the casing is located at the crossing of the main axis and an axis of the external frame. Also, the center of gravity of the external frame is on its axis. The stabilizing system, enabled via contacts at the casing and at the external support frame, performs interframe correction to provide orthogonality of the main axis and an axis of the external support frame.

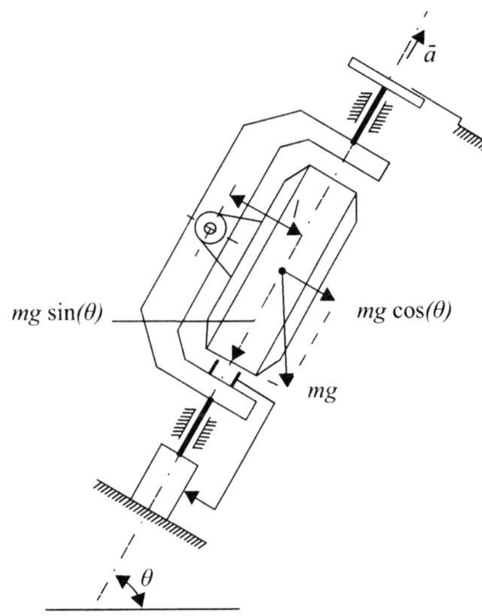


**Figure 6.7.** The gyroscopic integrator for linear acceleration (GILA).

When a vehicle moves with acceleration $a$ and pitch angle $\theta$ a moment $M$ operates around a case axis equal to:

$$M = m\,[a + g\sin(\theta)]l \tag{6.11}$$

where $m$ is the mass of the gyro unit, $mg$ is the gravitational force, and $l$ is the distance between the casing support hinge $a$ and the center of gravity of the gyro unit.

If the main axis is always orthogonal to an axis of the external frame, then a force $mg\cos(\theta)$ and inertial forces arising from cross-sectional accelerations do not create a moment relevant to an axis of the gyro casing, and it will exhibit precession around an axis of the external frame with angular rate $\omega$ where

$$\omega = \dot{\alpha} = \frac{ml[a + g\sin(\theta)] + M^*}{H}$$

from which

$$\alpha - \alpha_0 = \int_0^t \frac{ml[a + g\sin(\theta)] + M^*}{H} d\tau \tag{6.12}$$

where $\alpha_0$ is the initial value of the angle at which the gyro turns about an axis of the external frame, and $M^*$ is the moment directed along an axis of the internal support frame arising due to unconsidered factors.

According to Davis and Ledgerwood (1961), when $(m, l, H) = const$ then:

$$\alpha - \alpha_0 = \frac{ml}{H}\int_0^t [a(\tau) + g\sin(\theta(\tau))]d\tau + \frac{1}{H}\int_0^t M^* d\tau = \frac{ml}{H}[(V - V_0) + \int_0^t g\sin(\theta(\tau))d\tau] + \frac{1}{H}\int_0^t M^* d\tau.$$

For a ballistic missile $V_0 = 0$ and taking $\alpha_0 = 0$, it is possible to obtain a basic relation describing GILA operation:

$$\alpha = \frac{ml}{H}[(V + \int_0^t g\sin(\theta(\tau))d\tau)] + \frac{1}{H}\int_0^t M^* d\tau \tag{6.13}$$

Thus, from the integral of $M^*/H$, the angle at which a gyro turns around an external frame axis, within about a degree of accuracy, is proportional to the flight speed plus some value representing the integral of a projection of the terrestrial gravitational acceleration on the missile longitudinal axis.

### 6.4.2 SOURCES OF GILA ERRORS

1. The GILA transfer factor $k = m \cdot l/H$. Therefore, deviations of values $l$ and $H$ from the desired ones (mainly due to temperature changes) directly influence the measurement.
2. The moment $M^*$ is caused basically by the frictional moments in the axes of the internal supporting frame and the current collector contacts (slip rings) that manage the operation of the stabilizing system and so forth, to produce an absolute linear time-dependent measurement error of: $\Delta V = tM^* / (kH)$.
3. During calculation of the amendment of the integral from a component $g\sin(\theta)$, deviations in the real values of $\theta$ from the prescribed values can appear. These are caused by drifts in the angular stabilization system and a nonparallel GILA sensitivity axis with the longitudinal axis of the vehicle. Furthermore, the actual active phase time of a vehicle trajectory can differ from the designated one because of deviations in the engine thrust, missile mass, and so forth.
4. The source of an error can be a conic movement (so-called scanning) of a missile.
5. The moment of friction in an external frame axis can also be a source of error.

For a ballistic missile, the accuracy in velocity as measured by a GILA should not be worse than 0.01%, which is rather high. Therefore, along with standards based on the minimization of the enumerated errors, the ultimate calibration of the instrument is conducted directly before launch with the missile in a vertical position.

GILA applications in control systems for space vehicles become simpler when there is no necessity for the evaluation of, and the correction registration for, modifications in the flight trajectory program. For example, before thruster initialization there is the possibility of defining GILA zero drift during weightlessness.

## 6.5 CONTACTLESS SUSPENSION GYROS

### *6.5.1 INTRODUCTION*

Gyroscopes with contactless rotor suspensions are built around a (usually spherical) rotor that is suspended by aerodynamic, aerostatic, magnetic, or electrostatic forces. At the present time, however, only gyroscopes with electrostatically suspended rotors are used in aerospace technology for high-accuracy autonomous attitude references because other types are not competitive with them in accuracy, reliability, or availability. Therefore, research on their development has largely ceased.

### *6.5.2 THE ELECTROSTATIC GYROSCOPE (ESG)*

The ESG is a device in which a spinning spherical rotor is suspended in an evacuated spherical vacuum chamber by electrostatic forces. By virtue of being a free gyroscope, it is the most accurate of all available sensitive elements used at the present time (Peshekhonov 2003). Its principle of operation is based on the well-known property of electric field lines surrounding a conductor to form normals to the surface of that conductor. Hence, the interaction of such a field with a perfectly spherical conductor does not cause any disturbing moments, which makes possible an ideal gyroscopic suspension. Hence, high drift stability can be achieved because a relatively small number of physical factors lead to errors. This high stability allows the development of adequate deterministic drift models, so making possible algorithmic compensation for real drifts (Anfinogenov et al. 1992; Martynenko 1988).

The general problem to be solved in developing an ESG can be formulated as follows: to provide levitation for a perfectly spherical conducting ball in a vacuum in the absence of an electric field, to provide stable spinning, and to read out information about the angular position of its axis of rotation relative to the gyroscope case.

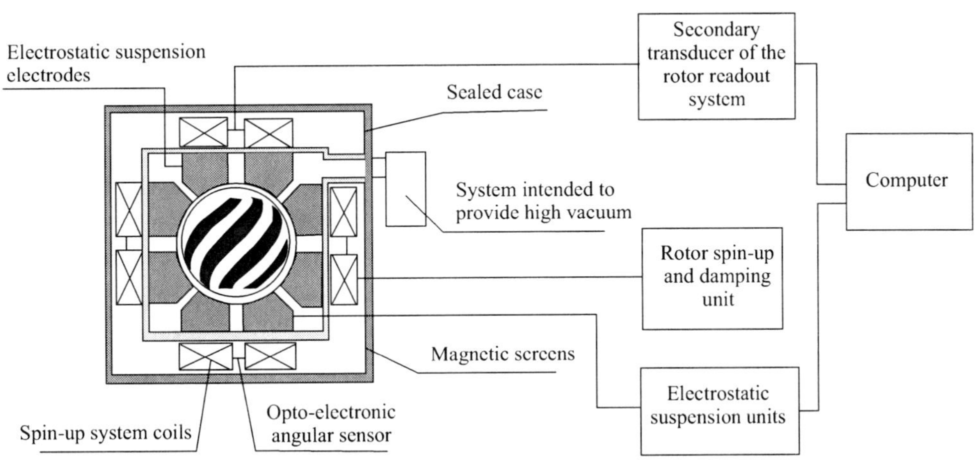


**Figure 6.8.** A Single Degree of Freedom (SDF) gyro system.

The accuracy of such a gyroscope in terms of drift parameter stability depends on a relatively small number of disturbing factors: rotor nonsphericity, stability of the rotor center-of-mass relative to the geometrical center, smallness and homogeneity of its surface resistance, stability of the electrostatic suspension parameters, and the efficiency of the magnetic shielding.

Among the major drawbacks of the ESG are low overload capacity (no more than 10–15 g) and low reliability should the rotor levitation support system cut off when the rotor is spinning at high speed. However, the ESG shows considerable promise for spacecraft due to a favorable combination of physical properties and operating conditions, as follows.

First, the basic disturbing moments that result in drift are caused by errors in rotor nonsphericity, which are proportional to the force load on the suspension. When inertial loads are absent, these moments are close to zero and do not cause systematic ESG drift, and for the same reason any temperature strain in the gyro rotor does not influence the drift.

Second, the natural rotation of the gyro case together with the orbital rotation of the spacecraft about the rotor's angular momentum vector (which is fixed in space) leads to self-compensation of most of the disturbing moments associated with that case, mainly caused by the residual magnetic field inside the suspension cavity.

Thirdly, attitude determination in inertial space using the ESG does not require numerical integration of Poisson differential equations and does not impose any restrictions connected with integration, processing speed, or computer memory. Hence, a definite advantage of the ESG as a free gyro is its property of maintaining inertial direction even in the case of a failure in the data processing channel. Furthermore, the ESG is free of drift components that depend on the gyro case angular velocity, unlike rate gyros, in which the error is caused by instability of the scale factor.

Information about the possibility of creating an ESG and practical methods for its development date back to the 1970s. Practical work on designing and using the ESG has taken place in the United States (Elwell 1973–1974; Everitt 1973, 1978, 1989; Fairbank 1961; Hendricks 1985; McLeod 1979; Nordsieck 1961; Schiff 1960); in Russia (Anfinogenov et al. 1987, 1991; Landau 1993; Orlov 1967); and in France (Leger 1984; Mathey 1967).

There are two types of ESG differing from each other by rotor construction: either a hollow or a solid rotor, both usually made of beryllium, this being a light and strong material that is stable with temperature changes. Both can be used in gimbaled and strapdown systems. Hollow-rotor ESGs for gimbaled systems afford high-accuracy performance, and under conditions of autocompensation by gyro case rotation and algorithmic compensation after preliminary starting calibration, the residual drift may not exceed $10^{-4}$–$10^{-5}$ deg $h^{-1}$. Strapdown attitude reference systems represent one of the prime ESG applications and considerable work has been performed in this area (Anshakov 2002; Dmitriev et al. 1995; Duncan 1973; Elwell 1973–1974; Gurevich 2001; Gusinsky 1998; Landau 2000, 2007, 2010; Nordsieck 1961; Schmidt 1979; Somov et al. 1999).

The advantages of the solid-rotor ESG are as follows: high temperature and temporal stability of the rotor shape, relatively low control voltages in the electrostatic suspension that allow the use of simpler and more reliable circuits in electronic units, and high stability of the ESG drift model coefficients from run to run.

Current ESG technology has developed from using hollow rotors of 38–50 mm diameter to gyroscopes with solid rotors of 10 mm diameter; and from gyroscopes for high-accuracy gimbaled systems to those for strapdown systems.

Practical implementation of ESG technology has involved the development of a number of systems, units, and assemblies taking into account their interactions, and for determining the viability of operation under different operating conditions. In fact, entire complexes for such research, design, manufacture, and testing have been built.

The basic work that must be carried out in designing an ESG for widespread applications involves the following:

1. Development of a gyro drift model for any mode of operation
2. Development of a fabrication technology, and metrological provision for rotor manufacture
3. Development of an electrostatic rotor suspension system, including the servo system that provides rotor levitation within the electrode system
4. Development of a system for reading out data about the rotor angular position relative to the gyro case, and the development of an error model for this system
5. Development of a system for spinning-up, nutation, oscillation damping, and stabilization of the rotor spinning rate
6. Development of a system for the provision, maintenance, and control of the vacuum in the gyroscope
7. Development of the gyro magnetic shielding system
8. Development of methods, aids, and software for testing, checking, and calibration of the gyroscope, and
9. Identification of the developed error model parameters.

The development of the ESG for strapdown systems poses fundamentally new problems, the most complicated of them being as follows:

1. Development of a strapdown ESG drift model for subsequent algorithmic error compensation
2. Development of a system for reading out data about the rotor angular position relative to the body-fixed frame within an unlimited range of angles, and including the development of an error model for this system, plus methods and aids for identifying the parameters of this model.

### 6.5.2.1 ESG Accuracy

Literature devoted to the theoretical analysis of ESG accuracy is extensive (Bryushkov 1978; Buravlev 1993; Dzhashitov et al. 1996; Gubarenko et al., 1994; Landau et al. 1996; Levin et al. 1999; Martynenko 1982, 1988, 1992, 1993; Zavgorodnii et al. 1994; Zhang 1992; ); and a general ESG drift model has been developed for a gyroscope with orthogonally arranged electrostatic suspension axes.

The main moments of forces acting on a spherical rotor in an electric field are caused by axial and radial rotor unbalance, nonsphericity of its surface, anisoelasticity of the suspension channels, and residual magnetic fields. The drift model, denoted by $\omega_1$, in the projection of one of the axes (e.g., Axis 1) of the case can be written as follows:

$$
\begin{aligned}
\omega_1 = {} & k_0\gamma_1\left[-\left(1-\gamma_1^2\right)\gamma_1^2+\gamma_2^4+\gamma_3^4\right]+k_1\left[-\left(1-\gamma_1^2\right)\,V_1+\gamma_1\gamma_2V_2+\gamma_1\gamma_3V_3\right] \\
& +k_2\gamma_1\left[-\left(1-\gamma_1^2\right)\,V_1^2+\gamma_2^2V_2^2+\gamma_3^2V_3^2\right]+k_3\gamma_1\left[-\left(1-\gamma_1^2\right)\,\gamma_1V_1+\gamma_2^3V_2+\gamma_3^3V_3\right] \\
& +k_4\gamma_1\left[-\left(1-\gamma_1^2\right)\,\gamma_1^2V_1^2+\gamma_2^4V_2^2+\gamma_3^4V_3^2\right]+\gamma_1\left(\mu_{12}\gamma_2^2-\mu_{31}\gamma_3^2\right)+\gamma_2\gamma_3v_{23} \\
& +\left(h_1\gamma_1+h_2\gamma_2+h_3\gamma_3\right)\left\{\frac{\alpha''}{H}\left(h_3\gamma_2-h_2\gamma_3\right)+\frac{\alpha'}{H}\left[h_1-\gamma_1\left(h_1\gamma_1+h_2\gamma_2+h_3\gamma_3\right)\right]\right\},
\end{aligned}
$$

where $\gamma_1$, $\gamma_2$, $\gamma_3$ are the direction cosines that characterize the rotor spin axis attitude relative to the case and $V_1$, $V_2$, $V_3$ are the relative control voltages on the suspension electrodes.

Here, $V_i = \frac{V_{ci}}{V_0}$ $(i = 1,2,3)$, where: $V_0$ is a constant reference voltage in the suspension, and $V_{ci}$ is a control voltage along the $i$-axis.

$h_i$ are the projections of the magnetic flux density on the axis of the case,

$\alpha'$, $\alpha''$ are the real and imaginary parts of the rotor magnetic polarizability coefficient as determined experimentally; and

$\mu_{ij}$ and $v_{ij}(i, j = \overline{1,3})$ are the coefficients that define the conservative and dissipative parts of the moment caused by interaction of the anisoelastic suspension with the radially unbalanced rotor, respectively.

$\gamma_i$, $V_i$ are the measured values and $k_n$, $\mu_{ij}$, $v_{ij}$, $h_i$ are the coefficients that must be identified during the tests with different ESG initial orientations.

The drift projections relative to the two other axes are derived by circular permutation of the indices.

One of the most important areas of work lies in the development of methods and techniques for the precision calibration of gyros from run to run (Andrews 1973), and in the referencing of the gyro measurement axes to the basic axes of the vehicle. The solution of these problems is particularly important when ESGs are used in attitude control systems for Earth remote sensing satellites (Emelyantsev et al. 2004, 2005; Landau et al. 2007, 2010).

### *6.5.2.2 The ESG Rotor*

The rotor is the basic element of the gyroscope and must meet numerous requirements, the most crucial being as follows:

1. The rotor surface must have the best possible spherical shape over the nominal angular velocity and operating temperature ranges.
2. The rotor surface must be uniformly electroconductive and have minimal electrical resistance.
3. Mass distribution in the rotor must provide minimal gyro drift.

These requirements are met using the state-of-the-art methods and techniques for the measurement and subsequent elimination of the rotor unbalance.

4. The rotor must have a contrast pattern on its surface for optoelectronic data readout.
5. The rotor surface must have a minimal coefficient of friction with the component parts of the housing which the rotor touches in the off-state.
6. The geometric parameters of the rotor along with its mass distribution must be stable and must not change under the effect of thermal fields during the whole service period (operation, storage, and transportation).
7. The rotor must not contain any magnetic material.

### *6.5.2.3 The Rotor Electrostatic Suspension*

The design of the rotor electrostatic suspension is a governing factor determining the development strategy of the gyro as a whole. This problem, in addition to the appropriate description and estimation of the electrostatic suspension parameters, are the subjects of many publications and patents (Atkinson 1969, 1972, 1974; Clavelloux 1970; Gubarenko et al. 1994; Klinchuch 1972) The main technical requirements for the suspension are reliability, maximal overload capability, low energy consumption, and high rotor positioning stability under static and dynamic overloads. It must also be capable of providing passive or active stabilization of the rotor rotation and charge stabilities.

When choosing control voltages, a DC suspension may be preferential. Despite some advantages of AC suspensions (relative simplicity of circuitry and lower probability of charge accumulation on the rotor), the DC system increases suspension overload capability for the same control voltage levels which is of crucial importance.

The electronic suspension servo system is rather sophisticated, and its design should allow for the following specific features:

1. Positive feedback intrinsic to the electrostatic suspension as conditioned both by the electrostatic interaction of charged bodies and by the geometry of the gap in the suspension
2. The necessity of using relatively high control voltages in the output stages of the suspension servo system.

In designing the electronic unit for the suspension servo system it is essential to provide:

1. a minimal statistical error in the rotor suspension;
2. the introduction of a loop for stabilizing the rotor rotation speed;
3. the possibility of rotor position control in order to align its center with the geometric center of the electrode sphere or the supporting elements of the case;
4. the possibility of accurate adjustment of each of the suspension channels to provide equi-elasticity and spatial stabilization of the rotor rotation speed;
5. a suspension overload capability taking proper account of the performance requirements (for the majority of real gyro applications this value does not exceed 10–15 g).

### *6.5.2.4 Angular Rotor Position Readout*

Readout of data about the rotor angular position relative to the gyro case-fixed frame is one of the most complicated development problems (Boltinghouse 1978; Dyugurov 1997). In fact, it is the key problem for a strapdown gyro system. The requirements of this system are as follows:

1. The readout system sensitivities for both gimbaled and strapdown ESGs must be only a few arcs. These requirements also determine the level of noise immunity of the readout system.
2. The readout system must not produce any gyro rotor perturbations.

The operation of any readout system is based on the use of the rotor deterministic parameters that depend on their geometric positions relative to the axis of the maximum moment of inertia. The ESG uses both measurement of the relation between the amplitude of the spinning spherical rotor beat frequency caused by rotor radial unbalance and the pulse-width angle (capacitive reading), and measurement of the relation between the linear velocity of a rotor point and the pulse-width angle. The first method was used for solid-rotor ESGs (Boltinghouse 1978; Orlov 1967) and hollow-rotor ESGs (Leger 1984), the advantages of this type of readout system being comparatively simple construction and technology. However, the accuracy of this system is limited because in addition to the beat-frequency signal information, the signals from the suspension have inclusions caused by dynamic disturbances of the base on which the gyro is placed. For this reason the development of a rotor angular position readout system using optical principles seems to be more promising (Dyugurov 1997). In this case the main advantages—noise immunity and no disturbing effects on the rotor—are worth further complications in the technology and eventual construction.

An important prerequisite for applying optical methods is the possibility of using the rotor surface as a high-quality spherical mirror and hence as an optical sensor element in the readout system. In practice, autocollimation principles cannot be used in the ESG sensor because the rotor for a strapdown system must be of spherical form.

Present-day optical data readout systems are based on radiant flux modulation with the use of a contrast pattern on the reflective surface of the rotor. Of practical interest are raster readouts of four types: phase-pulse, pulse-width, pulse-frequency, and code systems.

Analysis of various designs of optical data readout systems has shown that a phase-pulse raster optoelectronic data readout system is preferential. Its principle of operation is based on the comparison of modulation signal phases of the radiant flux reflected from the surface of a rotating rotor bearing a special raster pattern that is read out by two diametrically placed sensors (Figure 6.9). Three pairs of such sensors placed along the orthogonal axes that form the coordinate frame of the gyro case provide data about the angular position of the rotor relative to that gyro case.

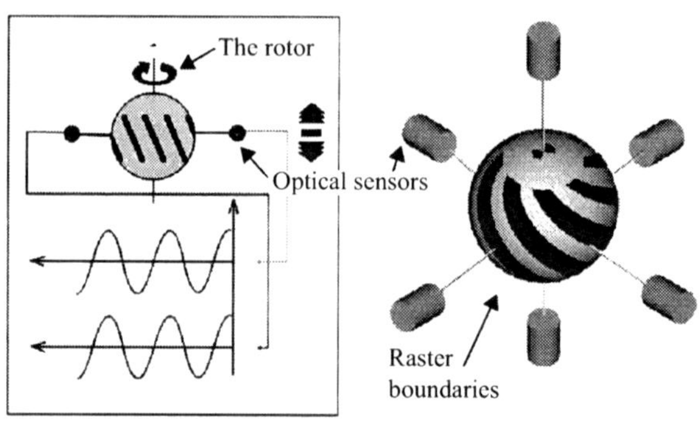


**Figure 6.9.** Feedback contour technique.

Generally, a data readout system for each axis consists of the following functional units and assemblies:

1. An optical channel with a rotor raster pattern
2. An optoelectronic device (sensor) including an illuminator, a device for forming light fluxes (generated by the illuminator and reflected from the rotor), and a photodetector
3. A signal preamplifier for each sensor
4. An electronic unit for converting the data into a form ready for subsequent use. For a gyro intended for a gimbaled system it is usually an amplified analog signal for the servo stabilization system; for a strapdown gyro it is a code for subsequent use in the computing device of the orientation system.

The main problem of providing high accuracy in the optoelectronic system for reading out the angular position of the ESG rotor in strapdown systems reduces to the development of methods and techniques for determining systematic errors in the readout system and designing algorithms for error compensation for the gyroscope operating under real conditions with the required response. In this case the raster form is of considerable importance. For the simplest design and technological solutions, two types of raster patterns, differing in their forms of raster boundaries, are worth consideration:

1. Raster boundaries formed by the lines of a great circle inclined toward the equatorial plane, and
2. Raster boundaries formed by loxodromes.

In the first case the technology and facilities for applying the pattern and controlling its form are simpler; but the algorithm for calculating the phase difference $\Delta\varphi$ as a function of the pulse-width angle is of rather complicated:

$$\Delta\varphi = 2N \arcsin\left(\frac{\tan\lambda}{\tan\alpha}\right) \tag{6.14}$$

where $N$ is the number of raster lines,
$\alpha$ is the inclination angle of the raster boundary to the equatorial plane, and
$\lambda$ is the pulse-width angle.

A loxodromic raster (seen in Figure 6.9) is preferential for devices where the requirements for the readout system response are more stringent, this being determined by data processing time.

### 6.5.3 CONCLUSION

The ESG is the most promising type of gyroscope for autonomous spacecraft attitude reference systems, and its further development—which presents considerable challenges—depends on the advent of new materials and even more sophisticated technologies. Results providing evidence in support of amazing ESG accuracies in space applications have been obtained in the United States via the Relativity Gyroscope Experiment aimed at the verification of Albert Einstein's theory of general relativity.

## 6.6 THE FIBER OPTIC GYRO (FOG)

All optical gyros are based on the Sagnac Effect (Sagnac 1913), which states that an optical path length difference is experienced by light beams propagating along opposite directions in a rotating frame. In fiber optic gyroscopes these two counterpropagating waves propagate within a closed fiber coil, and when this coil rotates the resultant phase difference is proportional to the rotation rate.

Fiber optic gyroscopes are desirable devices for many navigation and guidance applications because, being solid state devices, they have several major advantages including light weight, long working lifetimes, no moving parts, and operate using low voltage power. There are two main designs of FOG, the first being the *Interferometric Fiber Optic Gyro* (IFOG) and the second being the *Resonator Fiber Optic Gyro* (RFOG).

### 6.6.1 THE INTERFEROMETRIC FIBER OPTIC GYRO (IFOG)

Over the last 30 years IFOG research and development has evolved from a promising experiment to an industrial device used for many applications, and its principles and modes of application are described below.

#### 6.6.1.1 The Basic IFOG Scheme and the Sagnac Effect

The basic scheme of the IFOG is illustrated in Figure 6.10 (Lefevre 1993). It is a passive Sagnac interferometer where a fiber optic coupler is employed to split the light from a light-emitting diode (LED) into two counterpropagating waves, clockwise (CW) and counterclockwise (CCW), in the fiber coil and to recombine the waves after propagation at a photodetector (PD). The phase difference is cumulative over a long fiber coil and is expressed by Looez-Higuera (2002) as

$$\varphi_s = \frac{8\pi SN}{\lambda \cdot c}\Omega_p \qquad (6.15)$$

where $S$ is the area enclosed by the each fiber loop, $N$ is the number of loops of the fiber coil, $\Omega_p = \Omega \cdot \cos\psi$ and is the component of the angular velocity perpendicular to the plane of the optical path ($\psi$ being inclination angle), and $\lambda$ and $c$ are the wavelength and the light speed, respectively.

For ideal fibers and components, the current $I$ generated at the photodetector will be

$$I = I_0\left(1 - \cos\varphi_s\right) \qquad (6.16)$$

where $I_0 = \sigma\frac{P}{2}$. Here, σ is the photodetector responsivity and $P$ is the power coupled into the input fiber.

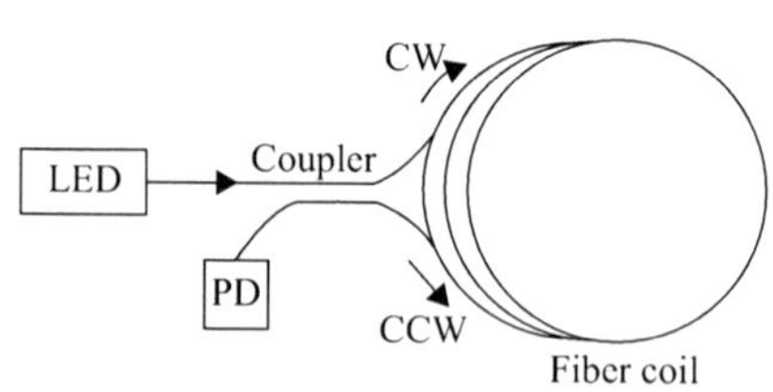


**Figure 6.10.** The IFOG basic scheme.

To achieve a sensitivity that approaches the quantum limit, it is necessary to eliminate all the sources of nonreciprocity other than those induced by the Sagnac Effect in the propagation of the CW and CCW waves (Ulrich 1976). If the counterpropagating waves accumulate a nonreciprocal phase shift due to the quality of the optical components or because of ambient-induced disturbances, zero-point errors and fluctuations will occur and completely mask the Sagnac phase shift (Ulrich 1980). Actually, special polarization-maintaining fibers as well as optical filters and polarization controllers are employed in the FOG.

#### *6.6.1.2 Open-Loop Operation*

The output signal $S$ of the FOG results from the interference of the two waves and is given by

$$S = S_0 \cdot \sin(T \cdot \Omega) \tag{6.17}$$

where $\Omega$ is the rotation rate, $S_0$ is the fringe amplitude, and $T$ is a time constant such that $T{\cdot}\Omega$ is the phase difference between the two interfering waves.

This type of signal has potential drawbacks depending upon the application. For example, the output signal is a nonlinear function of the rotation rate and has a dynamic range limited by the sinusoidal waveform. Also, the output signal being in the form of an analog electric current, it is necessary to digitize it for processing. Thus, this type of gyroscope is difficult to use in inertial reference systems where the data from three gyroscopes and three accelerometers must be simultaneously analyzed. Furthermore, the stabilities of the quantities $S_0$ and $T$ are questionable. The fringe amplitude $S_0$ depends on the optical power of the source, the optical power loss through the system, and on the states of polarization of the interfering waves. However, these variations may possibly be determined through independent measurements and removed from the signal.

A typical setup of a practical FOG in all-fiber technology is shown in Figure 6.11, where a *phase modulator* (PM) is inserted in the fiber coil close to the fiber output so that a different phase delay is accumulated by the counterpropagating waves (Ezekiel 1982). Usually, the phase modulator is constructed by winding and cementing a few fiber turns on a short, hollow, piezoceramic tube (PZT) (Martini 1987).

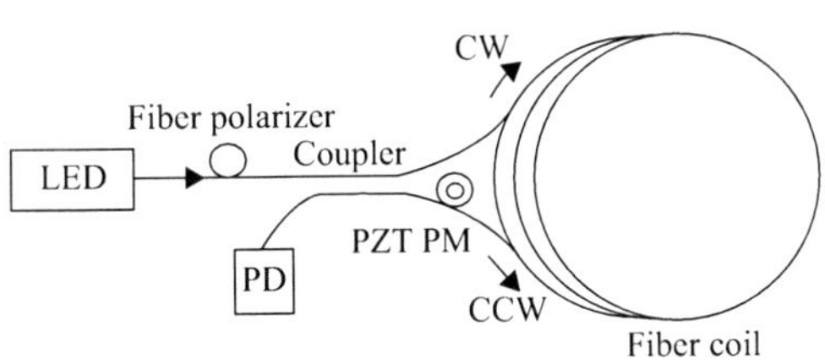


**Figure 6.11.** Open-loop IFOG configuration with piezo-ceramic phase modulator (PZT PM).

#### *6.6.1.3 Closed-Loop Operation*

In a closed-loop gyroscope a feedback mechanism maintains the open-loop signal at zero by compensating for the Sagnac phase shift within the sensing loop. This is achieved by canceling the Sagnac phase shift by adding a controlled phase delay, the measurement of which then reveals the rotation-rate information.

Typically, single-axis IFOGs use the so-called minimum configuration (Figure 6.12) that provides reciprocal optical paths for two beams counterpropagating in a fiber loop. The FOG consists of the one light source, one photodetector, one 1:1 fiber splitter (coupler) to divide the light into two parts, one ring interferometer to sense angular rate, plus printed circuit boards containing signal processing circuitry. The ring interferometer consists of a multifunction integrated optical chip (MIOC) and a polarization-maintaining (PM) fiber coil. The MIOC is a three-port integrated optical gyro chip providing three functions. Firstly, it polarizes the propagating light to reduce bias instability due to polarization nonreciprocity. Secondly, it splits the light into clockwise and counterclockwise waves, each with equal optical power, and recombines them using a Y-junction waveguide. Thirdly, using electro-optical phase modulation, it applies a biasing phase shift between the counterpropagating beams. PM fiber is used in order to reduce both the drift caused by polarization cross-coupling and that caused by the Earth's and other magnetic fields via the Faraday Effect.

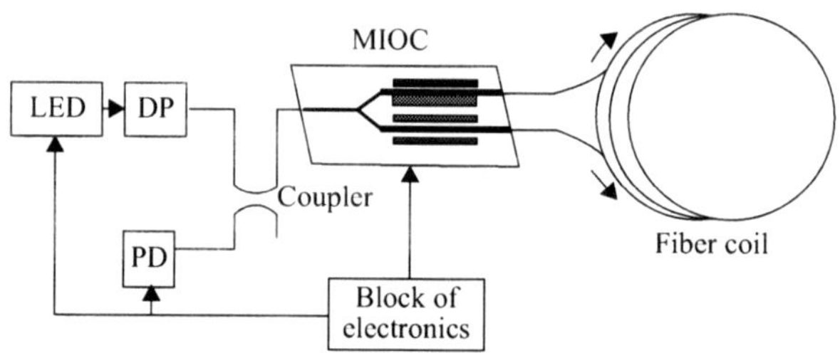


**Figure 6.12.** Minimum configuration of single-axis gyro. LED, light-emitting diode; DP, depolarizer; PD, photodetector; MIOC, multifunction integrated optical chip.

The dynamic range of the device is easily configurable via the length and diameter of the fiber loop; and the maximum measured rotation rate corresponding to the Sagnac phase shift will be equal to Ω.

$$\Omega_\pi = \frac{\lambda \cdot c}{2L \cdot D} \tag{6.18}$$

### 6.6.1.4 Fundamental Limitations

1. *Optical losses*
   The sensitivity of the IFOG is limited by shot noise that is proportional to the square root of the power, and decreases with fiber length. However, the Sagnac Effect increases with fiber length, so that these two competing effects define the length of the fiber for a given sensitivity.
2. *Thermal Noise*
   A time-dependent temperature gradient along the length of the fiber can introduce spurious phase shifts and is due to the temperature dependence of the refractive index of that fiber. To minimize this effect, fibers with smaller $dn/dT$ dependence should be used. Also, quadrupole winding such that equidistant points from the fiber center are physically close to each other strongly reduces this effect.
3. *Backscattering of light*
   Backscatter at the output–input couplers and the MIOC interfaces can interfere with the main beams, creating parasitic interferometers. The use of a special immersion cell to reduce the refractive index step, plus the use of tilted MIOC interfaces, can attenuate such backscatter effects.

4. *Optical Kerr Effect*
   The electric fields of the counterpropagating beams can cause changes in the refractive index that are nonreciprocal if the light is split into unequal parts (Frigo 1983). The nonreciprocity induced by the nonlinear Kerr Effect can be strongly reduced by using broad-band, low coherence, unpolarized optical sources or even with a simple 50% duty cycle modulation of the input optical power.
5. *Magneto-optical effect*
   The magneto-optical Faraday Effect is a nonreciprocal effect, which is potentially problematical by adding to the Sagnac Effect (Hotate 1987). However, this problem is now almost solved by the careful use of untwisted polarization-maintaining fibers as well as by using cases made from special materials such as Permalloy.

For an IFOG with perfect components (ideal splitter, no backscattering, etc.), the measurement limit is imposed by the shot noise in the light as measured by the photodetector (Lin 1979). The uncertainty, $\delta\Omega_\pi$, generated by the fluctuation in the light due to shot noise, can be expressed (Davis 1978) as

$$\delta\Omega_\pi = \frac{c}{L \cdot D} \frac{\lambda/2}{\left(n_p n_D \tau\right)^{\frac{1}{2}}} \tag{6.19}$$

where $n_p$ is the number of photons per second reaching the photodetector, $n_D$ is the detector quantum efficiency, and $\tau$ is an average time constant.

#### *6.6.1.5 The Multiple-Axis IFOG*

Complete *Inertial Measurement Units (IMUs)* for determining the motion of a vehicle in three-dimensional space have six sensors as a minimum without redundancy, normally three linear accelerometers and three single-axis gyros. Affecting the design parameters in the field of sensor packages for avionics, flight control, and missile guidance are volume, power consumption, and weight. The first and most obvious way to optimize a multiple-axis system is to use one common light source to supply all the FOGs, and fortunately, light sources available today are powerful enough to supply three FOGs in parallel. The electrical power consumption, volume, and cost of two light sources can thus be saved with a tolerable reduction in the optical signal level.

The next step in the improvement of multiple-axis systems is to use one common optical detector for the outputs of all the interferometers. Usually one (3×3) or two (1×1 and 1×2) fiber optical couplers distribute light to the three FOGs and gather their output on one detector. In such a configuration, frequency multiplexing or time multiplexing methods are used to separate the signals coming from the different axes.

#### *6.6.1.6 The Depolarized IFOG*

One of the greatest challenges facing the widespread use of IFOG technology is in the area of cost reduction, and one of the most critical areas where this is necessary to realize an attractive product cost is that of the sensing coil itself. A very promising approach is in the use of a standard single-mode fiber coil in conjunction with polarization randomization in the loop, that is, *depolarized technology* (Sanders 1996). This technology is particularly attractive for high-performance IFOGs where large amounts of polarization-maintaining fiber are used. Navigation-grade performance levels comparable to PM gyros have been obtained (Szafraniec 1995) by this method.

#### 6.6.1.7 Applications of the IFOG

Typical applications of open-loop and closed-loop IFOGs are shown in Table 6.3.

**Table 6.3.** Application of open-loop and closed-loop IFOGs

| IFOG type | Performance category | Applications |
|---|---|---|
| Open loop, PM fiber | 1–10 deg $h^{-1}$ | Aircraft attitude heading reference systems |
| Closed loop | 1–10 deg $h^{-1}$ | Tactical guidance |
| Closed loop, depolarized fiber | 0.01–0.003 deg $h^{-1}$ | Aircraft navigation, spacecraft, and land navigation |
| Closed loop, PM fiber | <0.001 deg $h^{-1}$ | Precise space applications |

In the United States, companies such as Honeywell and Northrop Grumman are the leaders in aerospace FOG development. JAE and Mitsubishi Precision in Japan, Sagem in France, and Optolink in Russia are also developing aerospace FOGs. Honeywell's open-loop IFOGs are used in *Attitude Heading Reference Systems* (AHRS) for the Dornier 328 passenger aircraft and as *Standby Attitude and Air Data Systems* (SAARU) for Boeing 777 commercial aircraft. Optolink's closed-loop three-axis gyro is working in the control systems of the landing module of the Russian manned transport spacecraft, Soyuz TMA, operating within the International Space Station program.

Insofar as the nonaerospace use of IFOGs are concerned, Hitachi Cable is possibly the leader, although US firms are also investigating these markets. Sagem (France) and Fizoptika (Russia) are also active players in this market.

### 6.6.2 THE RESONATOR FIBER OPTIC GYRO (RFOG)

Similarly with the IFOG and ring laser gyroscopes (RLGs), the resonator concept (Shupe 1981) utilizes the phase difference experienced between clockwise (CW) and counterclockwise (CCW) waves traveling around a closed path that is rotating with respect to an inertial frame—again the Sagnac Effect .

A typical RFOG (Strandjord 1991) consists of a recirculating passive optical cavity operating as follows. As the frequency is tuned such that an integral number of wavelengths fit inside the optical pathlengths of the ring, the input energy is absorbed into the recirculating loop and a sharp resonance dip appears at the output. In this method, light is extracted from the ring by the use of a fiber coupler giving resonance peaks in the transmission mode operation of the resonator (Figure 6.13(a)) or in the reflecting mode operation (Figure 6.13(b)). These resonances are tracked by servo loops (Ezekiel 1977), one that adjusts the laser frequency to the CCW resonance center and the other that adjusts a frequency shift $\Delta f$ to allow resonance tracking in the CW direction. This configuration operates as a narrow-band fiber ring resonator with a high $Q$-factor, and values above $Q = 100$ are quite practical. By comparison, the IFOG has an equivalent $Q$-factor of about 2. Since a high $Q$-factor enhances sensitivity, good performances can be obtained with a much shorter fiber length. This feature gives the RFOG a significant potential advantage over the IFOG for high-performance applications where the coil length of the IFOG (typically from 200 m to 2 km) becomes a significant portion of the device cost and size. Also, the shorter the coil length, the smaller will be the influence of thermal nonreciprocities on bias drift (Shupe 1981). In fact, it is possible to use single layer winding schemes that allow direct isothermal contact of the whole coil with a conducting bobbin.

The main technical problem for RFOG developers is in the precise determination of the CW and CCW resonance frequencies. For precise measurement of the (shot-noise-limited) rotation rate, a typical measurement accuracy should be less than $10^{-7}$ of the line width. Also, the resonance center can be precisely measured only if the resonance line shape is highly symmetrical. However, there are two factors that compromise this line shape symmetry. The first is optical scattering from the CW to the CCW direction (and vice versa). The second is the existence of a further polarization state in the ring (Meyer 1983). The propagation of light in this second state is supported by the fiber and most often excited by cross-coupling in the fiber optic coupler or by imperfect launching conditions at the input to the ring. The existence of this undesired light in the ring gives rise to a second resonance dip, which causes gyro bias errors. Another negative factor that reduces RFOG performance is the optical Kerr Effect. In fact, the sensitivity of RFOG depends on an optical power mismatch between CW and CCW waves (Sanders 1992).

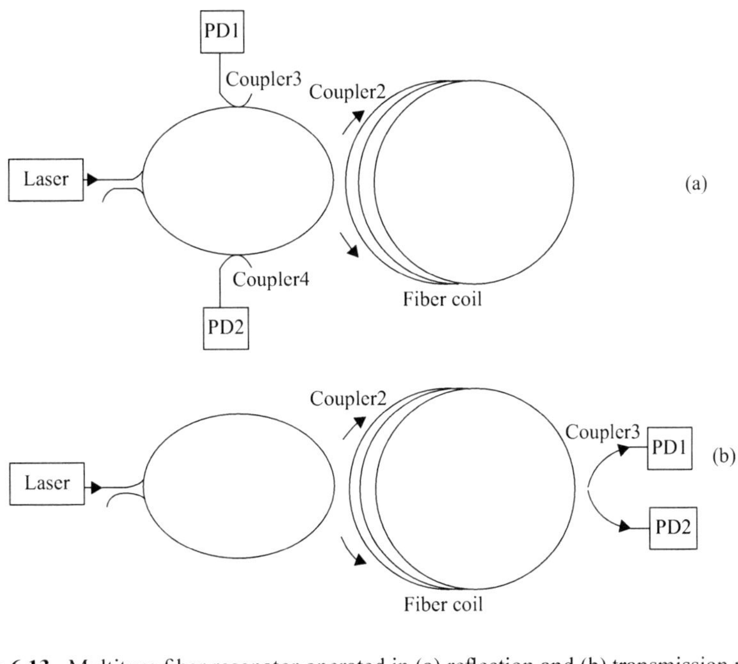


**Figure 6.13.** Multiturn fiber resonator operated in (a) reflection and (b) transmission modes.

In the case of the passive multiturn fiber resonator with perfect components (all above-mentioned effects being negligible), the fundamental limit is determined by the uncertainty in the measurement of $\Delta f$ due to the shot noise of the light at the photodetector (Lin 1977). In each direction of propagation, the shot noise of the light causes a frequency uncertainty in the measurement of the corresponding resonance frequency. The uncertainty in the measured rotation rate $\delta\Omega$ can be expressed as (Burns 1994):

$$\delta\Omega_{\pi} = \frac{\sqrt{2}\cdot\lambda nP}{4A}\frac{G}{\left(n_{\mathrm{p}} n_{\mathrm{D}} \tau\right)^{\frac{1}{2}}} \qquad (6.20)$$

where G is the cavity line width, $n$ is the refractive index, $P$ is the perimeter of the resonator, and $A$ is the area enclosed by the light path. The other parameters are as defined in Equation (6.5).

While the RFOG maintains the passive structure of the IFOG, stimulated Brillouin scattering is used in the design of RFOGs with active fiber ring resonators. This approach exhibits a high-performance capability mainly because of the reduced negative effects induced by nonreciprocity and the imperfect behavior of the components involved. To reduce the Kerr Effect and extend the dynamic range of a Brillouin fiber optic gyroscope, an intensity modulator is included in the optical loop that periodically attenuates the Brillouin light waves that counterpropagate in the optical loop so that they each propagate as quadrature waves. The use of quadrature wave modulation for the counterpropagating light wave reduces the cross-effect of the Brillouin waves to substantially the same magnitude as the (direct wave) self-effect so that the nonreciprocal Kerr Effect is substantially reduced or eliminated. In order to support the counterpropagating quadrature waves, the optical loop is pumped with light having frequency components selected to provide Brillouin light at the frequencies necessary to generate quadrature waves in the counterpropagating Brillouin waves.

The *ring fiber laser gyro* (RFLG) takes advantage of rare-Earth doped fibers as the active medium sustaining laser oscillation in the sensing coil. The configuration of the RFLG is similar to that of the RFOG (Figure 6.13), but in the RFLG the coupler is a *wavelength-division multiplexing* (WDM) fiber coupler arranged so as to cross the pump power at the pump wavelength $\lambda_{\mathrm{pump}}$ from the external laser diode, and to bar the oscillating field at $\lambda_{\mathrm{sign}}$ in the ring. In this design, very high ring $Q$-factors can be achieved and the expected performance of such an RFLG device is close to the performance of the helium–neon RLG.

## 6.7 THE RING LASER GYRO (RLG)

### 6.7.1 *INTRODUCTION*

The principle of ring laser gyro (RLG) operation is based on an effect discovered by Sagnac in 1913 during his examination of the properties of the multi-mirror ring optical resonator, also known as the passive Sagnac *Interferometer* (Sagnac 1913).

A basic layout for an ideal optical ring loop of radius $R$ is shown in Figure 6.14. It contains a mirror $A$ on which light from emitter $E$ falls and is split into two rays 1 and 2. These two rays are propagated clockwise and counterclockwise, respectively. This mirror rotates with an angular velocity $\Omega$ and a second position is also shown.

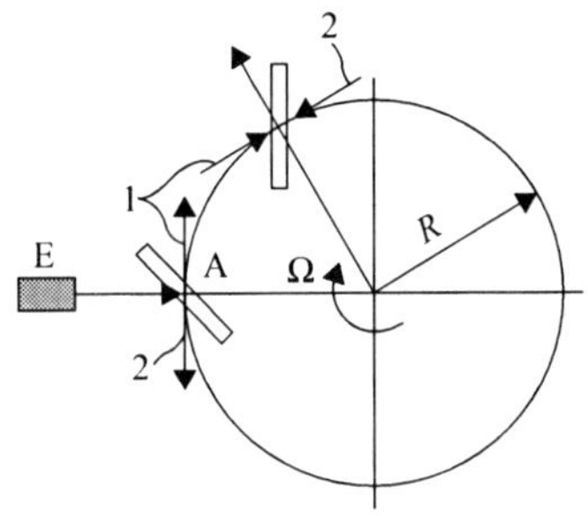


**Figure 6.14.** Schematic diagram of the ring resonator (the Sagnac passive interferometer): A, mirror; 1, 2, directions of ray propagation; E, emitter.

If the resonator is immobile, the time the light takes in passing through the closed loop is identical for both rays and is equal to $t = 2\pi R/c$, where $c = 3 \times 10^8$ m s$^{-1}$, the velocity of light.

When the resonator rotates with velocity $R\Omega$, clockwise light transiting its perimeter will take a time $t_1$ and counterclockwise light will take $t_2$ because the mirror moves away from the incident light ray (see the second position of the mirror in Figure 6.14). These times are defined by the equalities:

$$t_1 = 2\pi R/(c - R\Omega), \quad t_2 = 2\pi R/(c + R\Omega).$$

Taking into account that, $c^2 >> R^2\Omega^2$, the following approximate equality is true:

$$\Delta t = t_1 - t_2 \approx \frac{4\pi R^2}{c^2}\Omega = \frac{4S}{c^2}\Omega. \tag{6.21}$$

It is important to note that the value of $\Delta t$ is proportional to the area $S$ of the resonator loop.

Because the values of $\Delta t$ are extremely small, direct measurements of object rotation rate by means of passive Sagnac interferometers are possible only in fiber-optical gyros containing 500–1000 m of fiber wound on a spool so that the effective value of $S$ is increased significantly.

The Sagnac phase shift between the two counter-rotating rays will be:

$$\Delta\varphi = \nu\Delta t = \frac{8\pi fS}{c^2}\Omega \approx \frac{4S\nu}{c^2}\Omega = \frac{8\pi S}{\lambda c}\Omega \tag{6.22}$$

where $f$ is the frequency of the electromagnetic radiation, $\nu = 2\pi f$ is the angular frequency, and the wave length is $\lambda = c/f$

From the Sagnac phase Equation (6.8) there follows the possibility of measuring rotation rate using interference fringes.

## 6.7.2 PRINCIPLE OF OPERATION

The ring laser gyro (RLG) is representative of an *active interferometer* that utilizes a laser beam in a closed loop, the triangular form of which is shown in Figure 6.15. Here, the active environment is a closed-loop resonator having mirrors located at the vertices as follows: *A*—plane mirror, *B*—spherical mirror, *C*—semi-permeable mirror. Coherent laser waves are excited in the resonator, the generation requirement in the ring being that along the length L0 of its perimeter an integral number $n$ of wave lengths $\lambda$ gives $L_0 = n\lambda = nc/f$; or $(n = L_0/\lambda)$. Thus, the frequency of the generated radiation is $f = nc/L$.

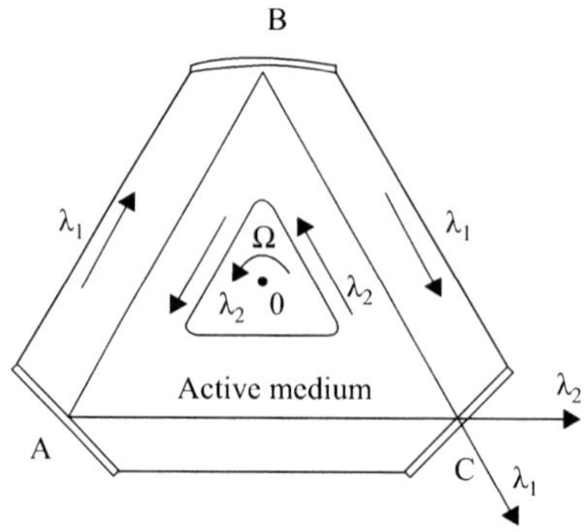


**Figure 6.15.** Schematic of the active interferometer.

During rotation of the resonator with angular velocity Ω around axis *O* there is a wavelength difference between the clockwise ($\lambda_1$) and anti-clockwise ($\lambda_2$) waves. This represents a path length difference of $\Delta L = 4S\Omega/c$ so that the frequency difference is

$$\Delta f = f_1 - f_2 = nc\left(1/L_1 - 1/L_2\right) = nc\left[1\Big/\left(L_0 - \frac{\Delta L}{2}\right) - 1\Big/\left(L_0 + \frac{\Delta L}{2}\right)\right] \approx \approx nc\frac{\Delta L}{L_0^2} = nc\frac{4S\Omega}{L_0^2 c} = \frac{L_0}{\lambda}\cdot\frac{4S\Omega}{L_0^2} = K_{MP}\Omega, \tag{6.23}$$

where $K_{MP} = 4S/L_0\lambda$ a resonator scale factor.

According to Equation (6.9), the change in the output frequency of the ring resonator is directly proportional to the angular rate of rotation of the base on which it is mounted. It should be noted that this change in frequency $\Delta f$ is a beat frequency that results from the superposition of two counter-rotating waves, and that it may be measured by combining the two waves exiting one of the mirrors at the cathode of a photo-receptor (Bychkov 1975).

The sinusoidal signal received at the photodetector is shaped to form square impulses for conversion to digital form, and the scale factor of Equation (6.10) used for the transition to the final measured angular velocity.

The RLG can be applied to the measurement of the actual rotation angle of the base on which it is installed, in which case its scale factor is defined by the number of impulses per unit angle of displacement.

### 6.7.3 *FREQUENCY CHARACTERISTICS AND MODE-LOCKING COUNTER-ROTATING WAVES*

The output characteristic of the RLG depends not only on the rotation rate of its base, but also on many processes taking place within the active environment of the resonator itself.

The relationship of the output impulse frequency at the photodetector to the RLG rotation is the basic frequency characteristic, and is depicted in Figure 6.16.

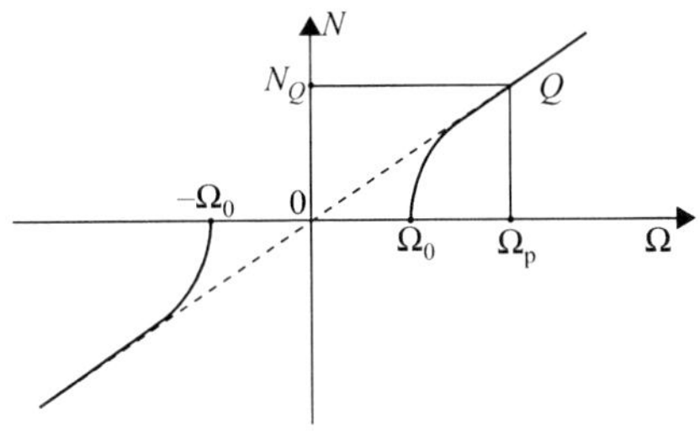


**Figure 6.16.** Frequency characteristic of the laser gyro.

In an ideal RLG such a characteristic would be simply a straight line as shown dashed in Figure 6.16. However, in real laser gyros there are unfortunately a variety of harmful stray effects leading to departures from this linear dependence. The most serious is the so-called "mode-locking" effect in the counter-rotating waves. The physics of mode-locking can be easily understood when it is realized that the mirrors, though all of the highest quality, nevertheless produce some dispersion. For example, because of this the clockwise ray, being dispersed by the mirrors "imposes" its phase through the active environment to the counterclockwise ray. The converse also occurs via the same mirrors. As a result of the gyro rotation, the ring laser generates counterwaves with the same frequency as would be the case for an immobile gyro even though, really, the laser gyro is rotating so that the frequencies of counterwaves should be different.

Thus, apparently insignificantly small dispersions can lead to a complete loss of sensitivity of the laser gyro to angular rotation in the so-called *lock-in zone* of angular velocities. In fact, a lock-in zone $2\Omega_0$ can be quite large. It should be noted that over a certain range of angular velocities this real frequency characteristic gradually departs from an ideal linear dependence so that as a first approximation, mode locking appears not in all angular velocity regions, but only near the origin. This is shown in Figure 6.16 as from $-\Omega_0$ to $+\Omega_0$.

### *6.7.4 THE ELIMINATION OF MODE-LOCKING IN COUNTER-ROTATING WAVES*

There are many ways to eliminate or diminish the locking of frequencies in counter-rotating waves. One of them is obvious—to minimize or even remove dispersion in the dielectric mirrors and other optical units in the resonator. Here it must be noted that this path is taken only when practically all other possibilities are either already exhausted, or will be exhausted in the near future. Today the reflection factor of laser mirrors is 99.9995 to 99.99995%, and each "nine" after the dot has been obtained after huge technological efforts and costs in the million dollar region.

Transfer of the operating point of an RLG to that part of the frequency characteristic which is practically linear (for example, to point $Q$ in Figure 6.16 at a rotation frequency of $\Omega p$) is another way of minimizing the harmful influence of mode-locking. This can obviously be achieved by rotating the RLG at a constant angular rate $\Omega p$ appropriate to this point $Q$.

Another means of mode-locking prevention is the method of *vibrational support*, the implementation of which consists of mechanically rocking the resonator around a sensitivity axis at a frequency Ωp. This may be achieved using an oscillating torsion-type elastic unit, but this type of mechanical vibrational support is inconvenient in realization and leads to complementary errors.

Transfer of the operating point to a linear region may also be achieved by superimposing a constant or variable magnetic field on the active environment of the laser, so constituting an "electronic support." The resultant "apparent rotation" of the laser gyro at a frequency $\Omega_p$ is based on the magneto-optical Zeeman Effect.

The Faraday Effect may also be used in the realization of an electronic support.

### 6.7.5 ERRORS

As with all gyros, laser gyro parameters are unstable in time, for example, because of thermally dynamic effects. The fundamentally desired parameter in a gyro is zero drift—that is, consistency in the frequency of an output signal (number of impulses from a reversible counter $N_Q$) at the operating point of the gyro. Modern laser gyros with helium–neon ring lasers are characterized by zero drifts of 0.005–0.001 deg $h^{-1}$ depending on their dimensions and construction.

The laser gyro is actually an integrating instrument because the reversible counter displays the total number of impulses over a certain period of time taking into account their sign (i.e., rotational sense). Hence the gyro measures the angle through which the vehicle has rotated around the given axis during this period irrespective how its angular velocity might have changed. (Angular velocity can obviously be calculated by angular differentiation.)

The time period mentioned above can range from fractions of a second to tens of minutes and is defined according to the particular application and required performance of the system. In cases when this time period is required to be small, the noise component of the random zero drift at the operating point must be considered. This is because over the short time periods during which the required information is gathered, the noise component of the zero drift can be perceived as a quickly changing rotation. Conversely, when the time period is longer, the influence of random drift is smaller.

Immediately after switch-on, it is to be expected that RLG parameters will change due to temperature rise. This is largely related to the fact that even a very small change in the perimeter length of the laser gyro resonator leads to a large change in optical frequency generation that will be perceived by the system as a vehicle rotation. So, for a triangular resonator with 13 cm sides, the necessary length maintenance precision of the resonator's perimeter over all expected climatic ranges should be not be worse than $\lambda/300$ that is, not worse than 20 Angstroms! In practice, such length "superstabilization" in the resonator's perimeter is realized at the expense of applying precision *piezocorrectors* that can modify the resonator perimeter length and are actuated via a feedback loop.

The influence of exterior magnetic fields on gyro zero drift is also very important. To reduce this influence the laser gyro is normally located within a multilayered magnetic screen that can lower the integrated magnetic field sensitivity to quite acceptable values of 0.5–1 grad $h^{-1}$ oersted$^{-1}$ (for drifts of the order 0.1° $h^{-1}$). However, there is still a variety of errors leading to linearity violations in the frequency characteristic of the laser gyro, such as zero drift in a frequency characteristic, or variations in counterwave frequencies relative to those of the forward waves, among others.

The cumulative operation of all these effects leads to nonlinearity in the frequency characteristic and thereby to dependence of the scale factor $K$ on the angular rate of rotation of the gyro. Thus, one of the major problems in the laser gyro is scale factor nonlinearity.

### *6.7.6 PERFORMANCE AND APPLICATION*

The performance of some first-generation RLGs with mechanical vibrating supports are shown in Table 6.4 (Aleshin, Veremeenko, and Chernomorskij 2006)

**Table 6.4.** Some first-generation RLG performances

| Characteristic, Dimensionality \ Model / Developer | LG-8028 | GG-1342 | ASLIG | SM-11 | BLG-1 | LG-1 | "Morion" |
|---|---|---|---|---|---|---|---|
| | Litton Inc. | Honeywell | Sperry | SRI "Polus" (Russia) | OKB "Temp" (Russia) | MIEA (Russia) | NPO "Astrophysics" (Russia) |
| Range of measured angular velocities, grad $s^{-1}$ | ±600 | ±800 | ±100 | — | ±100 | — | ±200 |
| Systematic error, grad $h^{-1}$ | 0.01–0.2 | 0.01 | 0.1 | 0.01 | 0.5–1.5 | 0.01 | 0.2–2.0 |
| Casual error, grad $h^{-1}$ | 0.003 | 0.003 | 0.05 | 0.003 | — | 0.003 | 0.03 |
| Stability of a scale factor | 5·10–6 | 5·10–6 | 5·10–5 | — | 10–5 | 5·10–6 | — |
| Power consumption, W | 3 | — | — | — | 6 | — | 9 |
| Perimeter, cm | 28 | 32 | — | 44 | 16 | 28 | 26 |
| Resource, $10^3$ hours | 15 | 20 | 30 | 10 | 0.5–1.0 | — | 5–7 |
| Mass, kg | 1.7 | 1.9 | 1.8 | — | 1.6 | 1.8 | — |
| Number of measuring axes | 1 | — | — | 1 | 3 | 1 | 3×2 |

RLGs with zero drifts between 0.005 and 0.01 grad $h^{-1}$ are rather large and demand a high technological capability in their manufacture. They are applied to autonomous navigation systems in vehicles for long-distance operation such as civil transcontinental airliners.

RLGs with zero drifts between 0.1 and 5 grad $h^{-1}$ are considerably smaller and also demand high, but not cutting-edge, techniques in their manufacture. These can be successfully used in integrated navigation systems along with other systems such as SNS.

RLGs with zero drifts of 1 to 10 grad $h^{-1}$ are the smallest and can be used in motion control systems for highly maneuverable vehicles having small operational ranges.

### *6.7.7 CONCLUSION*

Thanks to good drift, stability, and scale factor linearity characteristics over a broad range of measurements, along with good vibration and shock stabilities, the RLG is now the main gyroscopic instrument in INS. However, progress continues, and developments in Germany together with New Zealand, for example, include super-large-scale laser gyros with a quadrate side of 40 m in which perimeter rays are generated. The expected zero drift of such devices should achieve $\sim 10^{-7}$ grad $h^{-1}$.

## 6.8 DYNAMICALLY TUNED GYROS (DTG)

### 6.8.1 INTRODUCTION

The dynamically tuned gyro (DTG) is a kind of *rotor vibrating gyro* (RVG) in which the principle of dynamic adjustment is realized (Brozgul and Smirnov 1970, 1989; Craig 1972a,b).

Referring to Figure 6.17(a) the DTG consists fundamentally of a rotor (1) (symmetric or asymmetric) driven by a synchronous motor (4) via a shaft (3). This shaft is connected to the rotor via a pair of torsion bars (2) that allow a vibratory motion of the rotor to take place about axis $x$, which it does as a result of the gyroscopic moment produced by any angular motion of the base upon which the system is mounted. This vibratory motion provides a measure of the angular speed of the base rotation. Figure 6.17(b) shows how a gimbal may be installed via an inner pair of torsion bars oriented orthogonally to the outer pair, so allowing vibratory motion around axis $y$. This general scheme is usually named after its inventor, E. W. Howe, and has the following major attribute. Whereas the torsion bar stiffness is independent of the spin speed, the dynamic inertia arising from the gyro effect is proportional to the square of the spin speed, and this provides a negative spin stiffness. At a certain *tuning speed* these moments cancel out so that the ideal condition of zero gyro torque can be approached.

Dynamic (resonant) tuning of the RVG reduces interaction of the rotor with the base. On average, for one rotor turn at a small angle of base turn, the elastic support becomes "momentless," hence it is possible to use the DTG as an angular rate integrator.

Both a vibrating signal and the combination of a constant rotor deviation with its vibration at double the frequency of rotation can be used as an output signal. In the latter case the gyro is called a *vibratory-precessional* DTG (Burdess and Fox 1977, 1978a,b). Multipurpose DTGs are vibratory-precessional DTGs that simultaneously measure two mutually perpendicular components of angular speed and linear acceleration vectors about the general axes.

DTGs are successfully used as gyroscopic units for indicating stabilizers and strapdown INS. They successfully compete with gyros of other types, especially in systems for orientation and navigation of moderate accuracy.

### 6.8.2 KEY DIAGRAMS AND DYNAMIC TUNING

As has been explained, in all DTGs the rotor is mounted on a motor shaft, and the most widespread DTG basic support schemes are shown in Figure 6.17.

When torsion bars transmit small angular deviations to a rotor relative to a shaft around axes $x$ and $y$, elastic moments will appear within that rotor. In Figure 6.17(c) such a rotor has an asymmetrical form, in contrast to that of Figure 6.17(b). Another gyro scheme in which the rotor support about its shaft is also realized using torsion bars, but along with two frames, is shown in Figure 6.17(d). In this form of gyro (actually a development of the Howe scheme), if frame (5) is used instead of a ring, and in the equatorial plane of the rotor, a two-frame support design will result (Brozgul and Smirnov 1970; Ormandy and Maunder 1973). Figure 6.17 shows two such frames in detail.

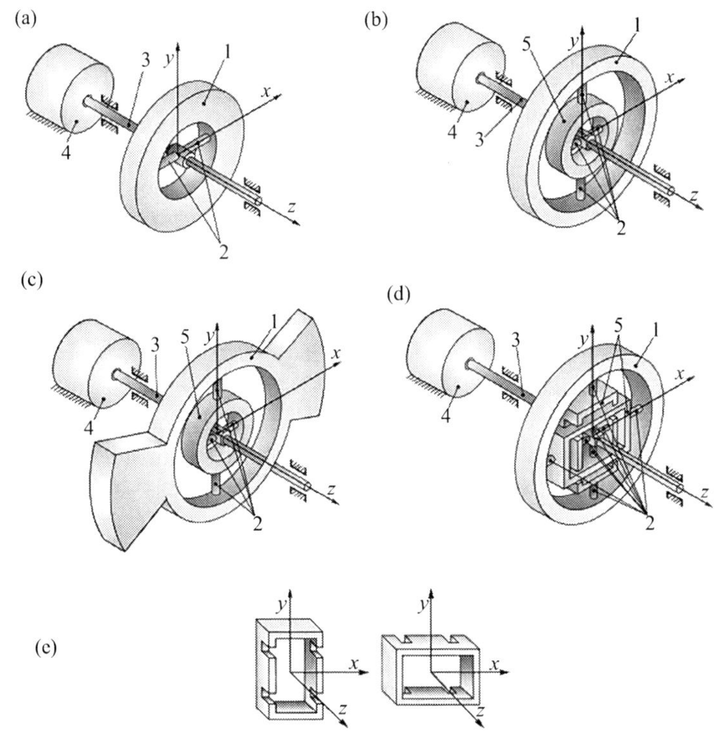


**Figure 6.17.** Basic schemes for the DTG. (a) without intermediate rotor support elements, (b) one-ring symmetric rotor support, (c) one-ring asymmetrical rotor support, (d) two-frame support, (e) frameworks.

For reduction of the elastic moments between rotor and shaft, the necessary dynamic adjustments consist of the careful selection of flexor, ring, and frame parameters, and of those of the rotor and its angular speed. At this speed the rotational rigidity of the torsion bars is compensated by the centrifugal moment of inertia of the ring, frames, and rotor (Brozgul and Smirnov 1970; Fox and Burdess 1980).

As a first approximation, the condition of dynamic adjustment for a gyro without intermediate rotor support elements appears as follows:

$$\Omega^2 = c\left(A + B - C\right)^{-1} \tag{6.24}$$

where $\Omega$ is angular speed of a rotor, $c$ is an angular rigidity factor for the torsion bars, and $A$, $B$, and $C$ are equatorial and axial moments of the rotor inertia (usually $A = B$).

A first approximation of the condition for dynamic adjustment for a gyro with a one-ring rotor support appears as follows:

$$\Omega^2 = 2c\left(A_1 + B_1 - C_1\right)^{-1} \tag{6.25}$$

where $c$ is an angular rigidity factor for the torsion bars relevant to the rotor support axes ($c_x = c_y = c$), and $A_1$, $B_1$, and $C_1$ are equatorial and axial moments of inertia of the ring (usually $A_1 = B_1$).

The condition of dynamic adjustment for a gyro with two-frame support under the condition of equality of the equatorial and axial moments of inertia of the first and second frames ($A_1 = A_2$, $B_1 = B_2$, and $C_1 = C_2$), and equality of the angular rigidity factors of each pair of torsion bars ($c_x = c_y = c$), is similar to the condition of dynamic adjustment of a gyro with one-ring support.

### 6.8.3 OPERATING MODES

A gyro using asymmetrical rotor support (as opposed to a ring) reacts to the angular speed of the base by producing a peak-modulated output vibration frequency. The rotor oscillates about a torsion bar twist axis at the rotor frequency, and after the end of a transient representing dynamic adjustment the rotor establishes a constant oscillation amplitude that does not depend on its rotation speed. Figure 6.18(a) represents this sequence and shows how the rotor oscillation reaches a constant amplitude after the end of the transient. This is the normal vibration operating mode:

$$a_{\mathrm{cm}} \approx C\,\omega/b \tag{6.26}$$

where $\omega$ is the angular speed of the base and $b$ is the rotor oscillation damping factor.

It is possible to realize this form of DTG with a damping moment close to zero ($b \approx 0$) and less than that for any of the other constructional methods. In this case, the oscillation amplitude of the rotor at its rotation frequency is proportional to any angle $\omega t$ made by the base in inertial space:

$$a \approx \frac{C}{2A}\omega t \cos \Omega t \tag{6.27}$$

where $\Omega$ is the rotor angular speed, and $A$ is the equatorial moment of the rotor inertia (see Section 6.8.2).

Another DTG operating mode can be characterized as *precessionally vibrating*, as in Figure 6.18(b). Here, it is obvious that the angle of rotor precession is limited by the greatest possible angle of the torsion bar's twist and, hence, over an unlimited operating time, the application of a feedback channel driven by the precession angle is required.

For a DTG with one-ring support, and at small angles of shaft (and hence of case) deviation, the movement of the rotor axis in a motionless system of coordinates has the character of precession around the axes connected to the case. In a system of coordinates rigidly aligned with the shaft, the rotor makes oscillations around the torsion bar axes with the rotor rotation frequency, so the precessionally vibrating operating mode (Figure 6.18(c)) is realized.

It is possible to show that it is convenient to measure rotor deviations in motionless coordinate systems, when such a gyro is often termed a *gyroflex*. If the DTG rotor is asymmetrical, the movement of its axis has a vibrating character in both moving and rotating systems of coordinates (Friedland and Hutton 1978; Maunder 1979).

Modifications of the DTG with a vibrating output are often termed *rotor vibrating gyros* (RVG) or *vibrorotors*. Independence of the accuracy characteristics from the rotor dimensions is a typical theoretical feature of the RVG (Raspopov and Savelyev 1979).

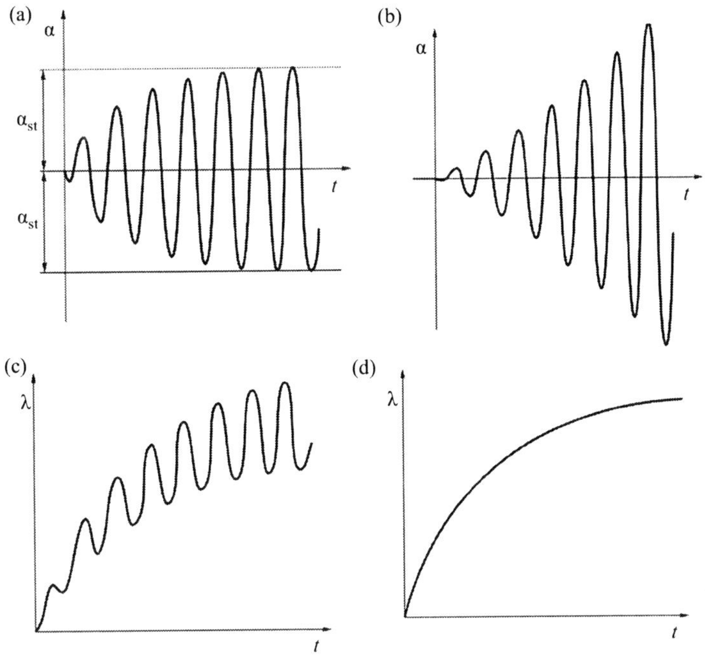


**Figure 6.18.** Reactions of the DTG to constant angular speed (operating modes).

Vibrorotors are two-component gyroscopic measuring instruments for angular speed or angle of turn depending on the operating mode (differentiating or integrating). They can measure the angular speed vector of a translational rotation lying in the plane of the rotor rotation; and such measurements can be combined using a single gyro.

In practice, the RVG with a symmetric rotor and two-frame support (or with alternative elastic support designs) has achieved the widest distribution. During dynamic adjustment, the performance of such RVGs approaches that of the *astatic gyro* (which has no error that depends on the excitation value, but only on its derivative) and they also are usually termed DTGs. When the base turns through a small angle $\lambda$, this angle appears equal to the angle between the axes of the motor shaft and the rotation of the rotor because of the astatic feature. Hence, projections of the moment vector of the motor $\bar{M}_{en}$ relevant to an axis of the rotor's own rotation $M_{en}\cos\lambda \approx M_{en}$ and on a perpendicular axis to it in an equatorial plane of the rotor $M_{en}\sin\lambda \approx M_{en}\lambda$ must be considered.

The first component of the motor moment is counterbalanced by the moment of forces resisting rotor rotation, which for a low pressure (or rarefied) gas environment may be considered as proportional to the speed of rotation. That is, $M_{en} = K_{if}\Omega$ where, $K_{if}$ is an integrating factor depending on the parameters of the gas environment and the form and character of the gas flow around the rotor as it rotates with a speed $\Omega$ and has a kinetic moment $H = C\Omega$.

The second component causes the gyro to drift toward a reduction in the angle λ with a speed:

$$\dot{\lambda} = -\frac{M_{en}\lambda}{H} = -\frac{K_{if}\Omega}{C\Omega}\lambda = -\frac{K_{if}}{C}\lambda \tag{6.28}$$

whence the possibility of using the DTG as an angular rate gauge follows. Actually, if the base turns at a constant angular rate $\omega = \dot{\lambda}$ this angle is given by

$$\lambda_{cm} = -\frac{C}{K_{if}}\omega \qquad (6.29)$$

That is, the rotor deviation from the equilibrium position is a measure of the angular speed of the base. Thus, a precession operating mode is realized (Figure 6.18(d)). It is obvious that the angular speed of the DTG drift disappears when $\lambda = 0$, so that it is necessary for the DTG to be used in closed systems, when a vector $\bar{H}$ is near to an axis of shaft rotation for all time, for example as in gyrostabilizers.

From the point of view of dynamic adjustment operation, the one-ring gyro (rotor vibrating gyro or RVG) is preferential. Furthermore, in comparison with two-ring gyros they are less labor intensive during manufacture, have simpler read-out schemes and signal transformations, and are less subject to the influence of zero signal drift in the angle gauge and feedback amplifier. The RVG can provide ranges of measurement from zero to ±180 deg $s^{-1}$ or even ±600 deg $s^{-1}$. Finally, the two-frame DTG is more accurate than the one ring.

### *6.8.4 DISTURBANCE MOMENTS DEPENDING ON EXTERNAL FACTORS AND INSTRUMENTAL ERRORS*

The DTG has the following attributes: the rotor is a symmetric body, the rotational rigidities of the torsion bars around the rotor axes are equal, and the design of two-ring supports are such that both intermediate rings (both frames) are identical (or very closely so) in terms of mass and inertial characteristics. The frames are positioned in planes of rotor rotation at angles of 90° to each other and to the main axis so that the minimum moments of inertia coincide with the torsional axes (or are parallel to them). However, even with these attributes, external influences and instrumental errors occur that cause disturbing moments (Maunder 1979; Pelpor, Matveev, and Arsenyev 1988; Vlasov and Filonov 1980; Zbrutskij and Pavlovsk 1981). These include the following:

1. Moments due to unequal linear support rigidity.
2. Moments due to an axial unbalance and failure to cross torsion bar axes, these being proportional to accelerations perpendicular to the motor shaft axis.
3. Moments due to axial oscillations in the shaft of frequency $\Omega$ are formed when $\omega=\Omega$. These are proportional to $\Omega^2$ and also to the rotor axial oscillation amplitude.
4. Moments due to axial oscillations of the shaft at a frequency $2\Omega$ are formed when $\omega=2\Omega$. These are also proportional to $\Omega^2$ and to the amplitudes of the rotor axial oscillation.
5. If axial oscillations are absent, and radial ones appear at a frequency of $2\Omega$, then constant components of moments are formed because of unequal support rigidity, the type of axial unbalance, and failure to cross axes.
6. Constant moments can also be generated by axial oscillations of frequency $\omega=2\Omega$ and radial oscillations of frequency $4\Omega$. These moments are proportional to vibro-acceleration amplitudes and to rotor axial oscillation amplitudes at a frequency $2\Omega$.
7. Constant moments will also arise at radial oscillations of frequency $2\Omega$ and axial oscillations at frequency $\omega=4\Omega$. These moments are proportional to vibro-acceleration amplitudes and to rotor axial oscillation amplitudes at a frequency $4\Omega$.

8. If there is an angular vibration in the gyro case, then along with periodic moments at frequencies $2\Omega$ and $4\Omega$ the constant moments around axes perpendicular to the axis of the motor shaft will appear, which become equal to zero if

$$A_1 - B_1 - C_1 = A_2 - B_2 - C_2$$

(defined in Section 6.8.2)

Thus, the influence of an angular vibration at double frequency is almost absent in a DTG with a two-ring (two-frame) support. However, a DTG with a one-ring suspension is not protected from angular vibrations at double the frequency of the rotor rotation.

### *6.8.5 MAGNETIC, AERODYNAMIC, AND THERMAL DISTURBANCE MOMENTS*

External and internal magnetic fields cause diversion moments similar to those that arise in mechanical gyros, and these are usually minimized by the application of various forms of screening. Research into such aerodynamic and thermal moments have presented difficult problems (Pelpor, Matveev, and Arsenyev 1988; Smith 1979. However, some practical improvements were made much earlier than when serious theoretical research began on the same problems. For example, the aerodynamic moments requiring *radial corrections* arose either because of deviation of the rotor rotational axis from that of the shaft axis in the cylindrical chamber, or when the rotor and the cylindrical chamber were not concentric because of the effect of a "gas wedge." Recommendations for the reduction of these moments were to completely remove the gas, to establish concentricity of the rotor and the chamber, and to increase the radial clearance between the rotor and its casing.

However, the realization of these recommendations does imply some essential contradictions. For example, it is impossible to remove the gas completely because of heat transfer requirements, and because the stability and accuracy characteristics worsen because of gas evolution from the lubricants and constructional parts of the gyro that rises with temperature. This situation can be alleviated somewhat by filling the gyro with hydrogen or helium at pressures of a few millimeters of mercury.

A large radial clearance between a rotor and a casing increases overall dimensions and worsens heat transfer. Hence, it is necessary to establish a compromise using this radial clearance between the rotor and its casing, and the eccentricity between the axes of the shaft and the casing, as parameters. Eccentricity also arises from aerodynamic moments produced by the elastic suspension rings and these are applied to a rotor from the inside. Such moments are exacerbated by any nonuniform gas stream from outside a rotor, which means that even apparently insignificant dents on a casing around a rotor can worsen its accuracy characteristics.

A serious problem is also represented by thermal DTG drift as the readiness (or warmup) time proceeds. It is known that such thermal processes are described by a set of differential equations of the first order whose solutions are represented by exponents. The stability (repeatability over time) of this drift therefore allows algorithmically based compensation by taking into account the gyro case temperature as measured by a temperature-sensitive element.

### 6.8.6 DESIGN, APPLICATION, TECHNICAL CHARACTERISTICS

Any DTG design operating as a part of a closed loop system contains a rotor with an elastic support, a drive motor, and angle and moment gauges. Also, a DTG may be thermostatically controlled (Pelpor, Matveev, and Arsenyev 1988).

Figure 6.19 shows the kinematic scheme of a DTG corresponding to Figure 6.17(d), and with a two-frame support. This represents the basic elements and units used in any DTG design. Here, rotor (1) is mounted on the right-hand end of shaft (7) via the two-frame support (3) within case (11) using ball-bearings (10) having preliminary axial tightness. The drive is a synchronous hysteresis motor having a rotor (9) installed on the left-hand end of shaft (7), and a stator (8) mounted on case (11). Induction gauges for (12) rotor angles are installed in the case and are switched in pairs corresponding to the differential scheme described above. A magnetoelectric moment gauge is formed by ring magnet (13) installed within the rotor, and coil (2) installed on the case. Screws (4) afford dynamic adjustment. The DTG has a tight cover (5) and cable entry grommets (6) for internal and external circuit connections.

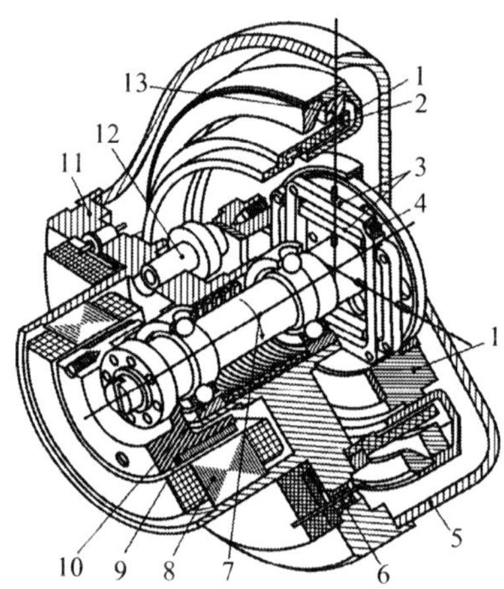


**Figure 6.19.** Kinematic scheme of a DTG with a two-frame support.

DTG design involves analysis of the influences of internal and external effects on drift caused by elastic support pliability, and leads to the following recommendations.

1. The DTG should be protected from external vibrations.
2. Measures must be taken to decrease inherent vibrations at the "dangerous" frequencies $0.5\Omega$, $1\Omega$, and $2\Omega$, caused by electromagnetic disturbances in the drive and ball-bearings.
3. It is necessary to design the elastic supports to be not only equally rigid, but also to provide zero pliability cross factors.
4. Gyro balancing technology during construction requires that any static unbalance in the gyro rotor is eliminated, rather than that of the full assembly of gyro rotor and drive rotor.
5. Along with gyro vibration insulation, a decrease in vibrating component drift can be achieved by raising the rigidity of the elastic supports in both axial and radial directions.
6. Vibration components of DTG drift depend essentially on ball bearing quality. During wear, their vibration levels increase, and hence, so do these concomitant vibration

components. Therefore, during service, a precision DTG system requires periodic updating of the coefficients in the mathematical error model on which compensation is based. Except for ball bearing deterioration, the accuracy characteristics over time are also defined by dimensional instability and the physico-mechanical characteristics of the materials used, along with the operating and storage conditions.

The first practical development of one-ring RVG support was started in the United States in 1958 by General Precision Inc. Another American firm, the Singer Corporation, developed a DTG for INS in its SKN-2400 for the SKRAM AGM-69 missile, which was widely applied in both military and civil fields. The Litton Corporation developed a serial INS, the LTN-72, for military and civil aircraft in which a DTG with two-frame support was used as the sensitive element. Teledyne Industries developed the DTG SDG-5 which was applied in the strapdown INS RIICS for fighter aircraft; and the DRIMS system for the Delta rocket in the standard inertial orientation block system DRIKU for space vehicles. This block was also applied to maintaining the exact orientation of onboard television cameras and transfer aerials on Voyagers 1 and 2, which began programs for the observation of Jupiter, Saturn, and other planets of the solar system in 1977 (Bahrami 1977; Engelder 1980).

The DTG is also designed in Germany by Litef and in France by Sagem to name but two. In Russia RVG-DNG are designed by design bureau Temp-Avia, the Ramensky instrument-making design bureau, and the Perm research-and-production instrument-making company amongst others. The technical characteristics of some RVG models with the possibility of dynamic adjustment will be found in Table 6.5 (Raspopov 2008).

### *6.8.7 CONCLUSION*

The history of the development and practical application of the dynamically tuned gyro is spread over about fifty years, but the technical characteristics reached by them do not yet match the possibilities suggested by the physical principles upon which they function. However, their advantages are being recognized by their increasing production volumes, especially where the accuracy of mission performance should be combined with small weight, dimension, power consumption, cost, and the preservation of stability characteristics.

Dynamically tuned gyros with two-frame supports reach average and above classes of accuracy and are used in INS for aerospace vehicles for multitudinous tasks. However, for vehicles with limited operating times, it is expedient to use gyros with one-ring support for control and navigation systems, including strapdown INS. It should also be noted that a number of designs can now be realized using silicon technology.

## 6.9 SOLID VIBRATING GYROS

### *6.9.1 INTRODUCTION*

A solid vibrating gyro employs two of the resonant vibration modes of an elastic body in its operation. These modes are uncoupled in the absence of rotation. The Coriolis forces resulting from rotation about a particular body-fixed axis (the sensitive axis) result in a coupling of the

**Table 6.5.** Performance characteristics of some versions of the RVG-DTG

| Designer | RVG-1M | GVK-6-1 | GVK-16 | GVK-17 | MG-4 | DNG-4 | MGL 80-3 | MGL 80-50 | K-273 |
|---|---|---|---|---|---|---|---|---|---|
| **Characteristic, Unit of measure** | **"Temp-Avia," (Russia)** | | **RIMDB (Russia)** | | | **PRPIMC (Russia)** | **SAGEM (France)** | | **Litef (Germany)** |
| Range of measured angular rates, grad $s^{-1}$ | ±150 | ±160 | ±200 | ±120 | ±60 | ±90 | ±3 | ±50 | ±200 |
| Casual drift (from run to run), grad $h^{-1}$ | ±5.0 | ±0.01 | ±(0.1–1.0) | ±(0.05–0.2) | ±0.2 | ±0.5 | ±0.36 | ±2.16 | ±0.3 |
| Drift, not depending on acceleration, grad $h^{-1}$ | — | ±3 | ±50 | ±25 | ±25 | ±50 | — | — | — |
| Drift, proportional to acceleration, grad $h^{-1}$ | — | ±1 | ±25 | ±15 | ±25 | ±6 | — | — | — |
| Voltage; V, kHz; motor; angle gauge; heating | — | 18(11);0.48<br>2,5; 19,2<br>115; 0.4 | 18(11);0.48<br>2.5; 19.2<br>36; 0.4 | 18(11);0.48<br>2.5; 19.2<br>36;0.4 | 15(11);0.48<br>2.5; 19.2<br>115;0.4 | 7,4;0,5<br>2,5; 32 | — | — | — |
| Dimensions, mm | Ø25×30 | Ø54×45 | Ø32×31.5 | Ø39×33 | Ø42×46 | — | Ø21×25 | Ø21×25 | Ø38×49 |
| Mass, g | 50 | 265 | 85 | 125 | 200 | 80 | 35 | 35 | 250 |

two modes. The natural frequencies of the modes are usually adjusted to be equal or nearly equal to maximize the resulting rotation-induced energy transfer between the modes. Figure A.2(a)–A.2(d) of Annex A of IEEE (IEEE Std 1431-2004, IEEE Standard Specification Format Guide and Test Procedure for Coriolis Vibratory Gyros, IEEE Aerospace and Electronic Systems Society, 20 December 2004) contains illustrations of the physical structure and the nature of the vibration modes of most of the solid vibrating gyros of current interest. Such gyros are undergoing intense development throughout the inertial industry because they lend themselves naturally to miniaturization in contradistinction to optical gyros, ring laser gyros, and fiber optic gyros.

The tuning-fork gyro is an example of a solid vibrating gyro in which the vibration modes utilized are dissimilar. The tuning-fork mode (the tines moving together and apart) is continuously excited. When the fork rotates about its axis of symmetry, the Coriolis forces acting on the vibrating tines induce a torsional oscillation of the fork about this axis.

Devices based on dissimilar modes of vibration like the tuning-fork gyro are often operated in the open-loop mode where one of the modes is excited to a prescribed amplitude. The damping and/or the difference in the natural frequencies of the two modes is increased to the extent that, in the presence of a constant rotation rate, the oscillation amplitude in the second mode (the readout mode) reaches a steady-state value proportional to the input rate in a time short enough to be consistent with the gyro's bandwidth requirements.

Solid vibrating gyros with similar vibration modes include those whose sensitive element is an axisymmetric body. Examples are the vibrating string, the vibrating rectangular (square) beam, the vibrating cylindrical shell, and the Hemispherical Resonator Gyro (HRG) in which the sensitive element is a hemispherical shell. The dynamical behavior of either the Foucault pendulum, or equivalently of a mass constrained to oscillate about the origin in the $x$–$y$ plane, reflects the behavior of any of these gyros.

### 6.9.2 *DYNAMIC BEHAVIOR OF THE IDEAL SOLID VIBRATING GYRO*

To every dynamic state of a solid vibrating gyro there corresponds an equivalent pendulum orbit. Pendulum orbits may therefore be used to represent the properties of solid-vibrating-gyro dynamic states diagrammatically (Zhuravlev 1997). This is shown (in Lynch 1998) (which is reproduced as Annex B of IEEE) to be true even for the dissimilar-mode solid vibrating gyros if the variables chosen to characterize the oscillations are normalized in such a way that their squares are equal to the amounts of energy in the respective modes.

The most general pendulum orbit is an ellipse. If the pendulum suspension is symmetrical (or, equivalently, the spring suspension system supporting the point mass in the $x$–$y$ plane develops a central force only) and damping is neglected, the orbit is stationary. One way of viewing the effect of rotation on the orbit of the point mass is to recognize that the orbit remains fixed with respect to inertial space as the springs supporting the proof mass rotate about the axis of symmetry in any way whatsoever (neglecting relativistic effects). This result is derived by Lynch (1998) directly from Newton's equations written in the rotating frame.

When viewed from the standpoint of inertial space, the elliptical orbit remains stationary. When viewed from a coordinate frame fixed in the vehicle, the orbit is seen to remain elliptical without change of the lengths of the semi-major and semi-minor axes, but to be rotating at the (instantaneous) input rotation rate and in the opposite sense. Thus, a measurement of the change in the angle between the spring-support system and the ellipse principal axes gives (the negative of) the change of the spring-support orientation with respect to inertial space. The ideal Foucault pendulum is thus a perfect integrating gyro.

This picture must be modified in order to represent the behavior of an actual solid vibrating gyro. Most gyros, even axisymmetric ones, do not have resonant modes in which all of the modal motion lies in a plane perpendicular to the sensitive axis. As a result, not all of the vibrating mass elements experience the maximum Coriolis force. The consequence of this fact is that, although the representative orbit continues to be a stationary ellipse, it no longer remains fixed with respect to inertial space but rotates in the same direction as the support frame at a lower rate.

When viewed from the support frame, it is still seen to be rotating in the opposite sense of the input rotation but at only a fraction of the inertial input rate. This fraction is called the *angular-gain factor*. The inclusion of an angular-gain factor does not alter the character of the ideal solid vibrating gyro as a perfect integrating gyro. A simple but informative equation reflects this behavior if an $x$–$y$ coordinate system is established in the plane of the spring-support system. If $\theta$ is the angle between one of the principal axes of the ellipse and the $x$-axis, $\Omega$ is the input angular rate (about the $z$-axis), and $k$ is the angular-gain factor, the equation for the time rate of change of $\theta$ is

$$\dot{\theta} = -k\Omega. \tag{6.30}$$

### 6.9.3 OPERATING MODES OF THE SOLID VIBRATING GYRO

Deviation of the dynamic behavior of the solid vibrating gyro from the ideal arises from the presence of damping and from a mismatch in the natural frequencies of the two modes of oscillation. The effects of these disturbances are described below. External forces must be applied continuously to compensate their influence and here it will be assumed that such compensating forces are in place so that the behavior of the gyro is close to ideal.

The open-loop mode of operation of solid vibrating gyros has been described briefly in Section 6.9.1, and a more complete description and accompanying analysis can be found in Lynch (1998). In this mode of operation, the gyro acts as a rate gyro, not a rate-integrating gyro.

In the whole-angle mode of operation, the orientation of the ellipse is simply measured periodically. The negative of the differences of successive measurements divided by the angular gain factor then provides an estimate of the changes in the spring-suspension orientation (about the sensitive axis) with respect to inertial space over the time intervals between measurements. (The operation of the HRG in whole-angle mode was first discussed in Lynch [1973].)

In the force-rebalance mode of operation, additional forces are applied continuously to maintain the ellipse in its initial orientation with respect to the support. The magnitude of the force required to thus restrain the ellipse is proportional to the input angular rate and provides a measure of it. (The operation of the HRG in force-rebalance mode was first discussed in Lynch [1972].) In both the whole-angle and the force-rebalance modes of operation, the gyro acts as a rate-integrating gyro.

### 6.9.4 THE NONIDEAL SOLID VIBRATING GYRO

If the system of linear springs constraining the mass to oscillate in the $x$–$y$ plane is not perfectly symmetric, there will exist in general two (orthogonal) directions in which the mass can oscillate back and forth along a line (a degenerate form of an elliptical orbit) at a constant frequency of oscillation. The frequencies of oscillation along these directions, $\omega_1$ and $\omega_2$, will be slightly different. To understand the effect this mismatch in natural frequencies has on pendulum orbits (in the absence of rotation), consider a case in which the principal axes coincide with the $x$ and the $y$ axes. If a linear orbit of peak amplitude $a$ is initially established that makes the angle $\theta$ with respect to the $x$-axis, its component along $x$ (of amplitude $a \cos\theta$) oscillates at one of the frequencies, say $\omega_2$, while the component along the $y$-axis (of amplitude $a \sin\theta$) oscillates at the other, $\omega_1$. Because of their difference in frequency, the two components that were initially in phase oscillate with a phase difference that grows linearly at a rate proportional to the difference frequency $\omega_1 - \omega_2$. As a result, the initially linear orbit direction migrates toward the $x$-axis and the orbit grows increasingly elliptical. When the ellipse axis reaches the $x$-axis, the ellipse has grown to its maximum ellipticity, after which point it begins to grow less elliptical again. It reestablishes itself as a linear orbit when it reaches the angle $-\theta$ (the $x$ and $y$ components that were initially in phase are now 180° out of phase). The phase difference continues to grow until, at the end of one difference frequency period, the initial linear orbit is reestablished. (The dependence of the pendulum variables on the difference frequency and the time can be simply obtained using Equation 31 of Lynch [1995].)

When this oscillation in the directions of the principal axes and of the ellipticity of the elliptical orbit are superimposed on the effects of the rotation-induced Coriolis forces (which cause any elliptical orbit to rotate), a simple estimate of the input rotation rate in terms of the measured orbit characteristics becomes difficult to identify. The simplest approach (used with all of the high-performance similar-mode gyros) is to introduce a closed-loop system in which the ellipticity of the orbit is measured and additional forces are introduced to drive it to null. It is shown in Lynch (1995) that maintaining the orbit linear is sufficient to reestablish the ideal rate-integrating gyro behavior (in the absence of damping).

The closed-loop that maintains the orbit linear is called the *quadrature-control loop* for the following reason: When the orbit is not quite linear, the oscillation component represented by the semi-minor axis of the ellipse must be detected and nulled. This component is oscillating in phase quadrature with the main component (represented by the semi-major ellipse axis).

The other effect that results in nonideal behavior of the solid vibrating gyro is damping of the oscillation modes. It is obvious that, in the presence of damping, additional forces will have to be exerted on the oscillating mass to sustain the oscillation amplitude against the damping losses. There is another important effect that is referred to as *differential-damping drift*. Damping of the vibrations of the gyro arise from a number of sources that are specific to the given gyro design. In general, the distribution of the damping forces is asymmetric. If a quadrature-control loop is in place to maintain the orbit linear, there will be one particular oscillation direction in which the damping is a minimum and another direction orthogonal to the first in which the damping is a maximum. (These directions do not usually coincide with the principal axes corresponding to oscillation frequencies $\omega_1$ and $\omega_2$.) For oscillations in either of these principal damping directions, the orbits remain linear and the sole effect of the damping is to reduce the oscillation amplitude. The time it takes damping forces to reduce an oscillation's amplitude to $1/e$ of its initial value is called the *damping time constant*. The maximum-damping time constant is termed $\tau_1$ and the minimum-damping time constant is $\tau_2$. If $a$ is the oscillation amplitude, the rate at which the amplitude decreases when the linear orbit is in the maximum-damping direction is then $a/\tau_1$ and the rate at which the amplitude decreases when the orbit is in the minimum-damping direction is $a/\tau_2$.

To understand the effect a mismatch in damping time constants has on the linear pendulum orbits under consideration (in the absence of rotation), consider a case in which the principal damping axes coincide with the $x$ and the $y$ axes. If a linear orbit of peak amplitude $a$ is initially established that makes the angle $\theta$ with respect to the $x$-axis, its component along $x$ (of amplitude $a$ cos$\theta$) decays at the rate $a$ cos$q/\tau_1$, while the component along $y$ (of amplitude $a$ sin$\theta$) decays at the rate $a$ sin$\theta/\tau_2$. If the amplitude $a$ is maintained constant by an amplitude-control loop, the following equations can be inferred:

$$\begin{aligned} -a\sin\theta\cdot\dot{\theta} &= -a\cos\theta/\tau_1 \\ a\cos\theta\cdot\dot{\theta} &= -a\sin\theta/\tau_2 \end{aligned} \tag{6.31}$$

The solution for the time rate of change of $\theta$ is

$$\dot{\theta}=\frac{1}{2}\sin 2\theta\left(\frac{1}{\tau_1}-\frac{1}{\tau_2}\right) \tag{6.32}$$

This equation shows that whenever the linear orbit is not aligned with one of the principal damping directions it will migrate toward the minimum-damping direction (in the current example, the $y$-axis which corresponds to $\theta = \pi/2$) at a rate proportional to $1/\tau_1 - 1/\tau_2$.

If Equations (6.12) and (6.13) are combined, an equation results that gives the instantaneous time rate of change of the linear orbit direction due to the influences of both inertial rate and differential damping.

$$\dot{\theta}=-k\Omega+\frac{1}{2}\left(\frac{1}{\tau_1}-\frac{1}{\tau_2}\right)\sin 2\theta \tag{6.33}$$

This equation may be regarded as the zeroth-order error model of any solid vibrating gyro. The error model is completed by including the effects of misalignment and gain-mismatch of the pickoffs and forcers, nonlinear effects, and offsets in the various feedback-control loops.

An offset in the quadrature-control loop leads to an incomplete nulling of the ellipse semi-minor axis (quadrature amplitude, $\theta$). It is shown in Lynch that the existence of a nonzero quadrature amplitude leads to an additional term in the error model (for the case in which the principal axes are aligned with the $x$ and $y$ axes)

$$\dot{\theta}=(\omega_1-\omega_2)\frac{q}{a}\cos 2\theta \tag{6.34}$$

If the above equations are generalized by letting $\theta_\omega$ represent the azimuth of the $\omega_2$ principal axis and $\theta_t$ represent the azimuth of the $\tau_1$ principal axis, both effects can be included in the error model by writing it in the following form:

$$\dot{\theta}=-k\Omega+\frac{1}{2}\left(\frac{1}{\tau_1}-\frac{1}{\tau_2}\right)\sin 2(\theta-\theta_\tau)+(\omega_1-\omega_2)\frac{q}{a}\cos 2(\theta-\theta_\omega) \tag{6.35}$$

In the higher-performance axisymmetric devices, the drift terms represented by Equation (6.16) are usually the largest that must be calibrated and removed from the rate estimate. An approach that allows the identification and thus the removal of these terms was pointed out by Loper (1984). If the amplitude of the linear orbit is driven to zero when the orbit azimuth angle is $\theta$ (i.e., the gyro is turned off) and then the orbit is reestablished at angle $\theta + \pi/2$, both the differential-damping drift and the residual-quadrature drift terms in the $\dot{\theta}$ equation change sign. By alternating the orbit azimuth between $\theta$ and $\theta + \pi/2$ (for any value of $\theta$), an estimate of the contribution from these drift terms can be obtained and a continuous calibration of their effects achieved. (This approach requires a redundant gyro or gyros at the system level so that information is not lost during the off time of each of the gyros.) This approach is particularly simple for gyros operating in the force-rebalance mode where it entails alternating the vibration modes employed as the driven and readout modes. In this form, it is applicable to gyros operating in the open-loop mode. This approach has recently been investigated at NASA's Jet Propulsion Laboratory for the continuous tuning and calibration of micromechanical vibratory gyros (Hayworth 2003).

### *6.9.5 CONTROL OF THE SOLID VIBRATING GYRO*

This following is a summary of what has been stated above about the control loops required for the operation of the solid vibrating gyro in its various operating modes.

In the open-loop operating mode, the only loop required is an amplitude-control loop to maintain the amplitude of the driven mode at its prescribed value. For more details about the design choices possible for the open-loop solid-vibrating gyro, see Lynch (1998).

In the whole-angle operating mode, two loops are required: an amplitude-control loop to maintain the equivalent-pendulum semi-major axis length at its prescribed value, and a quadrature-control loop to drive the semi-minor axis length to zero.

In the force-rebalance operating mode, one additional loop is required: a force-rebalance loop to drive any in-phase motion detected along the readout-axis direction to zero.

In the high-performance solid vibrating gyros operating in either whole-angle or force-rebalance modes, an additional loop is usually included: a phase-locked loop that locks the timing signals from a reference-phase generator (or clock) to the phase of the semi-major axis oscillations. These timing signals are then available for use in demodulating the readout signals and in remodulating the control signals developed by the various loops.

### *6.9.6 AXISYMMETRIC-SHELL GYROS*

The effect of rotation about the axis of symmetry on the flexural vibration modes of axisymmetric shells (shells of revolution) was first reported by Bryan (1890). The circular ring is the simplest axisymmetric shell and is the easiest in which to visualize the nature of the flexural modes. In the lowest-order flexural (bending) mode, which is the mode usually employed in the design of a gyro, the circular ring deforms into an elliptical shape, returns to a circle, then deforms to an ellipse again but with the semi-major and semi-minor axes interchanged, and finally returns to a circle to complete one cycle of the oscillation. (The oscillation frequency depends on the specifics of the design, but for most axisymmetric-shell gyros currently in existence, it is between 1 and 10 kHz.) These ellipses are not to be confused with the ellipses representing equivalent pendulum orbits. They are the actual shapes the deformed body takes during its oscillation cycle.

These alternating deformations form a standing elastic wave on the ring. (The lowest-order flexural modes of cylindrical and hemispherical shells have a more complicated behavior, but any cross section of those shells perpendicular to the axis of symmetry exhibits similar standing waves.) The standing waves of the lowest-order flexural mode have four nodes and four antinodes. It is these standing waves that are the analog of the linear pendulum orbits.

The Coriolis forces acting on the individual vibrating elements of the ring, when it rotates about its axis with respect to inertial space, cause the standing wave to rotate with respect to the ring. It is shown in Bryan (1890) that the rotation rate of the standing wave is less than the rotation rate of the ring so that, for an observer stationary with respect to the ring (rotating with it), the standing wave appears to rotate in the opposite sense of the inertial rotation. If $\theta$ represents the azimuth angle one of the antinodes (or line of antinodes in the case of a shell) makes with respect to a coordinate system fixed in the ring or shell, its time rate of change satisfies Equation (6.11) with an angular-gain factor $k$ determined by the geometrical shape of the shell (typically a number between 0.25 and 0.4).

The fact that the most general pendulum orbit is an ellipse with the semi-major and semi-minor axis components oscillating 90° out of phase has its analog for the flexural shell modes. The most general (lowest-order-mode) flexural shell vibration consists of a standing wave of the type described above upon which is superimposed a second such standing wave whose antinodes and nodes coincide with the nodes and antinodes (respectively) of the first wave, and which is oscillating 90° out of phase with the first standing wave. The first wave is called the *principal wave* and the second the *quadrature wave,* and the control problem of the axisymmetric-shell gyro may be simply expressed in terms of these waves (compare Section 6.10.5). Forces must be exerted on the shell in such a way as to maintain the amplitude of the principal wave at a prescribed value and to drive the quadrature wave to null. To operate the gyro in the force-rebalance mode, additional forces must be applied to hold the principal wave at a prescribed azimuth.

It should be noted that, unlike the pendulum where the principal and quadrature linear orbits are 90° apart in azimuth, the antinodes of the principal and quadrature waves in axisymmetric-shell gyros operating in the lowest-order flexural mode are separated by only 45°. The principal frequency and principal time-damping axes are likewise 45° apart. As a result, the error-model equation takes the form:

$$2\dot{\theta} = -k\Omega + \frac{1}{2}\left(\frac{1}{\tau_1} - \frac{1}{\tau_2}\right)\sin 4(\theta - \theta_\tau) + (\omega_1 - \omega_2)\frac{q}{a}\cos 4(\theta - \theta_\omega) \tag{6.36}$$

If this is to be taken as the standard form of the error equation of the axisymmetric-shell gyros, the angular-gain factor must be identified as $k/2$.

### 6.9.7 THE HRG—HISTORY AND CURRENT STATUS

The Hemispherical Resonator Gyro (HRG) is the only solid vibrating gyro to have met (in fact, exceeded) aircraft-navigation performance requirements. The remainder of Section 6.9 will therefore be devoted to the discussion of its history, its current status, and its design characteristics.

Development of the HRG began at the Boston Laboratory of the AC Spark Plug Division of the General Motors Corporation in the mid-1960s under the direction of the author. The effort was transferred to the Milwaukee, Wisconsin Operations in 1969 and then to the Santa Barbara, California, Operations in 1972, both of the Delco Electronics Corporation of General Motors. Delco's Inertial Division was purchased by Litton Guidance and Control Systems (now part of the Northrop Grumman Corporation) in 1996 and the effort transferred to the Woodland Hills, California, Operations in 2002. Litton/Northrop Grumman has been producing HRG-based systems for satellite guidance and deep space missions for more than a decade. The many successful space missions guided by HRG-based systems include Cassini, NEAR (Near Earth Asteroid Rendezvous), Messenger, and Deep Impact.

HRG development began in the Soviet Union in the mid 1980s with a comprehensive analytical study of axisymmetric-shell gyros at the Russian Academy of Science's Institute of Problems in Mechanics in Moscow (Zhuravlev 1985). Semi-independent HRG developments were then carried out at the Moscow Institute of Electromechanics and Automatics (now a division of the Aviapribor Corporation), at the Ramenskoye Instrument Design Bureau in Ramenskoye, Moscow region (now the Ramenskoye Design Company), at Elektromechanika in Miass, Chelyabinsk region, and at the Arsenal Corporation in Kiev, Ukraine. Funding for HRG development became severely limited after 1991, but it has continued in the former Soviet Union at the Moscow Institute of Electromechanics and Automatics (Izmailov 1999), at the Ramenskoye Design Company (Djandjgava 1998), at the Research and Production Enterprise Medicon of Miass (Bodunov 2001—Medicon is a private firm continuing the HRG development begun at Elektromechanika), and at Arsenal and Lileya, Ltd. of Kiev (Yatsenko 2000—Lileya is a private firm continuing the HRG development begun at Arsenal).

In France, Sagem Defense Security (now part of The Safran Group) has completed the development of an HRG for missile guidance (Taverna 2004). There is additional HRG development in other parts of the world, but there have as yet been no reports of these activities in the open literature.

### *6.9.8 HRG DESIGN CHARACTERISTICS*

The principal features of the Delco/Litton HRG design are presented in Loper (1990). This US patent contains descriptions of (1) the essential materials and processes involved in the fabrication of the Delco/Litton HRG, (2) specific approaches to using the readout and forcer electrodes to identify the equivalent-pendulum parameters and to affect the control of the gyro, (3) block diagrams of the various control loops, and (4) a discussion of the essential error sources including nonlinear effects. As such, it should be regarded as the basic technical reference for high-performance-HRG manufacture and use. Although the original design dates from the early 1980s, most of the design elements are unchanged from that time.

The three basic assemblies that make up the HRG are the hemispherical resonator, the pickoff housing, and the forcer housing. The material usually chosen for the resonator is fused silica because of its material stability, low coefficient of thermal expansion, and, most particularly, its low internal damping. HRG resonators, with natural frequencies between 2 and 8 kHz, routinely achieve $Q$s in excess of $10^7$ ($Q = \omega\tau/2$). The pickoff and forcer housings are usually made of the same material to eliminate problems caused by mismatches in thermal expansion coefficients. The three parts are bonded together in such a way that the pickoff and forcer housing surfaces on which the pickoff and forcer electrodes are plated are separated from the metallized resonator surfaces by small gaps. The electrodes and the resonator thus form capacitors that can be utilized to read out and force the resonator's flexural motion capacitively. (Capacitive readout and forcing were chosen in order to disturb the resonator motion as little as possible.) There are thirty-one tines that extend beyond the equator of the hemispherical-shell resonator that are utilized in the balancing process. Small amounts of mass are removed from selected tines to reduce $\omega_1 - \omega_2$, and to reduce the vibration-induced gyro drift. (See Lynch (1987) for a discussion of the vibration sensitivity of the HRG.)

The use of capacitive forcing and readout minimizes the disturbances to the resonator but introduces an additional problem: electrical crosstalk among the forcers, pickoffs, and their connections. The effects of this crosstalk must be minimized if aircraft-navigation performance levels are to be achieved. The HRG design presented in Loper (1990) accomplishes this minimization by separating the frequencies at which the various loops operate. Voltages varying at twice the resonant frequency are applied to a ring-forcer electrode (an electrode in the form of a ring in the circumferential direction) by the amplitude-control loop. The desired amplitude is maintained through the controlled excitation of a parametric resonance. (The use of a ring electrode to parametrically excite the HRG was first discussed in Lynch [1973].) A discussion of the phenomenon of parametric resonance is given in Landau (1976). DC voltages are distributed on a set of discrete-forcer electrodes by the quadrature-control loop to drive the quadrature wave to null. (This use of DC voltages to affect the nulling of the quadrature wave was first discussed in Loper [1979].) In the whole-angle operating mode of the HRG there are no other forcing voltages required and therefore no spurious signals varying at the resonator natural frequency to be picked by the readout circuits. In the force-rebalance operating mode, voltages varying at the resonator natural frequency are applied to specific discrete-forcer electrodes by the force-rebalance loop to maintain the output of the "orthogonal" readout channel at null. Although there are some crosstalk problems in this arrangement, they are minimal.

The use of parametric excitation with Micro Electrical Mechanical Systems (MEMS) sensors is currently under investigation (Zhang 2002). It appears that both parametric excitation and DC quadrature control should be eminently suited to MEMS gyros because of the scaling laws satisfied by solid vibrating gyros. For example, the force exerted on the HRG resonator when the DC voltage $V_R$ is placed on the resonator and the voltages $+V_F$ and $-V_F$ varying at the resonator natural frequency are placed on selected discrete-forcer electrodes is proportional to $V_R V_F A/d^2$, where $A$ is the electrode area and $d$ is the electrode-resonator gap length. The scaling laws for parametric excitation and DC quadrature control contain an additional factor $w/d$, where $w$ is the radial deflection of the shell at the equator at an antinode. These quantities are independent of scale. The oscillating mass they are acting on is, however, proportional to the cube of the linear dimension of the gyro. If the HRG resonator diameter were reduced from 30mm to 15mm and all other dimensions were scaled proportionally, the use of the same voltages would result in an increase of force authority of the various groups of forcers by a factor of eight. The hemispherical geometry of the HRG makes the possibility of reducing its size limited. These same scaling considerations, however, apply to most solid vibrating gyros. It is therefore quite possible that, although micromechanical solid vibrating gyros will have much larger natural frequency mismatches and much smaller damping time constants than the HRG, the additional force authority possible at the smaller dimensions will allow both their parametric excitation and DC quadrature control.

### 6.9.9 ADDITIONAL HRG REFERENCES

In addition to Zhuravlev (1985), there are at least two book-length treatments of HRG theory and design and at least one in preparation. The first is *Introduction to Theory of Vibratory Gyroscopes* (Egarmin 1993). It is based on work first carried out at the Institute for Problems in Mechanics in Moscow. It treats the dynamic equations of an imperfect HRG, the principal drift sources in the HRG, and the control of the oscillations in the HRG.

The other book is *Design of the Solid-State Wave Gyro* (Matveev, Lipatnikov, and Alekhin 1998). The authors describe their book as a manual addressing the general problems of HRG design including selection of the calculation model, device error analysis, recommendations for determining the features of the gyro itself, of resonator control systems, data acquisition and processing systems, gyro balancing, and testing.

The reference Loper (1984) contains a description of the basic HRG software model used to fit whole-angle HRG gyro data as a function of readout angle, temperature, and drive voltage. Other useful technical papers include Loper (1986), which discusses the projected scaling of HRG performance with size, and Zhuravlev (1995), which contains a fairly complete treatment of the electrical model of the HRG including the effect of circuit resistances. Loper (1986) also contains an improved version of the software model that allows a separation of the contribution to the differential damping drift that arises from physical sources from the contribution that arises from nonuniformities in the negative damping produced by the parametric excitation via the ring electrode, the so-called Parametric Force Drift, or PFD.

## 6.10 MICROMECHANICAL GYROS

### 6.10.1 INTRODUCTION

MEMS built on single chips include Micro-Mechanical Gyros (MMG), and as now, silicon was the basic material for manufacturing them (Petersen 1982). Within such a device is a sensitive element that responds to angular acceleration along with (some of) the electronics necessary for its operation. This sensitive element is comprised of a flexure part and an inertial mass (IM). When the reference frame angular speed changes, the energy of the induced (or *primary*) oscillations of the inertial mass, along with the flexure assembly (or *resonator*), is transformed into *secondary* oscillation energy that contains information about the angular rate to be measured.

This transformation is carried out owing to the influence on the resonator of Coriolis forces or moments of inertia due to rotation of the resonator with respect to the reference-frame angular speed, the vector of which is perpendicular to the vector of the angular momentum (or moment of momentum) that arises via translational or rotary primary fluctuations in the IM (Barbor 1996; Bryzek and Petersen 1994; Kranz and Fedder 1997; Weinberg 1995).

Primary oscillations can be regarded as the *drive mode* (DM), or mode of motion, along the excitation coordinate, and secondary oscillations, or the *sensitivity mode* (SM), as motion along the output signal coordinate.

Linear-linear gyros (LL-type), rotary-rotary gyros (RR-type), and linear-rotary gyros (LR-type) are distinguished by the form of the inertial mass motion in the DM and the SM. In LL-gyros the inertial mass in the DM and the SM makes a translational motion, in RR-gyros a rotary motion, and in LR (or RL) gyros various combinations of translational and rotary motion.

MMGs also include *fork* and *wave* types. The distinctive feature of the fork MMG is the presence of rod structures (or *legs*). It is essential that the inertial masses of these legs are distributed regularly along their lengths and that one end of each is free to move. When a translational angular speed occurs, the vector of which is perpendicular to the angular momentum vectors of the elementary masses distributed along the legs, Coriolis forces of inertia arise which generate secondary oscillations in these legs (Loveday and Rogers 1998).

The distinctive attributes of wave MMGs is the presence of a resonator having a ring form fastened to the case via elastic support elements, or having the shape of a rod whose fastening to the case does not interfere with its longitudinal and lateral oscillations. The first type is usually called a *ring MMG*, and the second, a *rod MMG*. In both cases inertness of standing waves raised in the ring or the rod is used. These waves precess with translational angular speed whose vector is perpendicular to the planes of a vibrating ring or directed along a vibrating rod. This is known as the Bryan Effect (Ayazi and Najafi 1998; Zhuravlyov and Klimov 1985).

Each kind of MMG is characterized by a set of classification attributes, the most important of which are number of measuring axes (a one- or two-component MMG), the number of inertial masses (a single or multimass MMG), the type of inertial mass suspension (contact, contactless, internal, or external), the inertial mass moving in one plane or in various planes, the type of inertial mass drive in the DM (electrostatic, magnetoelectric, etc.), and the type of signal pick transducer up in the SM.

Further typical characteristics of MMGs include the measurement range, the sensitivity, the pass-band, the scale factor and its stability, and the cross sensitivity, noise, and temperature stability of the characteristics and various other operational parameters.

The MMG can operate in direct and compensative transformation modes of measuring data takeoff.

### 6.10.2 OPERATING PRINCIPLES

The main part of an MMG that determines the functionality is the *sensitive element* (SE), which may be a suspended inertial mass with a drive providing the mode of motion. In the presence of a translational angular speed, and owing to the Coriolis acceleration and the corresponding inertial forces, secondary (SM) oscillations are generated. On this basis, MMGs are sometimes regarded as devices for measuring Coriolis acceleration.

The single-mass MMG of the LL and RR-types, and also ring, fork, and rod MMGs, are the most simply realized versions and have the widest distribution.

#### 6.10.2.1 Linear-Linear (LL-type) Gyros

The basic schematic of the LL-type of MMG is shown in Figure 6.20. It consists of an IM (1) (of mass *m*) and a support with elastic elements (3, 4) fixed to the base (5). Elements (2) provide integrity and rigidity in places where the elastic coupling elements interface with the IM. A drive of any physical nature (electrostatic, electromagnetic, etc.) is not shown in the figure, but is functionally part of the SE.

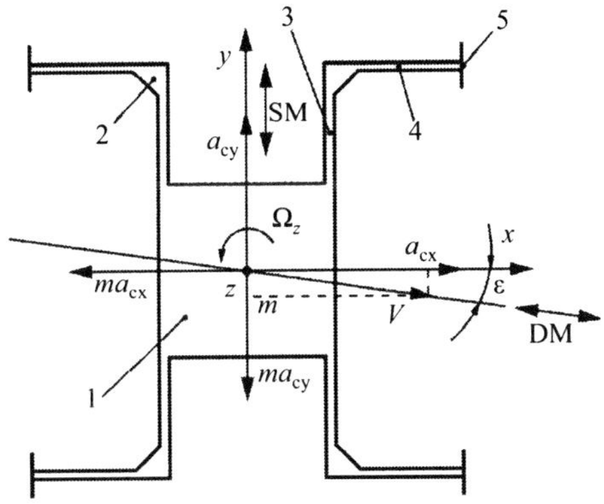


**Figure 6.20.** Basic scheme of single-mass sensitive element (SE) of the LL-type of MMG.

Generally, the drive develops a force $F_0 \sin pt$ (where $F_0$ and $p$ are the amplitude and frequency) which is directed at some small angle ε (sin ε ≈ ε and cos ε ≈ 1) to the $x$-axis and produces IM oscillations $x = x_0 \sin pt$ (where $x_0$ is the amplitude). The instantaneous vector of the linear velocity of the inertial mass (IM) in the drive mode (DM) has projections on to the $x$ and $y$ axes as follows:

$$V_x = V\cos\varepsilon \approx V;\; V_y = -V\sin\varepsilon \approx -V\varepsilon \tag{6.37}$$

For a translational angular speed $\Omega_z$ whose instantaneous vector is directed in the positive direction of the $z$-axis, Coriolis accelerations arise along axes $x$ and $y$:

$$a_{cx} = 2\Omega_z V\varepsilon \;\; a_{cy} = 2\Omega_z V \tag{6.38}$$

That is, the IM reacts to the action of the Coriolis inertial force $ma_{cx}$ along the $x$-axis, and $ma_{cy}$ along the $y$-axis.

If inertial and damping forces are included along with the elastic forces applied to the IM, the equations of motion in this elementary case appear as follows:

$$\left.\begin{aligned} m\ddot{x} + b_x\dot{x} + G_x x &= F_0 \sin pt - 2mV\Omega_z\varepsilon; \\ m\ddot{y} + b_y\dot{y} + G_y y &= -\left(F_0 \sin pt\right)\varepsilon - 2mV\Omega_z, \end{aligned}\right\} \tag{6.39}$$

where $b_x$, $b_y$ are the damping factors of the IM in the direction of the corresponding axes, and $G_x$, $G_y$ are the rigidities of the elastic supports, also in the directions of the corresponding axes.

The first equation of system (6.39) describes the DM and the second the SM, from which it follows that the IM movement along the $y$-axis under the action of Coriolis forces is deformed by the projection of a drive force along the same axis. This results in a measuring error in the MMG. Neglecting the influence of this distortion, the established mode of motion of the IM in the SM is determined by the following expression:

$$y = -\frac{2mV\Omega_z}{G_y} \tag{6.40}$$

from which it follows that the amplitude of the secondary IM oscillations is proportional to the angular rate of the base, so that the MMG actually provides a measurement of that base angular rate. From Equation (6.40), the necessity of maintaining a constant rate $\dot{x} = V$ in the DM also follows, which is a difficult task.

Also to be considered is a quadrature measurement error in $\Omega_z$. Because $V = \dot{x} = x_0 p \cos pt$, the Coriolis acceleration will be $a_{cy} = 2\Omega_z x_0 p \cos pt$. In the DM, the projection of the IM acceleration on to the $y$-axis is equal to $a_y = \ddot{x}\varepsilon = -\varepsilon x_0 p^2 \cos pt = \varepsilon x_0 p^2 \cos(pt + 90)$. The IM reacts to this component of acceleration in addition to $a_{cy}$, which results in a measurement error called the *quadrature error* because there is a phase lag of 90° between accelerations $a_y$ and $a_{cy}$.

The relationship between the acceleration amplitudes is

$$\frac{a_y}{a_{cy}} = \frac{\varepsilon p}{2\Omega_z} \tag{6.41}$$

from which it follows that if the orders of the accelerations are identical, the accuracy of the observed perpendicularity between the directions of the DM and the SM should be very high. For example, at $\Omega_z = 0.05$ rad s$^{-1}$, $p = 10^5$ rad s$^{-1}$, and $a_y = a_{cy} = 1$ the allowable value of the angle e = $10^{-6}$ rad, but usually this cannot be realized in practice.

The quadrature signal frequency is equal to the drive frequency, and this makes noise filtration difficult. However, because of the 90° phase shift, the quadrature signal can be partially excluded with the help of a phase-sensitive detector. The efficiency of such filtration depends on how precise the phase ratio can be, as allowed by the relevant electronics.

In an MMG with a single-mass SE, it is difficult to separate the useful signal caused by the base Coriolis acceleration from a signal caused by a linear acceleration whose vector has a component along an axis of secondary oscillation, that is, an output axis.

#### *6.10.2.2 Rotary-Rotary (RR-type) Gyro Principles*

In the RR-type MMG, the motion of the IM in the DM and the SM has a rotary character. The combination of the rotary motion of the IM in the DM, and its translational rotary motion in the SM, results in the generation of an inertial Coriolis force moment called the *gyroscopic moment.*

In Figure 6.21, the IM (rotor) (1) is connected to the base (3) via elastic support elements (2) that have low torsional rigidity around the $y$-axis and also, via bending, around the $z$-axis. The rigidity of these elastic elements is much greater around the $x$-axis, which means that oscillations of the rotor around both axes $z$ and $y$ are dominant.

The drive provides rotor oscillatory motion with a speed $\dot{\gamma} = \Omega$ around the $z$-axis in such a manner that during each first half-period, the kinetic moment $H_1$ is directed to the positive side of the $z$-axis, and during second half, $H_2$ is directed to the negative side. The equality $H_1 = H_2 = H = J_\gamma \Omega_\gamma$ ($J_\gamma$ being the axial moment of rotor inertia) is maintained rather precisely, and is actually a primary oscillation mode, that is, a DM.

On the occurrence of a base translational speed $\omega$ (this being precession in the rotor), there is a gyroscopic moment $M_{G1}$ for the first half-cycle of the DM and $M_{G2}$ for the second half-cycle of the DM ($M_{G1} = M_{G2} = H\omega$). So, the periodically changing direction of the gyroscopic

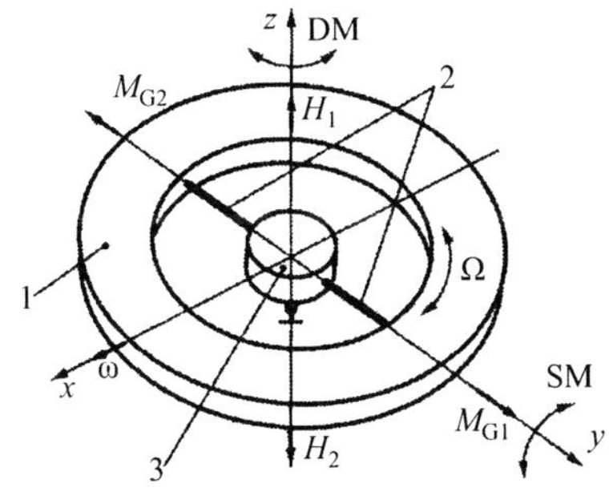


**Figure 6.21.** Kinematic scheme of the SE in the RR-type MMG.

moment causes oscillations of the rotor around the $y$-axis. This is the mode of secondary oscillation, that is, the SM.

The drive moment $M_0 \sin pt$ (where $M_0$, $p$ are its amplitude and frequency) overcomes the inertial moment, the damping moment, and the moments of the elastic suspension forces around the $z$-axis. The gyroscopic moment overcomes the similar moments of the forces around the $y$-axis. Hence, the elementary equations of the rotor motion may be written as follows:

$$\left.\begin{aligned} J_\gamma \ddot{\gamma} + b_\gamma \dot{\gamma} + G_\gamma \gamma &= M_0 \sin pt; \\ J_\alpha \ddot{\alpha} + b_\alpha \dot{\alpha} + G_\alpha \alpha &= H\omega, \end{aligned}\right\} \tag{6.42}$$

where $J_\gamma$, $J_\alpha$ are the axial and equatorial moments of the rotor inertia,
$b_\gamma$, $b_\alpha$ are damping factors of the rotor in the DM and the SM,
$G_\gamma$, $G_\alpha$ are the rigidities of the elastic support elements around the $z$ and $y$ axes, and
$\gamma$, $\alpha$ are angles of rotor fluctuation in the DM and the SM.

For an established SM, it follows that from the second equation of system (6.42) the angle of a rotor turn around the $y$-axis is:

$$\alpha = \frac{H\omega}{G_\alpha}, \tag{6.43}$$

and it is this that contains information on the angular rate of rotation of the base upon which the gyro is mounted. Hence, a DMG of the RR-type is also a gauge for angular rate measurement.

The RR-type of MMG also exhibits a quadrature error.

The equations for the sensitive element's (SE) motion in a single-component MMG are similar to those of system (6.42). Then, from expression (6.43) it follows that the measurement accuracy of the base translational speeds depend initially on the stability and predictability of the parameters that determine the values $H$ and $G_\alpha$.

If in the scheme of Figure 6.21 the rotor is provided with elastic elements along the $x$-axis, the gyro would become a two-component instrument capable of measuring also the angular rate of the base whose vector is directed along the $y$- axis.

The SE motion of a two-component MMG is also described by the system equations (6.42), to which it is necessary to add the equation of rotor motion around the second measuring axis.

### 6.10.2.3 Fork and Rod Gyro Principles

The principle of sensitive element (SE) operation of a fork gyro is illustrated in Figure 6.22.

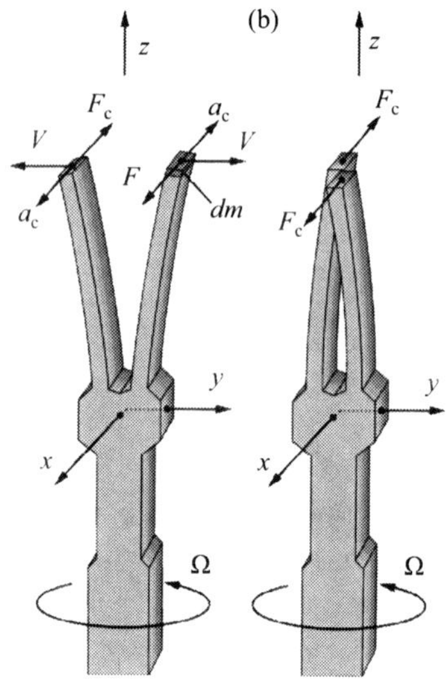


**Figure 6.22.** Principle of the SE operation of a fork gyro.

Here, the branches of the SE, made from a monocrystal, oscillate in antiphase motion in the *zy* plane. Each branch element, of mass *dm*, moves with a linear speed *V*. For a SE rotation of speed W around the z-axis, in the corresponding vector direction there is a Coriolis acceleration $a_c = 2V\Omega$ and a corresponding inertial force $F_c = 2V\Omega dm$. The Coriolis inertial forces are summed by the masses of each of each of the branches and result in their bending in a plane *xz*.

The SE representing a dual tuning fork is shown in Figure 6.23. The action of the Coriolis forces of inertia $F_c$ in response to the movement of the legs by the excitation forces $F_d$ and the measured speed Ω, is similar is to that of Figure 6.22. The dual SE allows for a reduced interaction between the vibrating elements.

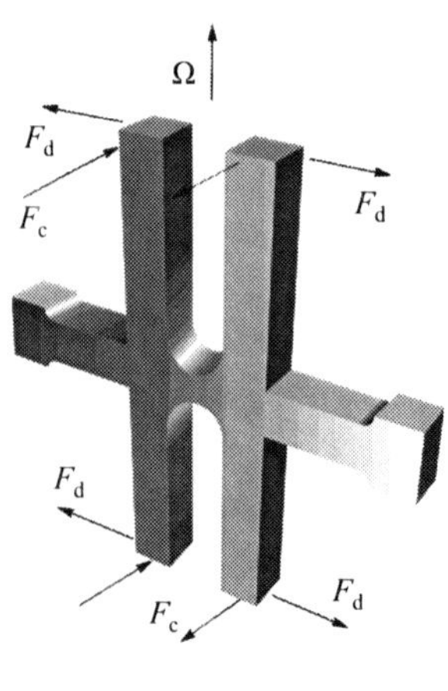


**Figure 6.23.** The sensitive elements (SE) of a dual tuning fork.

Rod gyros have a core as the sensitive element (SE) in which a lateral wave of deformation is generated. This then reacts to angular rate because secondary oscillations of the rod serve as the measurement source. This mechanism indicates why they are sometimes called *wave gyros*.

Figure 6.24 shows the structure of such a gyro, which consists of a rod (2) in a case (1) and carries piezoelectric elements (3,4,5,6) installed on the sides of this rod, shown in cross-section underneath. This rod cross-section can actually take any other form, for example, that of an equilateral triangle.

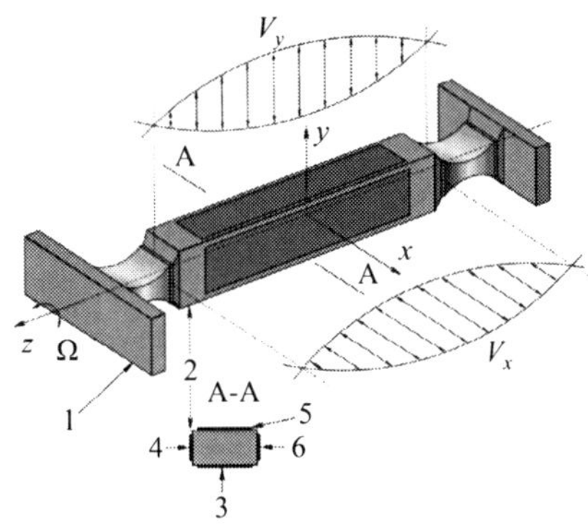


**Figure 6.24.** The sensitive element (SE) of a vibrating rod (or wave) gyro.

It will be seen that both ends of the rod are reduced in cross-section: these can be regarded as elastic connections to the case. The pair of piezoelements (3–5) serves to provide oscillation excitation at the fundamental in the plane $yz$, which results in each element of the rod acquiring a linear speed $V_y$. If the structure is rotated around the $z$-axis with angular speed Ω, the Coriolis forces of inertia cause the rod to oscillate in the $xz$ plane with linear speed $V_x$ in each element.

In the pair of piezoelements (4–6) only one serves to measure these oscillations, and another piezoelement can be included for damping purposes. With the help of an electronic circuit, the oscillation amplitude proportional to the angular speed W is measured, and a phase indication of the direction of rotation around the $z$-axis is also determined.

### *6.10.2.4 Ring Gyro Principles*

The wave solid-state gyro (SSG), also called the *hemispherical resonator gyro*, is a form of ring MMG (RMMG). The operating principle is based on using the inert properties of elastic waves treated as radial oscillations of the second mode in hemispherical, cylindrical, or ring resonators.

The effect of elastic wave inertness in rotating axisymmetric bodies was described theoretically and confirmed experimentally by G. H. Bryan in 1890. He showed that during the rotation of a vibrating shell, and as a result of the action of inertial Coriolis forces, there is a splitting of the basic mode of oscillation frequency due to bending in its walls that results in the precession of a stationary wave relative to both the shell and the inertial space.

The ring resonator can be thought of as a ring that has been cut from the circumference of a hemispherical or a cylindrical resonator, and the precession of a stationary wave in such a rotating resonator is illustrated in Figure 6.25. Stationary waves in two mutually perpendicular directions $x$, $y$ are formed in this elastic ring resonator, and are characterized by antinodes and nodes.

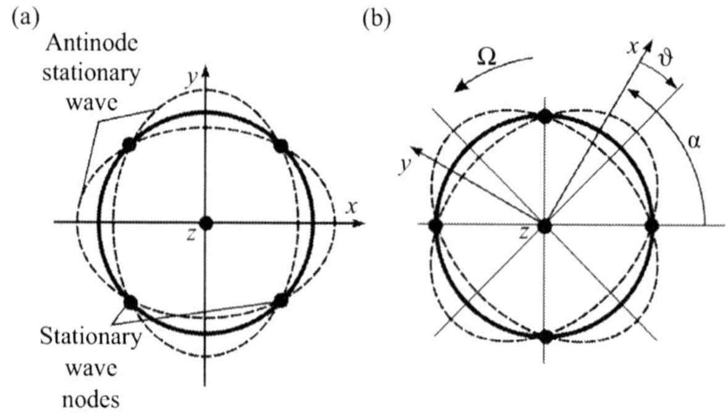


**Figure 6.25.** Precession of a stationary wave in a rotating resonator.

During rotation of the resonator around an axis of sensitivity that is perpendicular to the plane of the ring resonator, and is on an axis of its symmetry, the waveform of the stationary wave lags behind the rotating resonator. That is, the nodes and antinodes precess relative to the resonator by an angle $\vartheta$ and in inertial space by an angle $\alpha$.

The operating mode of the SSG as an angular rate gauge, or as an integrating gyro (IG), depends on how the resonator oscillation excitation is produced. For positional excitation using a pair of opposing electrodes, the pressure changes with a frequency that is half that of the basic mode. At $\Omega = 0$, the orientation of the stationary waves in the resonator is constant and is determined by the positions of the excitation electrodes, or in other words $\vartheta = \vartheta_0$ and the standing wave is locked to the resonator. When the base rotates, the orientation of the stationary wave relative to the resonator is determined by $\vartheta = \vartheta_0 - \frac{4KQ}{\omega_0}\Omega$, where $\omega_0$ is a resonant oscillation frequency, $K$ is the scale factor of the resonator, $Q = \frac{1}{2\xi}$ is the resonator selectivity, and $\xi$ is the damping factor of the resonator material (Zhuravlyov and Klimov 1985).

The SSG-IG mode requires parametric excitation of the resonator, which is surrounded by a ring electrode to form a cylindrical capacitor. This capacitor is then excited at an amplitude and frequency close to the natural frequency of the resonator. The orientation of the stationary wave of the resonator may then be determined using the following expression (Zhuravlyov and Klimov 1985):

$$\vartheta = \vartheta_0 - K\int_0^t \Omega(\tau)d\tau \qquad (6.44)$$

The technical realization of the IG mode for the RMMG presents significant difficulties connected with the accommodation of the excitation electrode system for producing oscillations in the resonator, which has elastic supports that limit the available space.

These ring resonator elastic supports can take various forms and may be placed either outside or inside the ring resonator. However, the methods of fixing these elastic elements both to the resonator and to elements of the case are of considerable importance. The elastic supports and the ring resonator form a vibro-structure that should have identical rigidity in radial directions throughout 45°. This can in fact be done by using up to eight elastic elements. Thus, it is possible to achieve a minimal discrepancy (splitting) of the vibro-structure oscillation frequencies in the directions of the antinodes of two stationary waves (Figure 6.25). This is important, as various kinds of two waves do not allow determination of the position of the resonator wave (Ayazi and Najafi 1998; Vavilov 2003; Zhuravlyov and Klimov 1985).

### *6.10.3 ADJUSTMENT OF OSCILLATION MODES IN GYROS OF THE LL AND RR TYPES*

The oscillatory systems of the MMG can be characterized by the resonant frequencies of the forced oscillations in DM, the frequency of the informative oscillations in the SM, and a quality (or damping) factor.

Neglecting the effect of the damping quality on the characteristic of the mechanical oscillator, another quality parameter may be used: the ratio of the oscillation amplitude at resonance to the pass-band of the measured signal frequencies.

Because the Coriolis acceleration is a narrow-band signal having frequencies centered on the DM frequency, the mechanical SM amplification factor can be adjusted to conform (or approximately conform) to the resonant frequencies of the DM and the SM.

There are several ways of adjusting the resonant frequencies for motion and sensitivity modes (Figure 6.26).

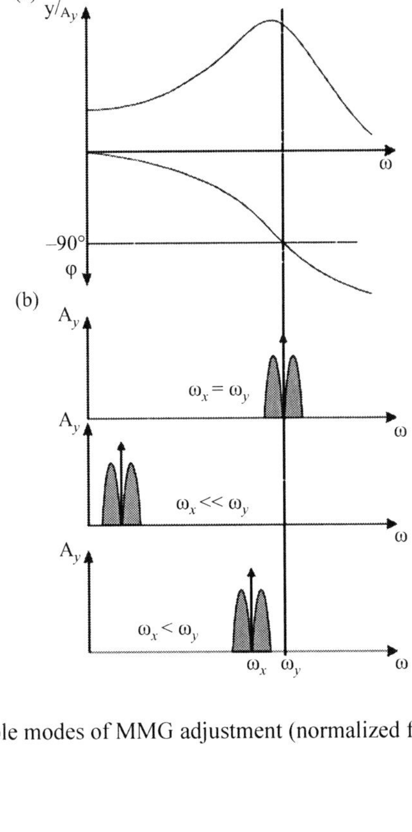


**Figure 6.26.** Possible modes of MMG adjustment (normalized frequency responses).

The ratio of amplitude $y$ of the DM oscillation along an informative axis to amplitude $Ay$ of the Coriolis acceleration (normalized frequency response), and the phases $\varphi$ of these oscillations, are shown in Figure 6.26(a) as functions of frequency. Possible modes of MMG adjustment are shown in Figure 6.26(b).

When the resonant frequencies in the DM and SM coincide ($\omega_x = \omega_y$), the MMG mechanical amplification factor is very high, but the pass-band of the input (measured speed) signal decreases and there is a phase shift at the output. If the frequency of the forced oscillation is much less than the sensitivity mode frequency ($\omega_x << \omega_y$), the gyro exhibits a wide pass-band and has no phase shift. Finally, if the resonant frequencies do not coincide within a small segment (for example, 10%), the mechanical amplification factor will be large and the pass-band will decrease insignificantly in comparison with the first case.

For MMG dynamics the ratio of frequencies is much more important than their absolute values.

Adjustment of the DM frequency $\omega_x$ is carried out by selecting the frequency of the power source; whereas adjustment of the SM frequency is achieved by changing the support rigidity in the direction of these oscillations via the electrostatic forces produced by the adjustment electrodes.

To illustrate the principle of electrostatic adjustment, an IM is shown in Figure 6.27, where the fixing points (or anchors) (3) are connected to elastic elements (1, 2) between which the rigid IM is supported. Fixed electrodes (4) and mobile electrodes (5) are parts of the electrostatic adjustment system included in the electronic unit that provides the electrostatic force.

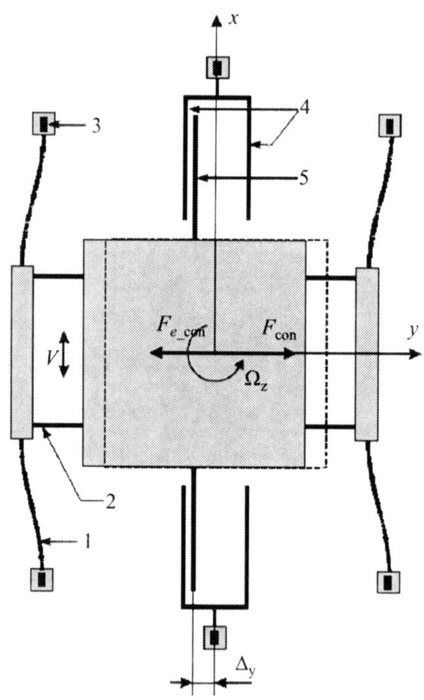


**Figure 6.27.** Adjustment of the sensitivity mode (SM) of an MMG.

Assume a lateral IM displacement (to the left of the $x$-axis in Figure 6.27) of value $\Delta y$ and speed $V$ caused by a rotation of the base with a speed $\Omega_z$. An elastic force $F_{con}$ in the support elements is directed against this displacement $\Delta y$ and is augmented by an electrostatic force $F_{e\_con}$ produced by the adjustment electrodes. The total force applied to the IM is therefore

$F_{\text{con}} - F_{e\_\text{con}} < F_{\text{con}}$ and hence both the rigidity of the suspension and the resonant frequency of the informative oscillations decrease.

Generally, the electrodes form $n$ pairs of capacitors, so that the resonant frequency for the sensitivity mode (SM) in the direction of the $y$-axis is determined by the formula:

$$\omega_{0y} = \sqrt{2\frac{\left(2G_{1y} - \sum_{i=1}^{n} \frac{C_{oi}}{\Delta y_{oi}^2} U_o^2\right)}{m}}, \tag{6.45}$$

where $C_{oi}$, $\Delta y_{oi}$ are the capacitances of the $i$-th pairs of capacitors and the gap between the mobile and motionless electrodes, $U_0$ is the voltage on the capacitors, $G_{1y}$ is the rigidity of the elastic support element, and $m$ is the IM mass.

From Equation (6.45) it follows that by changing the parameters of the electrostatic "comb" and voltage it is possible to change the oscillation frequency of the IM in DM, so achieving the necessary detuning of frequencies in relation to the primary resonant frequency of the IM in the drive mode (DM).

Detailed research of MMG dynamics is presented in a number of works (Barbor 1996; Evstifeev 2002, 2004; Geiger 1997, 1998; Kycherkov 2002; Lestev, Popova, Pjatyshev, and Raspopov 2007; Severov, Ponomarev, and Panferov 1998, 2003).

### 6.10.4 DESIGN, APPLICATION, AND PERFORMANCE

Many enterprises and organizations around the world, for example, Bosch, MS Lab., Draper Lab., Epson, Analog Devices, and HSG-IMIT, are engaged in MMG development and production.

#### 6.10.4.1 Gyros of the LL- and RR-types

One of the most famous developers of LL-type MMG is the American firm Analog Devices, which has developed and released its ADXRS series. This type of gyroscope consists of an integrated single-crystal silicon microcircuit that contains all the necessary electronic components for producing an appropriate output signal. In the center of the microcircuit are two SE micromechanical structures shown in Figure 6.28(a) (Vlasenko 2003) whose principle of operation corresponds to Figure 6.20. The SE for each of these structures is established in motion by electrostatic drive electrodes at the edges. Signal pick-up is carried out by similar structures to determine the angular speed of the case, this being perpendicular to the crystal plane. In such microstructures, the directions of the IM oscillations in the modes of motion and sensitivity (DM and SM) are mutually perpendicular, which makes it possible to avoid the influence of constant and vibro-accelerations on the gyro output signal.

The high-frequency pick-off signal obtained from the capacitive displacement electrodes is amplified and demodulated to result in an output signal voltage proportional to the angular speed of the encapsulation. This microcircuit also includes structures for temperature error compensation and for calibration, and also a precision power source unit.

The thickness of the mechanical structure is 6 microns and since there is no vacuum inside the encapsulation, the resonator quality factor relevant to the axes of motion in both DM and SM is good (typically ~ 45). The MMG therefore operates in a direct measurement mode because the frequency detuning results in a value of about 300 Hz. (The natural frequency of the support oscillation without damping is about 15 kHz.)

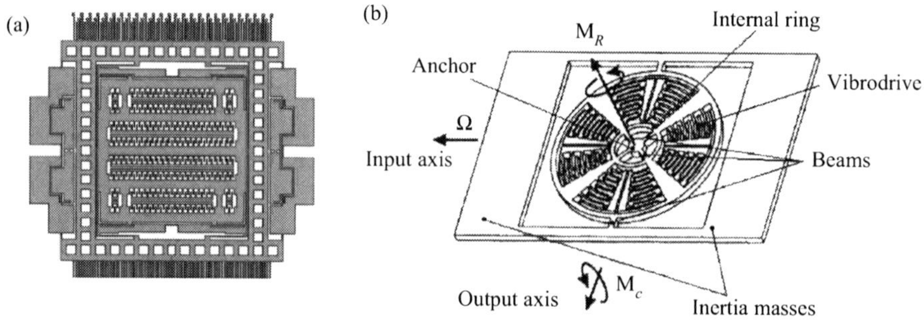


**Figure 6.28.** MMG structures. (a) LL structure (USA Patent 6,122,961 dated 26.09.2009, reproduced by permission of Analog Devices, Inc.; Norwood, Mass.) (b) RR structure by HSG-IMIT (from Fell [2001] and Geiger [1998]).

In the SM, the IM's movement is in the order of $10^{-16}$ m, and a high-sensitivity capacitor system for signal pick-up allows the measurement of such tiny movements within a limiting error of about $16 \times 10^{-19}$ m. The mass of the MMG is less than 1 g, and the size of the ceramic encapsulation is 7 × 7 × 3 mm. The random drift is about $0.3° \text{ s}^{-1}$, and the range of measurement for the ADXRS series is up to $300° \text{ s}^{-1}$.

ADXRS gyroscopes have stable output signals at accelerations up to 2000 g and can, for example, be used within integrated INS with GPS for the stabilization systems of some aerospace vehicles, among many other applications.

Development of RR-type gyros has been carried out by the Draper Laboratory at MIT along with experts at Robert Bosch GmbH and the HSG-IMIT institute in Germany.

The kinematic scheme of one variant of the RR-gyro by HSG-IMIT is shown in Figure 6.28(b) (Geiger et al. 1998). Here, an inner ring constitutes the IM and its outer rim is connected to the inner rim of an external ring via elastic elements as shown in Figure 6.28(b). This ring is realized on a silicon substrate that also carries a vibrodrive (or vibro-actuator), and also capacitive pick-offs for output signal acquisition. The vibro-actuator produces a rotating moment $M_R$ that interacts with the Coriolis force $M_c$ that appears due to any angular speed Ω of the case (and hence the substrate to which the MMG is anchored).

There is a vacuum inside this gyro encapsulation of 0.01 mBar, and the rotor oscillation frequency (that is, of the inertial mass in the DM) is 1420 Hz. The drift speed is about $65° \text{ h}^{-1}$ over a passband of 50 Hz and the measurable speeds range up to $200° \text{ s}^{-1}$.

One of the first microstructures that realized the principle of the LR-gyro was also developed in MIT's Draper Laboratory.

### *6.10.4.2 Fork and Rod Gyros*

Fork-sensitive elements (Figure 6.22) are used in such gyros as the DRZ from Temic (Daimler-Benz-Konzern) and the QRZ from Systron Donner (BEI Electronics, Inc).

Daimler-Benz offers the Temic gauge for angular speed measurement. This gyro has external dimensions of 63 × 47 × 35 mm and is made using hybrid technology for the fork-type sensitive element. The excitation and signal readout elements for pick-up from the vibrating parts are executed using silicon technology, and the service electronics appears in traditional printed circuit form. The metal case of the device is, of course, airtight. The output is in analog form, and the measured speeds range up to 75° $s^{-1}$. The basic applications of this gyro are in various control and fault monitoring systems.

The GiroChip™ is a Systron Donner product where the use of piezoelectric material has essentially simplified the design and has provided temperature stability and a long service life. The sensitive element, together with the electronics, is built into a rigid case, and when the device is energized at a constant voltage, it produces a high-quality analog output signal over a wide passband. The range of measured speeds is 50–1000° $s^{-1}$. This angular speed gauge has a wide sphere of application, including integrated INS with GPS, antennae stabilization, and various control systems.

The BEI GiroChip™ HORIZON angular speed gauge is produced using a similar sensitive element.

Systron Donner has also developed the QRS 11 quartz rate sensor gyro, the SE of which is made from monolithic quartz and is representative of the dual tuning fork structure of Figure 6.23. The mass of the QRS 11 is 60 g, its dimensions are 40 × 16 mm, and the passband is no greater than 60 Hz.

The complete *inertial module* (or *inertial block*) offers expanded application opportunities in comparison with single devices such as the GiroChip™, especially for navigational systems. One example is the Motin Pack™ that contains three gyros for angular speed and three accelerometers.

The French SAGEM SA firm has developed a gyro of macroscopic dimensions named the Quapason™ that uses a quartz resonator consisting of four rectangular-section rods on a common base connected to the case through a vibro-insulated leg (Ganrya, Fantonbi, and Karon 2004). Two models of these gyros have been developed with sizes 28 × 60 mm and 15 × 30 mm. The resonator can operate using different electronic circuits to provide measurement modes for either turn angle or angular rate. The passband is 100 Hz and the nominal measurement speed is up to 250° $s^{-1}$. This gyro finds applications in antennae stabilization and in optical systems for visual line stabilization, and so forth.

The Japanese firm, Murata, produces two modifications of piezoelectric vibrating gyros, ENV, and ENC. The rod-type sensitive element (Figure 6.24) of this type of gyro is essentially a prism suspended on extensions and having a cross-section in the form of an equilateral triangle, on the sides of which are piezo-elements for excitation (in the first mode of flexural oscillation of the prism), and also for acquiring the output signals. The prism is made from Elinvar, which exhibits an almost zero temperature coefficient of the modulus of elasticity that has allowed a major reduction in the temperature dependence of the gyro characteristics. Two gauges carry out excitation (at a primary oscillation frequency of about 25 kHz) and measurement of secondary oscillations, and a third serves to generate a feedback signal.

The Gyrostar ENV-05 D-02 has dimensions 18 × 30 × 41 mm and a mass of 50 g, and exhibits a measurement range up to ±90° $s^{-1}$.

### *6.10.4.3. Ring Gyros*

In the late 1990s, Silicon Sensing Systems, a joint venture between the Plymouth, UK arm of UTC Aerospace Systems (formerly part of BAe Systems) and Sumitomo Precision Products of Japan) developed the SGH01 inductive MEMS gyro as shown in Figure 6.29 (Hopkin 1997).

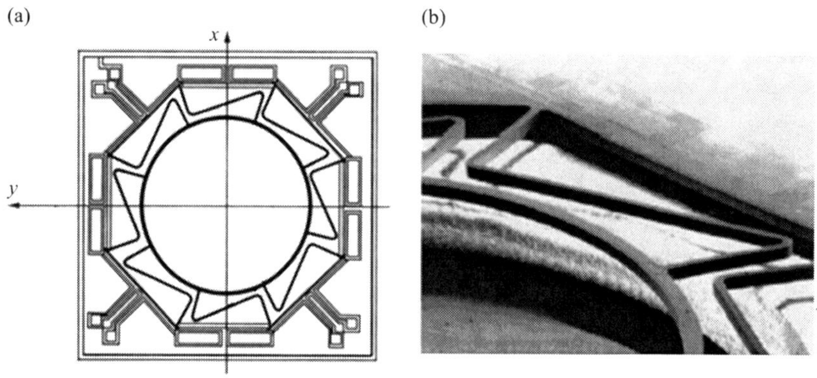


**Figure 6.29.** (a) Basic SE scheme of a microgyro with external ring resonator support and (b) a fragment of the microstructure.

Here, the resonator is a silicon ring having a diameter of 6 mm, width of 120 micron, and height of 100 microns mounted on a $10 \times 10$ mm glass supporting structure by eight elastic support elements. A metal film is deposited on the surface of the structure to provide drive and pick-off transducers, each of which spans one-eighth of the ring.

When mounted in a permanent magnetic field, the drive transducers generate a 2θ standing wave of frequency 14 kHz by passing an alternating current through the ring. The Coriolis effect causes the standing wave to move relative to the fixed support structure when the gyro rotates. This can be measured using the pick-off electronics, which allow the rate of rotation to be calculated.

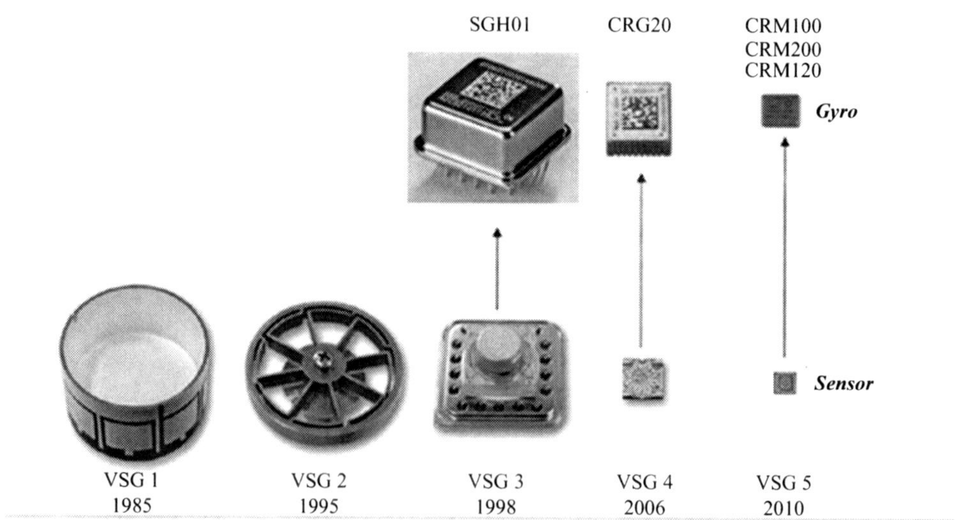


**Figure 6.30.** Evolution of vibrating ring sensors developed by UTC, Plymouth, UK. (Courtesy of UTC Aerospace Systems).

In a vacuum-sealed package, its uncompensated bias is <3° $s^{-1}$ and it exhibits excellent vibration and shock resistance up to 20,000 g. Initially it was destined for automotive and commercial applications but, with the use of innovative temperature compensation techniques, a bias performance of <100° $h^{-1}$ became attainable, making its performance appropriate for high grade commercial, military, and aerospace requirements.

Further developments of these MEMS gyros have led to the smaller 3 mm ring CRG20 device with electrostatic drive and pickoff and the 2 mm piezoelectric CRM series of gyros, commercially known as Pinpoint®. The evolution of vibrating ring sensors developed by UTC Aerospace Systems is shown in Figure 6.30.

### *6.10.5 CONCLUSION*

Though micromechanical gyros have only a short development history, and were thought to be rather rough devices, they have recently made considerable progress toward better drift reduction. Gyros of this type are already used in military applications, and those developed by Analog Devices have reached a claimed drift not worse than 10° $h^{-1}$, which places them into the middle level of accuracy. Even with a drift level of more than 100° $h^{-1}$ it is possible to use gyros of this type for solving some problems in the stabilization and control of highly maneuverable aerospace vehicles.

In addition to the above examples, there are numerous designs already declared in patents and also in various stages of development, but not yet released for commercial production, and so are absent from this review.

## REFERENCES

Aleshin B. S., K. K. Veremeenko, and A. I. Chernomorskij. 2006. *Orientation and Navigation of Moving Objects: Modern Information Technologies* (p. 424). Moscow: Fizmatlit. (In Russian.)

Alexeev K. B., and G. G. Bebenin. 1974. *Spacecraft Control*. Moscow: Mashinostroenie. (In Russian.)

Anfinogenov, A. S. 2002. et al. "Potentialities of an electrostatic gyro rotor with different structures of suspension." Materials of the 23rd Conference in memory of N. N. Ostryakov: St. Petersburg. (In Russian.)

Atkinson, J. L. 1974. Force Decoupler for Electrostatic Gyroscope Suspension System. US Patent No. 3, 785, 709.

Ayazi, F., H. H. Chen, F. Kocer, G. He, K. Najafi. 2000. "A high aspect-ratio polysilicon vibrating ring gyroscope." Solid State Sensor and Actuator Workshop, Hilton Head Island, South Carolina, June 4–8.

Ayazi F., and K. Najafi. 1998. "Design and fabrication of a high-performance polysilicon vibration ring gyroscope." Eleventh ASME International Workshop on Micro Electro Mechanical Systems. Heidelberg, Germany, January 25–29.

Ayazi F., and K. Najafi. 2000. "High aspect-ratio dry-release poly-silicon MEMS technology for inertial-grade microgyroscopes." Position Location and Navigation Symposium, San Diego, CA. pp. 304–8.

Bahrami, R. 1977. "Inertial attitude control of voyager spacecraft using dry tuned rotor gyro." AIAA Guidance and Control Conference, Hollywood, Florida, August 8–10.

Barbor N. et. al. 1996. "Micro-electromechanical instrument and systems development at Draper Laboratory." 3rd SPb International Conference of Integrated Navigation Systems. SPb.: CSPI "Electropribor"–Part 1. pp. 3–10.

Blanchard, R. L. 1978. "High accuracy calibration of electrostatic gyro strapdown navigation system." Jr. AIAA, Guidance and Control, Conference, Palo Alto, August 7–9, pp 130–6.

Bodunov, B. P., S. B. Bodunov, and V. M. Lopatin. 2001. In The 8th Saint Petersburg International Conference on Integrated Navigation Systems, Saint Petersburg, Russia.

Boltinghouse, J.C., and J. L. Atkinson. 1978. Electrostatic Pickoff System for Ball Gyros of the Electrostatic Levitation Type. U.S. Patent 4,074,580.

Branets V.N., and I. P. Shmuglevskiy. 1992. *Introduction to the Theory of Strapdown INS*. Nauka. (In Russian.)

Brozgul L.I., and E. L. Smirnov. 1970. *Vibrating Gyroscopes* (p. 216). (In Russian.)

Brozgul, L. I. 1989. *Dynamically Tuned Gyroscopes* (p. 280). (In Russian.)

Broxmeyer, Ch. 1964. *Inertial Navigation Systems*. New York: McGraw-Hill.

Bryan, G. H. 1890. "On the beats in the vibrations of a revolving cylinder of bell." *Proceedings of the Cambridge Philosophical Society* VII: 101–11.

Bryushkov, V. G., and Yu. G. Martynenko. 1978. "Drifts of an unbalanced gyroscope in anisoelasticity suspension." Izv. AN SSSR, MTT, 26 (in Russian.)

Bryzek, J., K. Peterson, and W. McCulley. 1994. "Micromachines on the march." // IEEE Spectrum, pp. 20–31. DOI: 10.1109/6.278394.

Burdess, J. S. and C. H. J. Fox. 1978a. "The dynamic of a multigimbal Hooke's joint gyroscope." *Journal of The Mechanical Engineering Science* 20 (5): 255–62. DOI: 10.1243/JMES_JOUR_1978_020_045_02.

Burdess, J. S. and C. H. J. Fox. 1978b "The dynamic of an imperfect multigimbal Hooke's joint gyroscope." *Journal of The Mechanical Engineering Science* 20 (5): 263–9. DOI: 10.1243/JMES_JOUR_1978_020_046_02.

Burns W. K., ed. 1994. *Optical Fiber Rotation Sensing*. Waltham, MA: Academic Press, Inc.

Bychkov, S. I. 1975. *Laser Gyro*. Moscow: Sovetskoe Radio. (In Russian.)

Clavelloux, N., and Bournault, J. 1970. Electrostatic Suspension Arrangements of Gyroscope Rotors. U.S. Patent 3,496,780.

Craig R. J. G. 1972a. "Theory of operation of an elastically supported tuned gyroscope." *IEEE Transactions on Aerospace and Electronic System* 8 (3): 280–8. DOI: 10.1109/TAES.1972.309510.

Craig R. J. G. 1972b. "Theory of errors of a multigimbal elastically supported tuned gyroscope." *IEEE Transactions on Aerospace and Electronic System* 8 (3): 289–97. DOI: 10.1109/TAES.1972.309511.

Davis J. L., and S. Ezekiel. 1978. "Techniques for shot-noise-limited inertial rotation measurement using a multiturn fiber sagnac interferometer." *Proc. SPIE* 157 (131). DOI: 10.1117/12.965477.

Davis S. A., and B. K. Ledgerwood. 1961. *Electromechanical Components for Servomechanisms*. New York: McGraw-Hill.

Djandjgava, G. I., G. M. Vinogradov, and V. I. Lipatnikov. 1998. The 5th Saint-Petersburg International Conference on Integrated Navigation Systems, Saint Petersburg, Russia.

Duncan, R. R. 1973. A strapdown inertial navigator using miniature electrostatic gyros. ION National Aerospace Meeting, Ramada Inn., Washington, DC.

Dyugurov, S. M., B. Ye. Landau. 1997. "Comparative analysis of systems for reading out angular data of strapdown electrostatic gyroscopes." *Gyroscopy and Navigation* 2 (17): 34–8. (In Russian.)

Edwards, Jr. A. (1971-1972), "The state of strapdown inertial guidance and navigation." *Navigation* 18 (4): 386–401.

Egarmin, N. E., and V. E. Yurin. 1993. *Introduction to Theory of Vibratory Gyroscopes*. Moscow: Binom Company.

Engelder P. D. 1980. "DRIMS—A redundant strapdown IMU for booster guidance cnd Control." *Proceedings of The IEEE 1980 National Aerospace and Electronic Conference*. NAECON, Dayton. New York, pp. 330–7.

Everitt, C. W. F. 1989. "The gravity-probe B relativity gyroscope experiment: Development of the prototype flight instrument." *Advances in Space Research* 9: 29–38. DOI: 10.1016/0273-1177(89)90005-7.

Evstigneev, M. I. 2004. "The problems of MEMS gyros design". *Gyroscopy and Navigation* 1 (44): 27–39. (In Russian.)
Ezekiel, S., and H. J. Arditty. 1982. *Fiber-Optic Rotation Sensors and Related Technologies*. Berlin: Springer Verlag.
Ezekiel, S., and S. R. Balsamo. 1977. "Passive ring resonator laser gyroscope." *Applied Physics Letters* 30: 478. DOI: 10.1063/1.89455.
Fell, P. 2001. "Angular rate sensor," US Patent # 6282958 (2001). (Assignee: BAE Systems, PLS, Farnborough, UK).
Fernandez, M., and G. Macomber. 1962. *Inertial Guidance Engineering*. Prentice-Hall.
Fountain, J. R. 2004. "Features and overview of the vibrating structure of a silicon gyroscope." NATO RTO lecture series 232 Advances in Navigation sensors and Methods of Integration (SET 064). CSRI 'Electropribor, St. Petersburg, May 26–28.
Friedland, D., and M. F. Hutton. 1978. "Theory and error analysis of vibrating-member gyroscope." *IEEE Transactions on Automatic Control* 23 (4): 545–56. DOI: 10.1109/TAC.1978.1101785.
Frigo N. J., H. F. Taylor, L. Goldberg, J. F. Weller, S. C. Rashleigh. 1983. "Otical Kerr effect in fiber gyroscopes: Effects of nonmonochromatic sources." *Optics Letters* 8 (2): 119–21. DOI: 10.1364/OL.8.000119.
Fox, C. H. J., and J. S. Burdess. 1980. "The dynamic of an imperfect Hooke's joint gyroscope." *Transactions on The ASME. Journal of Applied Mechanics* 47 (1): 161–6. DOI: 10.1115/1.3153596.
Geiger, W. 1997. "Improved rate gyroscope designs designated for fabrication by modern deep silicon etching." // Symposium Gyro Technology, Germany. pp. 0–2.8.
Geiger, W. et. al. 1998. "A silicon rate gyroscope with decoupled driving and sensing mechanisms MARS." RR // Symposium Gyro Technology, Germany.
Gelb A., and A. Sutherland Jr. 1967. "Design of strapdown gyposcopes for a dynamic environment." Semi-annual Report TR-101-1, The Analytic Sciences Corp.
Gelb, A., and A. Sutherland, Jr. 1968a. AIAA Paper №68-830, AIAA Guidance, Control and Flight Dynamics Conference, August 12–14.
Gelb A., and A. A. Sutherland, Jr. 1968b. "Design of strapdown gyroscopes for a dynamic environment." Interim Scientific Report TR-101-2. The Analytic Sciences Corp.
Gilmore, J. P. 1967. "A non-orthogonal gyro configuration." MIT Instrumentation Laboratory report, T-472.
Gilmore, J. P., and R. A. MacKern. 1972. "A redundant strapdown inertial reference unit (SIRU)." *Journal of Spacecraft and Rockets* 9 (1): 39–47. DOI: 10.2514/3.61628.
Gubarenko, S. I., et al. 1994. "Synthesis of transfer function of the electrostatic gyro suspension servo system." *Gyroscopy and Navigation* 2 (5): 21. (In Russian.)
Gusinsky, V. Z., B. Ye. Landau, and V. G. Peshekhonov. 1998. "An electrostatic gyroscope in a spacecraft inertial strapdown attitude reference system." *Proceedings of the 3rd Chinese-Russian Symposium*: Beijing, p. 104.
Hayworth, K. 2003. In NASA Motion Control Tech Briefs, October, pp. 70–2.
Hopkin, I. 1997. "Performance and design of silicon micromachined gyro." Symposium gyro technology, Stuttgart, Germany, September 16–17.
Hotate, K., and K. Tabe. 1987. "Drift of an optical fiber gyroscope caused by the Faraday effect: Experiment." *IEEE Journal of Lightwave Technology* 5 (7): 997–1001. DOI: 10.1109/JLT.1987.1075591.
Hung, J. C. 1972. "A study of strapdown platform technology." Elec. Eng. Dept., University of Tennessee, Scientific Rept, S-23, Chapter 3.
Hung, J. C., and G. B. Doane. 1972. "Progress in strapdown technology." Presented at the 15th Guidance and Control Panel Symposium of NATO/AGARD, Florence, Italy.
Hung T.C., and B. J. Doran. 1972. "High-reliability strapdown platforms using two-degree-of-freedom gyros." *IEEE Transactions on Aerospace and Electronic Systems* 9 (2): 253–9. DOI: 10.1109/TAES.1973.309793.

IEEE Std 1431. 2004. IEEE Standard Specification Format Guide and Test Procedure for Coriolis Vibratory Gyros, IEEE Aerospace and Electronic Systems Society, 20 December 2004; Annex A, Design features of Coriolis Vibratory Gyros, pp. 53–5 and Annex B, Coriolis Vibratory Gyros, pp. 56–66.

Izmailov, E. A., M. M. Kolesnik, A. M. Osipov, and A. V. Akimov. 1999., In The 6th Saint Petersburg International Conference on Integrated Navigation Systems, Saint Petersburg, Russia, pp. 6-1–6-9.

Janrua, A., P. Fantonby, and J. M. Caron. 2004. "A cheap, small size inertial measuring module of middle accuracy with vibration sensor for tactical applications." *Gyroscopy and Navigation* 1 (44): 48–58. (In Russian.)

Klinchuch, J. F. 1972. Electrostatic Gyroscope Suspension System. U.S. Patent 3,697,143.

Kucherkov, S. G. 2002. "Integration features of a vibration MEMS gyro with resonance adjustment used as a sensor of angular velocity construction." *Gyroscopy and Navigation* 2: 12–18. (In Russian.)

Landau, B. E. 1993. "A Solid-rotor electrostatic gyroscope." *Gyroscopy and Navigation* 1: 6. (In Russian.)

Landau, L. D., and E. M. Lifshitz. 1976. *Mechanics of Course of Theoretical Physics* (3rd edition, Section 27, p. 80). Oxford: Pergamon Press.

Landau, B. E., et al. 2000. "A solid-rotor electrostatic gyroscope for strapdown navigation and attitude reference systems." *Gyroscopy and Navigation* 4 (31): 50. (In Russian.)

Landau, B.E., et al. 2001. "Spacecraft attitude reference system based on strapdown electrostatic solid-rotor gyroscopes." *Gyroscopy and Navigation* 3 (34): 63. (In Russian.)

Lefevre, H. 1993. *The Fiber-Optic Gyroscope*. Norwood, MA: Artech House.

Leger, P., and R. Bihan. 1984. "Le gyroscope suspension electrostatice." *Navigation* (Fr.), Avril, 126: 223–38.

Lestev, A. M., I. V. Popova, E. N. Piatishev, and V. Y. Raspopov. 1999. "Development and research of a micromechanical gyroscope." *Gyroscopy and Navigation* 2 (25): 3–10 (in Russian).

Lin, S., and T. G. Giallorenzi. 1977. *Applied Optics* 18 915. DOI: 10.1364/AO.18.000915.

Lin, S., and T. G. Giallorenzi. 1979. "Sensitivity analysis of the Sagnac-effect optical-fiber ring interferometer." Applied Optics 18 (6): 915–31. DOI: 10.1364/AO.18.000915.

Looez-Higuera, J. M. Ed. 2002. *Handbook of Optical fibre sensing Technology*. John Wiley & Sons, Ltd.

Loper, E. J., and D. D. Lynch. 1979. Sonic Vibrating Bell Gyro. U.S. Patent 4,157,041, June 5.

Loper, E. J., and D. D. Lynch. 1984. In Proceedings of the National Technical Meeting of The Institute of Navigation, San Diego, CA, USA, January 17–19.

Loper, E. J., and D. D. Lynch. 1990. Vibratory Rotation Sensor. U.S. Patent 4,951,508, August 28.

Loper, E. J., D. D. Lynch, and K. M. Stevenson. 1986. In Proceedings of the 1986 Position, Location and Navigation Symposium (PLANS '86), Las Vegas, NV, November 4–7.

Lynch, D. D. 1972. Bell Gyro and Improved Means for Operating Same. U.S. Patent 3,656,354, April 18.

Lynch, D. D. 1973. Rotating-Wave Rotation Detector and Method of Operating Same. U.S. Patent 3,719,074, March 6.

Lynch, D. D. 1987. In Proceedings of the 43rd Annual Meeting of The Institute of Navigation, Dayton, OH, USA, 23–25 June, pp. 34–7.

Lynch, D. D. 1995. In The 2nd Saint Petersburg International Conference on Gyroscopic Technology and Navigation, Saint Petersburg, Russia, pp. 26–34.

Lynch, D.D. 1998. In Symposium Gyro Technology Stuttgart, Germany, pp. 1.0–1.14.

Magnus, K. 1971. *Kreisel: Theorie und Anwendungen*. Berlin-Heidelberg-New York: Springer-Verlag.

Martini, G. J. 1987. "Analysis of a single-mode optical fibre piezoceramic phase modulater." *Optical and Quantum Electronics* 19 (3): 179–90. DOI: 10.1007/BF02030653.

Martynenko, Yu. G. 1988. *Motion of a Solid Body in Electrostatic and Magnetic Fields* (p. 368). Moscow: Nauka. (In Russian.)

Matthews, J. B., and G. R. Taylor. 1969. Development program for a general purpose computer and strapdown inertial measurement unit, AIAA Guidance, Control and Flight Mechanics Conference, AIAA Paper No. 69–850.

Matveev, V. A., V. I. Lipatnikov, A. V. Alekhin 1998. "Design of the solid-state wave gyro." Moscow State Technical University named after N.E. Bauman, Moscow. (In Russian.)

Maunder, L. 1979. "Dynamically tuned gyroscope." Proceedings of The First World Congress on Theory of Machines and Mechanism, Montreal, 1979. New York, pp. 470–3.

Meyer R.E., S. Ezekiel, D. W. Stone, and J. Tekippe. 1983. Passive fibre-optic ring resonance for rotation sensing." *Optics Letters* 8 (19): 644–6.

Ormandy, D., and L. Maunder. 1973. "Dynamics of the oscillogyro." *Journal of the Mechanical Engineering Science* 15 (3): 210–17. DOI: 10.1243/JMES_JOUR_1973_015_036_02.

Pelpor, D.S., V. A. Matveev, V. D. Arsenyev. 1988. Dynamically Tuned Gyroscopes. p. 262. (In Russian.)

Peshekhonov, V. G. 2003. "Gyroscopes of the early 21st century." *Giroskopiya i Navigatziya (Gyroscopy and Navigation)* 4 (43): 5. (In Russian.)

Petersen, Kurt E. 1982. "Silicon as a mechanical material." *IEEE* 70 (5): 420–57.

Loveday, Philip W., and Rogers, Crayd A. 1998. "Modification of Piezoelectric Vibratory Gyroscope Rezonator Parameters by Feedback Control." *IEEE Transactions on Ultrasonics, Ferroelectrics, and Frequency control* 45 (5): 1211–15. DOI: 10.1109/58.726445.

Pondrom, W. L. 1984. "Electrostatically suspended gyroscope." *IEEE Transactions on Aerospace and Electronic System* 20 (4): 422–4.

Putty, M. W., and K. Najabi. 1994. "A micromachined vibrating ring gyroscope." in Proc. Digest, Solid-state sensors and Actuators Workshop, Hilton Head, pp. 213–20.

Raspopov, V. I. 2008. Inertial sensitive elements. Part 1. *Gyroscopes—The Avionics World*, №1. p. 52–9. (In Russian.)

Raspopov V. Y., and V. V. Savelyev. 1979. *Constructive Schemes and Elements of the Theory of Vibrating Gyroscopes* (p. 76). Tula. (In Russian.)

Raspopov, V. Y. 2007. *Micromechanical Gyros* (p. 399). Moscow: Mashinostroenie. (In Russian.)

Raushenbah, B. V., and B. N. Tokar. 1974. *Spacecraft Attitude Control*. Nauka. (In Russian.)

Roantree, P. J., and N. J. Kormanik. 1966. A generalized design criterion for strapped-down inertial sensor loops. AIAA/JACC Guidance and Control Conference.

Sagnac, G. (1913) "L'éther lumineux demontré par l'effet du vent relative d'éther dans un interféromètre en rotation uniforme." *Comptes Rendus de l'Academie des Sciences* 95: 708–10.

Sanders, G. A. 1992. "Fiber optic sensors, critical reviews of optical science and technology." CR44: 133–59.

Sanders, G. A., B. Szafraniec, R.-Y. Liu, C. Laskoskie, and L. Strandjord. 1996. "Fiber optic gyros for space, marine, and aviation applications." *Proc. SPIE* 2837: 61. DOI: 10.1117/12.258208.

Savet, P. H. 1961. *Gyroscopes: Theory and Design*. New York: McGraw-Hill.

Scoville, A. E. 1968. A Comparison of Strapdown and Platform Inertial System Performance Limitations, Proceedings of the ION National Space Meeting.

Scoville, A. E., and J. Yamron. 1966. The mechanization of a strapdown inertial system based on time-modulated torquing. AIAA/JACC Guidance and Control Conference.

Severov, L. A., V. K. Ponomarev, A. I. Panferov. 1998. "Micromechanical gyros: Construction and characteristics, technologies and development methods." *Isvestiya Vusov—Priborostroenie* 1: 57–73. (In Russian.)

Severov, L. A., B. K. Ponomarev, A. I. Panferov et al. (2003), "Information characteristics of micromechanical gyroscopes." *Gyroscopy and Navigation* 1: 76–82. (In Russian.)

Shupe, D. M. 1981. "Fiber resonator gyroscope: Sensitivity and Thermal Nonreciprocity." *Applied Optics* 20 (2) : 286–9. DOI: 10.1364/AO.20.000286.

Smith, S. G. 1979. "Strapped-down system and gyroscope technology." *The Journal of Navigation* 32 (1): 91–101. DOI: 10.1017/S0373463300033154.

Strandjord, L. K., and G. A. Sanders. 1991. "Resonator fiber optic gyro employing a polarization-rotating resonator." *Proc. SPIE* 1585: 163.

Szafraniec, B., J. Feth, R. Bergh, and J. Blake. 1995. *Proc. SPIE* 2510: 37. DOI: 10.1117/12.221716.

Taverna, M. A. 2004, July 19. *Aviation Week and Space Technology* p. 124.

Ulrich, R. 1980. "Fiber-optic rotation sensing with low drift." *Optics Letters* 5 (5): 173–5. DOI: 10.1364/OL.5.000173.

Ulrich, R., and M. Johnson. 1976. "Fiber-ring interferometer: Polarization analysis." *Optics Letters* 4 (5): 152–4. DOI: 10.1364/OL.4.000152.

Vavilov, V. D. 2003. *Integrated Gauges* (p. 503). Moscow: Nizhni Novgorod. (In Russian.)

Vlasenko, A. 2003. "Integrated gyroscopes IMEMC: An angular velocity sensor by analog devices." *Electronic Components* 2: 57–9. (In Russian.)

Vlasov, J. B., and O. M. Filonov. 1980. *Rotor Vibrating Gyroscopes in Navigation Systems* (p. 224). (In Russian.)

Weinberg M. et. al. 1995. "A micromachined comb drive tuning fork gyroscope for commercial applications." 2 S Pb. International Conference of Gyroscopic Technology and Navigation: CSPI "Electropribor"– Part 2. pp. 79–87.

Yatsenko, Y. A., S. F. Petrenko, V. V. Chikovani. 2000. In The 7th Saint Petersburg International Conference on Integrated Navigation Systems, Saint Petersburg, Russia, pp. 199–201.

Yemelyantzev, G. I. et al. 2004. In-flight Calibration of Strapdown Spacecraft Attitude Reference System by the Data from Stellar Sensor Proceedings of the 11th International Conference: St. Petersburg, p. 127. (In Russian.)

Zbrutskij, A. V., and M. A. Pavlovsk. 1981. "Dynamically tuned gyroscope in the conditions of spatial motions of the basis." *Mechanics of a Firm Body* 1: 16–26. (In Russian.)

Zhang, W., R. Baskaran, and K. L. Turner. 2002. "Effect of cubic nonlinearity in auto-parametrically amplified resonant MEMS mass sensor." *Sensors and Actuators A* 102: 139–50. DOI: 10.1016/S0924-4247(02)00299-6.

Zhuravlev, V. Ph. 1997. Controllable Foucault's pendulum as a model for a class of free gyroscopes. *Mechanics of Solids* 32: 21–8.

Zhuravlev, V. Ph., and D. M. Klimov. 1985a. *Hemispherical Resonator Gyro*. Moscow: Nauka. (In Russian.)

Zhuravlev, V. Ph., and D. M. Klimov. 1985b. *The Wave Solid-State Gyroscope*. Moscow: Nauka. (In Russian.)

Zhuravlev, V. Ph., and D. D. Lynch. 1995. "Electrical model of hemispherical resonator gyro." *Izvestiya Akademii Nauk—Mechanics of Solids* 5: 12–24.

CHAPTER 7

# Compasses

Konstantin K. Veremeenko
*Moscow Aviation Institute (State Technical University), Russia*

## 7.1 INTRODUCTION

The problem of determining a direction of movement or a direction to a desired point on the Earth has been known from very ancient times, and observation of the stars was probably the first approach to solving the problem. It is well known that to determine North may be accomplished by observing the North Star (Polaris) in the Northern Hemisphere, or in the Southern Hemisphere by observing two stars in the Southern Cross constellation (Crixa). Over the ages, mankind has sought and improved methods of solving this problem and has created special devices, to wit, compasses—instruments that point in the direction of the magnetic North Pole, which is near to the true North Pole but has a position that varies slowly with time. The directions of magnetic meridians, and hence geographic meridians, allow the determination of directional bearings on the ground.

In the Middle Ages, during the epoch of the great geographical discoveries, the magnetic compass was the main instrument for enabling sailors to remain on course in the open ocean. Later, other physical principles were used to create mechanical gyro-compasses, electronic radio-compasses (Aczel 2001; Gurney 2004), and the electronic magnetometers that now find wide usage in aerospace applications.

To define directions on the Earth some form of datum must be established. One of the simplest and most natural is the Earth's direction of spin. This direction has long been defined as being to the east, or anticlockwise when viewed from over the North Pole. The north–south direction corresponds to a line that is perpendicular to any east–west line. So, when one stands facing east, the North Pole is to the left and the South Pole to the right. These four directions—north (N), south (S), east (E), and west (W)—are the *Cardinal Points*. The four midway directions between north, east, south, and west are north-east (NE), south-east (SE), south-west (SW), and north-west (NW). These are the *Quadrantal Directions* (Figure 7.1) (Abanave 2005).

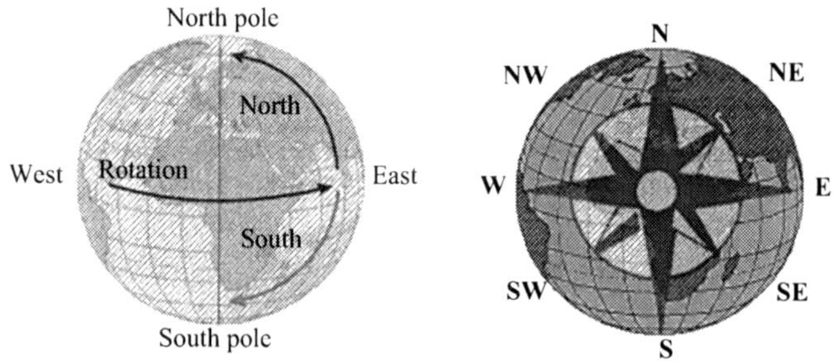


**Figure 7.1.** Earth's rotation, cardinal, and quadrantal points.

**Figure 7.2.** The Sexagesimal system.

This system of cardinal and quadrantal directions was widely used over a long period, but the development of civilization demanded the introduction of more exact ways of direction definition. New subdivisions were made and the *Sexagesimal system* was eventually suggested for measuring directions (Figure 7.2) with more precision.

In the sexagesimal system a full clockwise turn from north back to north through points east, south, and west is equal to 360°. North is defined as 0°, so consequently 90° corresponds to east, 180° corresponds to south, and 270° corresponds to west. Any angle on such a scale is referred to as an *azimuth*.

On a map, or on the Earth's surface, *true* azimuths are counted from the true northern directions of the geographic meridians (Figure 7.1), which are imaginary lines on the Earth's surface drawn from the true (or geographic) North Pole to the true geographic South Pole. *Magnetic azimuths* are measured from the northern direction of *magnetic meridians*. A magnetic meridian is a projection of a line pointing to the magnetic North Pole on the Earth's surface. Magnetic meridians represent complex curves that converge at the northern and southern magnetic poles.

The compass is an instrument for indicating the directions of the magnetic—and hence the geographic—meridian, and for measuring the angle in a horizontal plane between this direction and the longitudinal axis projection of the object upon which the compass is mounted. This angle, measured in degrees clockwise from north is named the *heading* and can be displayed as the true (geographic), the magnetic, or the gyro heading.

Today, compasses and magnetometers play a very important rôle in aerospace practice. Magnetic compasses themselves are now used mainly in aircraft for backup or reference navigation and trajectory control. In space applications magnetometers are widely used to measure the Earth's magnetic field for scientific and applied purposes.

In modern times, compasses are classified as magnetic, mechanical (gyro), radio, electronic, and celestial. In this chapter, attention is given to magnetic, electromagnetic, and electronic compasses, mechanical (gyro) compasses having been considered in Chapter 6.

## 7.2 MAGNETIC COMPASSES

### *7.2.1 BRIEF HISTORICAL SKETCH*

There is a rich store of experience in the use of aerospace applications of compasses and magnetometers stretching back over a century, and indeed the first use of a compass appeared more than two thousand years ago in China. The first compass consisted of a piece of magnetic iron ore (magnetite) in the form of a spoon with a thin shank and a spherical, carefully polished convex part. This "spoon" was balanced at the center of a polished bronze, copper, or wooden plate with 24 divisions marked on it. Thus, the spoon could easily rotate around its axis. After being given a slight perturbation, the "spoon" would settle down so that its shank pointed to the south. This was the most ancient device for direction definition (Figure 7.3), and was the progenitor of the magnetic needle.

It is known that in eleventh-century China a floating compass needle made of an artificial magnet existed. Usually it was made in the form of a small fish that was lowered into a vessel containing water. This freely-floating device was designed so that its head pointed south. Several versions of such compasses were devised by eleventh-century Chinese scientists, who actually performed considerable work on magnetic needle properties. One method for producing an artificial magnet was to magnetize a normal sewing needle with the help of a natural magnet and then to attach its center by means of wax to the top of a case via a freely hanging silk string. This compass was more precise than the floating version because it experienced a much smaller resistance to turning. Another design was even close to the modern version, a magnetized needle being pinned on to a stud.

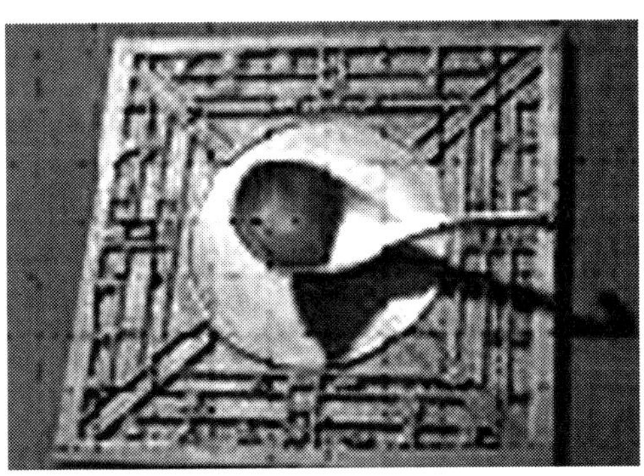

**Figure 7.3.** Traditional-style Chinese compass.

During such experimental work, it was learned that a compass needle did not actually point exactly to the south but exhibited some variation. Thus, magnetic variation was revealed and the ancient scientists learned to how to calculate this angle for various areas in China.

In the eleventh century many Chinese ships were equipped with floating compasses. Usually they were installed at both the prow and the stern of the ships so that in any weather captains could hold a correct heading, taking into account both readings. In the twelfth century similar kinds of Chinese compass were adopted by the Arabs.

At the beginning of the thirteenth century the "floating needle" became known to Europeans. Italian seamen were the first to adopt it from the Arabs, and from them the compass passed on to Spaniards, Portuguese, Frenchmen, and later to the Germans and English. At first the compass consisted of a magnetized needle and a lamina of wood floated in a water-filled vessel, and in due course the vessel was enclosed by glass to protect the float from the wind.

In the middle of the fourteenth century Europeans modernized the compass and placed a magnetic needle on a support in the middle of a paper circle—the *compass card.* Later, Italian Flavio Julio improved this design, having supplied it with a compass card divided into 16 parts with four divisions between each cardinal point. This simple adaptation represented a great step in compass improvement, and eventually the circle was divided into 32 equal sectors. During the sixteenth century, in order to reduce the influence of oscillations (for example, pitching and rolling at sea) gimbals were put into practice (Figure 7.4). A century later a mariner's compass was supplied with a rotating diametrical ruler that allowed more precise direction determination.

At the beginning of the twentieth century liquid compasses were created (Figure 7.5). In such compasses magnetic needles are located in a hermetically sealed flask containing a nonfreezing liquid. The needle is able to come to an instant stop after arriving in parallel with a magnetic meridian. Compasses of this kind were used by the early pilots undertaking long-distance flights at the beginning of aeronautics.

Being equipped with compasses, Spanish and Portuguese seamen ventured on long voyages at the end of fifteenth century. They left the sea coasts to which navigation had been referred over several millennia and embarked on voyages through the open ocean. For centuries this instrument remained one of the main means of navigation and, having undergone a number of changes, became one of the basic devices on board aircraft at the beginning of the twentieth century. It remains one of the obligatory devices in modern aviation instrument panels.

The invention of the compass resulted in a revolution not only in navigation but in allied fields—for example, it was the first instrument that allowed the accurate paving of a route. The

**Figure 7.4.** A gimbaled compass.

**Figure 7.5.** Pyser-SGI Francis Barker M73 black liquid prismatic compass (Mils Model).

importance of compasses is explained by their remarkable properties, which are in turn defined by the characteristics of the Earth's magnetic field.

### *7.2.2 THE EARTH'S MAGNETIC FIELD*

The Earth is a huge spherical magnet and like any classical dipole has two magnetic poles (Figure 7.6) which are not collocated with the geographic ones (The American Practical Navigator 2002; Williams 1992). The positions of the magnetic poles on the Earth's surface are not constant but exhibit small annual movements within the planet (Figure 7.7). This fact is obviously of major importance for measuring true headings using the magnetic compass.

Taking into account presence of two poles, it is possible to construct a grid of meridians relative to the magnetic north and south poles on which it is possible to define a magnetic north direction (Figure 7.8).

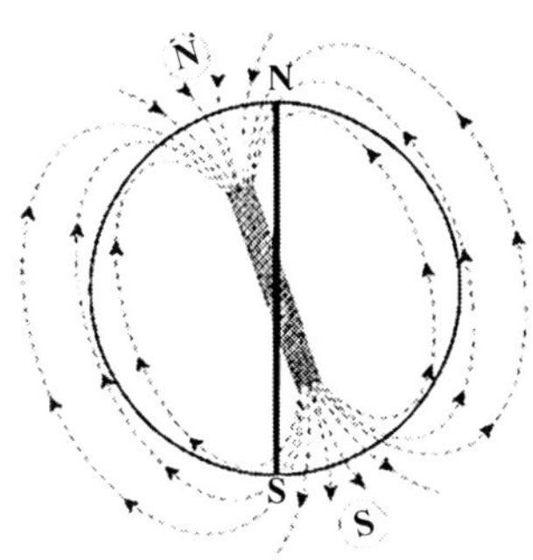


**Figure 7.6.** The Earth's magnetic poles.

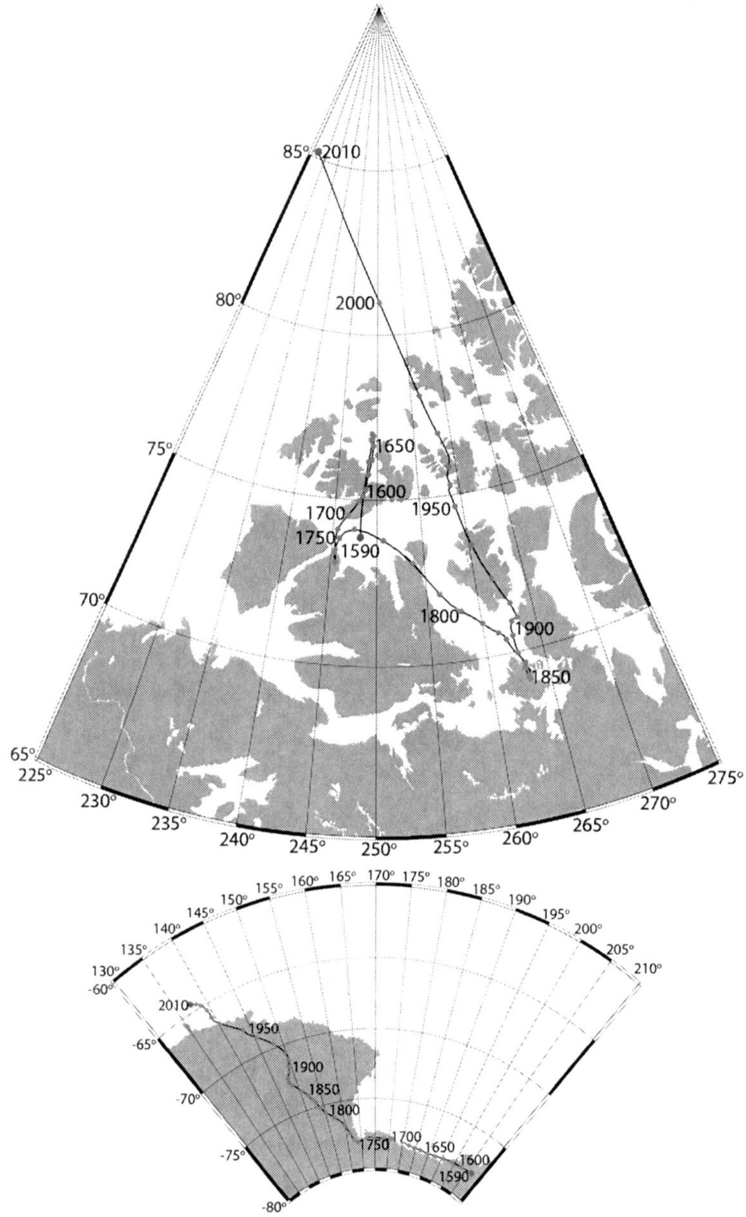


**Figure 7.7.** Magnetic poles movements. (a) North pole (b) South pole.

Figure 7.9 shows that magnetic and true headings cannot, in general, be equal, and that the magnetic needle is deflected from true north to form the angle of *magnetic declination.* Magnetic declination is considered positive if the northern end of a magnetic needle is deflected to the east from a geographical meridian and negative if to the west. Values of magnetic declination are available on magnetic compass cards and are used for the calculation of true headings using magnetic compass readings:

$$\psi_T = \psi_M + D \tag{7.1}$$

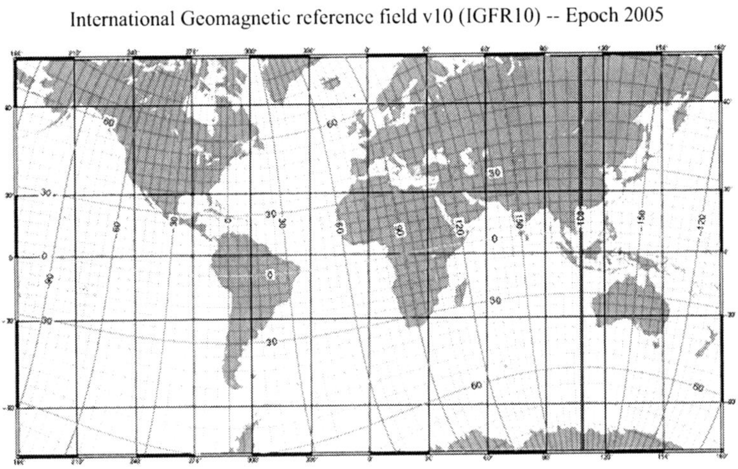


**Figure 7.8.** Geomagnetic coordinates.

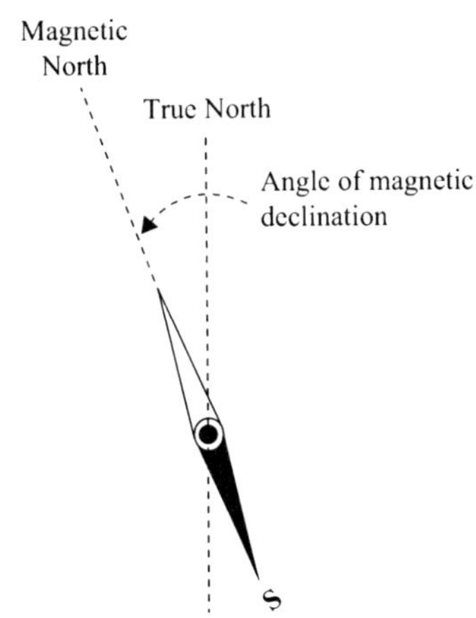


**Figure 7.9.** Magnetic declination.

where $\psi_T$ is true heading, $\psi_M$ is magnetic heading, and $D$ is magnetic declination. In aeronautical and other forms of navigation, the term *magnetic variation* is more often used than magnetic declination.

Actually, the structure of the Earth's magnetic field is much more complicated than the field of a simple dipole, being not perfectly uniform, but irregular. Heterogeneity in the distribution of magnetic mass in the body of the planet and external geomagnetic perturbations lead to deformations in the magnetic field of the dipole. That is why the magnetic field must be measured in many places to get a satisfactory picture of its distribution.

The Earth's magnetic field may be characterized by a vector of intensity $\boldsymbol{F}$. After many long-term observations, it has been ascertained that the Earth's magnetic field originates from three main sources so that:

$$\boldsymbol{F} = \boldsymbol{F}_0 + \boldsymbol{F}_1 + \boldsymbol{F}_e \tag{7.2}$$

The dominant component $\boldsymbol{F}_0$ is the so-called "main" or "core" field, generated by a hydrodynamic dynamo operating in the Earth's fluid outer core. The second contribution $\boldsymbol{F}_1$ comes from

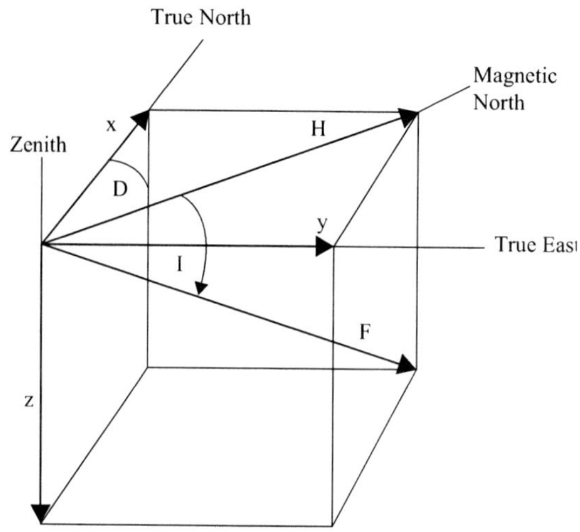


**Figure 7.10.** Magnetic field elements.

magnetized rocks in the Earth's lithosphere. Core and crustal fields together are denoted as "internal" fields because their sources are within the Earth. "External" fields $\boldsymbol{F}_{\mathbf{e}}$ are due to electric currents in the ionosphere and magnetosphere caused by the interaction of the Earth's main field with the sun. Finally, the time-varying external fields produce secondary, induced currents in the Earth's interior, which in turn cause a secondary, induced magnetic field. However, the largest part of the magnetic field at the Earth's surface comes from the sources inside the Earth.

The spatial orientation of the Earth's magnetic field intensity vector $\boldsymbol{F}$ is very important because a freely suspended compass needle aligns itself along this direction. Figure 7.10 shows how the earth's magnetic field ($\boldsymbol{F}$) is composed of a vertical ($\boldsymbol{Z}$) and a horizontal ($\boldsymbol{H}$) component.

The magnetic field is perfectly vertical at the magnetic poles (i.e., the horizontal intensity is zero) and perfectly horizontal at the magnetic equator (i.e., the vertical intensity is zero). Everywhere else there is a combination of vertical and horizontal components. The angle between the magnetic field intensity vector and the horizontal component is known as the *inclination I*, or *magnetic dip angle*. Magnetic declination $D$ is defined as the angle between the horizontal component of the magnetic field intensity vector and true north (the geographical meridian). A good compass should point in the direction of the horizontal component of the magnetic field where the compass is located.

An easy way to calculate magnetic declination and other components of the magnetic field is by reference to magnetic field mathematical models whose parameters are based on an analysis of magnetic observations over all or part of the entire world. The most common method used for producing global models is spherical harmonic analysis. The International Geomagnetic Reference Field (IGRF) and the World Magnetic Model (WMM) are the most commonly used models for navigational purposes (Barton 1997; Chapman and Bartels 1940; National Geophysical Data Center 2005; McLean et al. 2004, 2005). Because of the variability of the Earth's magnetic field these models are traditionally updated every five years. The models are global, but there are also models of the magnetic field over some specific geographical regions. Because the latter analyses were carried out over smaller regions, such models can reproduce smaller spatial variations in the magnetic field than can the IGRF. It is generally agreed that the IGRF achieves an overall accuracy of better than 1° in declination and the accuracy is better than this in densely surveyed areas such as Europe and North America, and worse in oceanic areas such as the South Pacific. The accuracy of all models is worse in the Arctic near the north magnetic pole. Nowadays, magnetic field models are used to calculate magnetic declination via computer programs wherein the user simply has to input data (latitude, longitude, year) to calculate the declination.

As long ago as in 1838 the famous German mathematician, astronomer, and physicist C.F. Gauss developed a method of representing the Earth's magnetic field. This method is based upon representing the Earth's scalar magnetic potential $V$ as a converging series, the terms of which are functions of latitude, longitude, and radial distance from the center of the Earth (Backus, Parker, and Constable 1996). Most modern models still use this method and in modern notation appear thus:

$$V = a\sum_{n=1}^{N_{max}}\left(\frac{a}{r}\right)^{n+1}\left[g_n^m \cos(m\varphi) + h_n^m \sin(m\varphi)\right]P_n^m(\theta), \tag{7.3}$$

where $\varphi$ refers to longitude, $\theta$ refers to latitude, $r$ is the radial distance from the center of the Earth, $a$ is a reference radius, $n$ is the degree of the term, and $m$ is the order of the term. It is now recommended that the World Geodetic System 1984 (WGS-84) datum and spheroid be used for coordinate transformations rather than the International Astronomical Union 1966 (IAU-66) spheroid previously recommended. Differences in output IGRF magnetic field values at the Earth's surface are less than 1 nT when this spheroid change is made. The parameter $a$ for the WGS-84 is $a = 6378.137$ km. The $P_n^m$ are called associated Legendre polynomials which look very much like distorted sine waves. The $g_n^m$ and $h_n^m$ are called Gauss coefficients which are determined through a least-squares analysis of a world-wide distribution of magnetic observations.

In theory the series goes to infinity but in practice some maximum degree, $N_{max}$ is chosen so that the series is able to reproduce the observed field to the desired resolution and accuracy. For example, for the IGRF, $N_{max} = 10$. To reproduce the field originating within the core of the Earth requires $N_{max} = 15$. To reproduce crustal anomalies visible in magnetic data at satellite altitudes requires $N_{max} = 80$.

The magnetic field components $X$, $Y$, and $Z$ (Figure 7.10) can be calculated from the scalar potential $V$ (Equation (7.3)) through the following partial derivatives (Davis 2004):

$$X = \frac{1}{r}\frac{\partial V}{\partial \theta} \qquad Y = \frac{1}{r\sin\theta}\frac{\partial V}{\partial \varphi} \qquad Z = \frac{\partial V}{\partial r}, \tag{7.4}$$

where the partial derivatives are calculated according to the following:

$$B_r = \frac{-\partial V}{\partial r} = \sum_{n=1}^{k}\left(\frac{a}{r}\right)^{n+2}(n+1)\sum_{m=0}^{n}(g_n^m \cos m\varphi + h_n^m \sin m\varphi)P_n^m(\theta),$$

$$B_\theta = \frac{-1}{r}\frac{\partial V}{\partial \theta} = -\sum_{n=1}^{k}\left(\frac{a}{r}\right)^{n+2}\sum_{m=0}^{n}(g_n^m \cos m\varphi + h_n^m \sin m\varphi)\frac{\partial P_n^m(\theta)}{\partial \theta},$$

$$B_\varphi = \frac{-1}{r\sin\theta}\frac{\partial V}{\partial \varphi} = \frac{-1}{\sin\theta}\sum_{n=1}^{k}\left(\frac{a}{r}\right)^{n+2}\sum_{m=0}^{n}m(-g_n^m \sin m\varphi + h_n^m \cos m\varphi)P_n^m(\theta). \tag{7.5}$$

The components of the Earth's magnetic field—declination, inclination, and total intensity—can be computed from the orthogonal components (Figure 7.10) using Equations (7.3)–(7.5) according to the equations:

$$D = \mathrm{arctg}(Y / X) = \mathrm{arctg}(X / Y), \quad I = \mathrm{arctg}\,(Z/H), \quad F = \sqrt{X^2 + Y^2 + Z^2}, \tag{7.6}$$

where $H = \sqrt{X^2 + Y^2}$.

The results of such calculations appear on maps and, in particular, lines of equal declination (*isogonic lines*) are extremely important for calculating true headings (Equation (7.1)) using magnetic compass readings. An example of a global map with lines of equal declination is shown in Figure 7.11.

### 7.2.3 MAGNETIC COMPASS DESIGN PRINCIPLES AND ERRORS

Modern aviation magnetic compasses are subdivided into combined and remote types. A combined compass has a sensitive system and a heading readout system combined in one case, and in a remote compass readings from the sensitive system are transferred to an indicator located some distance from it. Usually, *fluxgate compasses* are used as remote devices using magnetic field induction sensors. Such compasses will be described in Section 7.3.

The basic requirement of the modern aviation magnetic compass is the measurement of the horizontal component direction of the Earth's magnetic field (Figures 7.10 and 7.11) that defines the direction relative to magnetic north. In this sense the principle of the present day magnetic compass is no different from that of the ancient compasses. The advantages of contemporary magnetic compasses over ancient ones result from a better knowledge of the characteristics of Earth's magnetic field and the laws of physics that determine the behavior of the compass needle, and from greater precision due to better design and construction.

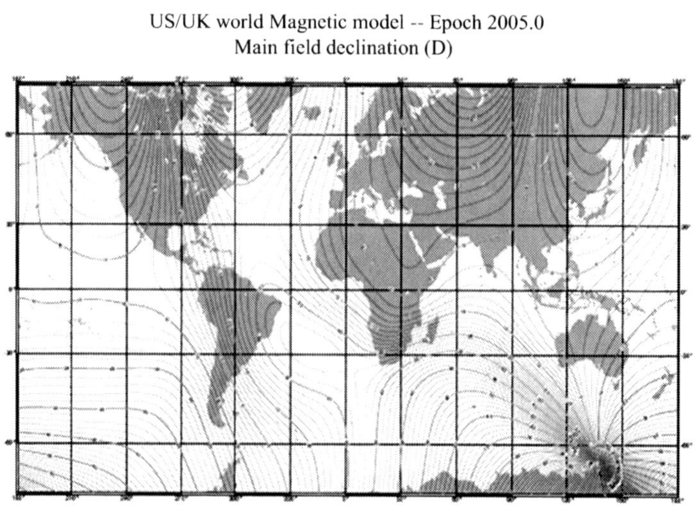


**Figure 7.11.** Main field declination.

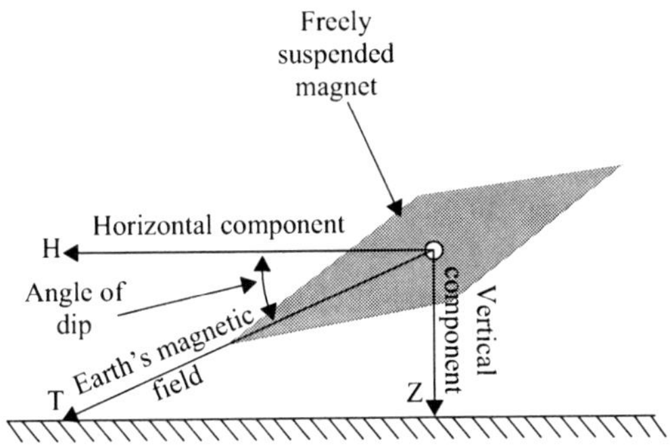


**Figure 7.12.** Compass needle positioning.

To achieve an accurate measurement of the horizontal component (***H***) of the Earth's magnetic field (Figure 7.12) the needle must be hung pendulously so that its weight offsets most of the effects of dip angle. It is this pendulous arrangement that becomes disturbed during turns and accelerations and leads to the well known *turning and acceleration errors* in aviation (see Figure 7.12).

Instead of a magnetized needle, the modern magnetic aeronavigation compass contains a special mobile magnetic system that can rotate in the horizontal plane to the direction of the horizontal component ***H*** of the Earth's magnetic field so that it displays the direction of the magnetic meridian via a scale. This scale is fixed to the mobile magnetic system and a course (or *lubber*) line is printed on the case window of the device. The frictional moments in the magnetic system support are decreased by means of a float rigidly fixed to the magnetic system and within a liquid that fills the case of the device. This liquid not only reduces the pressure of the magnetic system on the support, but also damps pendulum swings via viscous friction. In some devices a dry suspension bracket is used, and damping of pendulum oscillations is achieved by electric currents induced in the case by the movements of the magnetic system. A deviation unit (or *compensator*) is installed on the case of the compass that includes a system of permanent magnets whose fields compensate any permanent magnetic fields within the vehicle.

Under real working conditions such a compass will exhibit errors and deviations. The *absolute error* is defined as the difference between the compass reading and the true north direction; magnetic declination (variation) and inclination being the two main factors that affect its magnitude.

Magnetic declination can introduce large errors and the correction factor varies depending on where the compass is located, but by using a compensation table based on the Earth's magnetic field model (McLean et al. 2005) this error can be eliminated.

The inclination or magnetic dip angle might be a source of error if compasses are not able to accurately determine the horizontal component of the magnetic field. In high latitudes, and especially in the vicinity of the magnetic poles, the effect of this factor is much greater because the vertical component of the Earth's magnetic field can be larger than the horizontal component in these areas. To reduce the effects of inclination, special modifications are employed in modern compasses. Two main approaches are used for compensating such effects. One is to place the sensor in gimbals to position the magnetic element horizontally. A second approach is to keep the magnetic sensor fixed on the vehicle and to use a tilt sensor to measure the inclination of the sensor with respect to the Earth's horizon and then calculate corrections using a compensation algorithm.

*Relative errors* in magnetic compasses need to be considered over rather small areas of about 10–20 km when some reference direction is to be established. In this case, the effects of inclination will not play an important rôle because their influence will be approximately the same at all points in such small areas.

There are a few other factors that can affect the performance of a compass, one being local magnetic interference that may produce deviations in the Earth's magnetic field and consequently erroneous compass readings. Such interference could cause large errors and most of the effects are unpredictable. All sources of such deviations can be divided into natural and artificial (man-made). One of the main natural sources of deviation is solar activity, which can have a major impact on the Earth's magnetic field via solar magnetic storms. Artificial deviations include magnetic fields emanating from functioning electromagnetic devices and metal structures located near the compass.

Northern turning error is a result of the action of the vertical component of the Earth's magnetic field on the compass magnetic system when an aircraft turns on northern or southern courses. This causes the compass card to deviate from the plane of the magnetic meridian. The amount of turning error depends on the roll angle of the aircraft during the turn and the magnitude of the magnetic inclination at the specific location. Northern turning errors can be compensated by calculated corrections that are functions of roll angle, and of magnetic field and coordinate parameters. However, there is a simple rough practical method of compensating for this error. When performing a turn on a northern course, the turn should be completed when the compass indication is equal to the difference between the desired heading and the roll angle. For turns on a southern course, the turn should be completed when the compass reading is equal to the sum of the desired heading and the roll angle. On courses of 90° and 270° northern turning error is equal to zero because the vertical component coincides with the Earth's magnetic meridian plane. After a turn, the influence of the vertical component of the Earth's magnetic field ceases, and accurate compass readings are restored.

Compass deviation also arises as a result of the influence of magnetic fields within an aircraft and may result in a deflection of the compass card, which aligns itself along the so-called *compass meridian* at some angle from the magnetic meridian. Consequently, the angle between the magnetic and compass meridians becomes another compass deviation. Deviation is traditionally considered eastern (positive) $+\Delta_K$ if a compass meridian is to the right from a magnetic meridian, and western (negative) $-\Delta_K$ if a compass meridian is situated to the left from a magnetic meridian.

Modern aviation compasses designed according to statutory principles have deviations significantly dependent on the heading of the aircraft; a general view of the main error components is shown in Figure 7.13.

The magnitude and nature of deviation depends basically on an aircraft's magnetic field created by both soft and hard magnetic iron, and also by various sources of electric fields that generate semicircular and quarter deviations.

Semicircular deviations arise under the action of the magnetostatic field created by an aircraft's hard magnetic iron which, with a large coercive force, has the properties of a permanent magnet. When a vehicle heading is changing, the magnetostatic field caused by hard magnetic iron maintains a constant magnitude and direction over the three axes of the vehicle. Semicircular deviation is named so because when an aircraft turns by 360° it reaches a zero value twice, reaches a maximum value twice, and changes its sign twice ($\Delta_{KB}$, $\Delta_{KC}$, Figure 7.13).

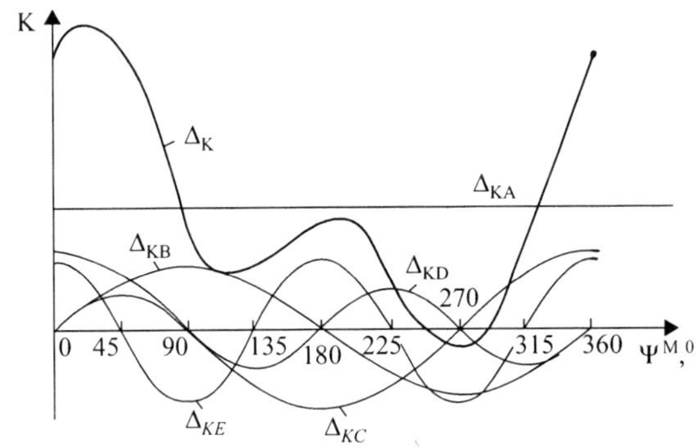


**Figure 7.13.** Graph of a deviation total value $\Delta_K$ and its components.

Semicircular deviation may be eliminated at the four cardinal points—0°, 90°, 180°, and 270° via a deviation compensator located in the bottom part of a magnetic compass.

Quadrantal deviation arises by the action of an alternating magnetic field created in the aircraft by soft magnetic iron having a small coercive force and which is therefore capable of being magnetized quickly and reversibly. The value and direction of the alternating magnetic field caused by soft magnetic iron are changed when the aircraft changes its heading with respect to a magnetic meridian. The term *quadrantal deviation* is explained by the fact that when an aircraft turns by 360° it reaches zero value four times, reaches a maximum four times, and changes its sign four times ($\Delta_{KD}$, $\Delta_{KE}$, Figure 7.13). Quadrantal deviation may be eliminated during a special procedure at the eight quadrantal Points—0°, 45°, 90°, 135°, 180°, 225°, 270°, and 315°, and is taken into account during a flight according to a special graph installed in a cockpit, or is compensated using a *quadrantal corrector*. Usually, residual quadrantal deviation will not be greater than ±5° after correction.

Inertial error arises when an aircraft turns because of centrifugal forces acting on a compass card, the southern end of which is made heavier in order to maintain its horizontal position. Long-term acceleration can take place during a flight, and this may deflect the compass magnetic system from a magnetic meridian plane that results in error accumulation in the relevant compass readings. Furthermore, during a turn, the compass card is displaced by the liquid and this phenomenon strongly distorts the readings. For all these reasons the use of a magnetic compass during a turn is inadvisable.

*Vibration error* arises under the influence of vibrating forces within an aircraft that force a compass magnetic system to oscillate relative to the lubber line. This error is reduced to a minimum by a rubber damping system.

### 7.2.4 EXAMPLES OF MAGNETIC COMPASSES STRUCTURES

In Figures 7.14–7.16, examples of aviation magnetic compasses of the combined type are shown. A liquid compass (pilotsweb.com) is shown in Figures 7.14 and 7.15, and a compass of the dry type (http://www.tghaviation.com) is shown in Figure 7.16.

The construction of the sensitive element in a liquid magnetic compass (Figure 7.14) consists of a float with two magnetic needles mounted on it and attached to a compass card bearing the letters N, E, S, and W for the cardinal headings, and numbers at 30-degree increments (without the last zero in each figure) marked on the same scale. Small divisions of the scale have no numbering and are equal to 5° each. The sealed chamber is filled with acid-free white kerosene and the float assembly is balanced on a pivot. The use of liquid reduces the frictional force in this pivot and decreases float oscillations. The compass has a glass window with a lubber line that serves as a reference behind which is the compass heading reading. Two compensating magnets are mounted on top of the compass and are adjustable so that the influence of the aircraft magnetic field can be corrected. Compass calibration is provided via two sets of screws marked by N-S and E-W letters. The external appearance of such a magnetic compass is shown in Figure 7.15.

A dry compass designed to replace the float/liquid type is shown in Figure 7.16. This is a precision aviation compass with a vertical card and its main advantage is that it has no liquid content that might leak. It uses eddy current damping instead of liquid damping, and the indicator consists of a vertical rotating dial about 2 in. in diameter that is again viewed via a window with a lubber line usually in the form of a small aircraft. The compass card rotates and presents all quadrants in their true relation to the line of flight. The heading is read at the 12 o'clock position at the nose of the small aircraft. Such a design requires no power for operation except for lighting, and as with any other magnetic compass it must be compensated using standard compensation procedures after installation.

Another variant of the liquid magnetic compass is shown in Figure 7.17, and in contrast to the previous examples it has a spherical glass face. This compass is mounted in the cockpit and is affixed via a universal fastening ring having a special rotary bracket that provides for elimination of the installation error. Inside the case (1) is an organic glass column (15) with an arm (16). A core (3) is pressed into a compass card cartridge (2) and this rests on the arm (16). The column (15) is isolated from vertical accelerations by a spring (10). There are holders on cartridge (2) in which two magnets (8) are located in cases in parallel with the of "N-S" line of the compass card. Compass card (18) is in the form of a truncated cone on which angle marks of 5° and figures at 30° appear. The heading readout again appears behind the lubber line (17). The plane passing through the lubber line and the center of the compass card coincides with a symmetry plane of the aircraft. A deviation compensator (11) is fixed at the bottom of the case and includes longitudinal platens (12) with cross magnets and also cross platens (13) also with longitudinal magnets. Screws for platens (12) and (13) can be operated via the extensions (14). The organic glass case is hermetically sealed with a round metal cover (7) and a ring (9). There is a hole for filling the compass with special a liquid (ligroin) and

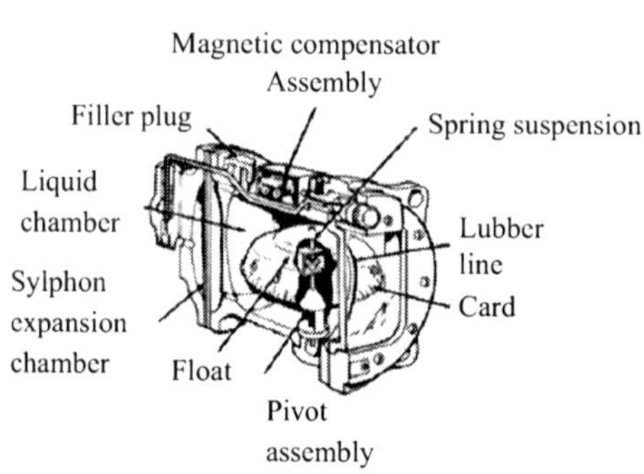


**Figure 7.14.** Section of an aircraft magnetic compass.

**Figure 7.15.** Physical configuration of an aircraft magnetic compass.

**Figure 7.16.** Precision vertical card aviation compass.

this is closed by the screw (5). Inside the upper part of the case is the withdrawing chamber, which is formed by a special partition (6) with a hole (4) to compensate temperature liquid volume changes.

## 7.3 FLUXGATE AND GYRO-MAGNETIC COMPASSES

### *7.3.1 FLUXGATE AND GYRO-MAGNETIC COMPASSES DESIGN PRINCIPLES*

By the end of the 1930s, electronic compasses had become common in aviation and are still widely used. One of the main components of such compasses is the induction sensor or *fluxgate magnetometer (flux valve)*; its operation is based on the saturation of magnetic materials.

An electromagnet consists of an iron core with a current-carrying coil wound around it. The magnetic field of the coil is strengthened by the iron core—a phenomenon that can be explained at atomic level. In nonmagnetized iron, the magnetic axes of the atoms are located chaotically, and the total magnetic field is practically zero. When a direct current flows in the coil a magnetic field is formed that forces many of the magnetic axes of the atoms to be aligned in the same direction, so creating a field considerably stronger than the magnetic field of the

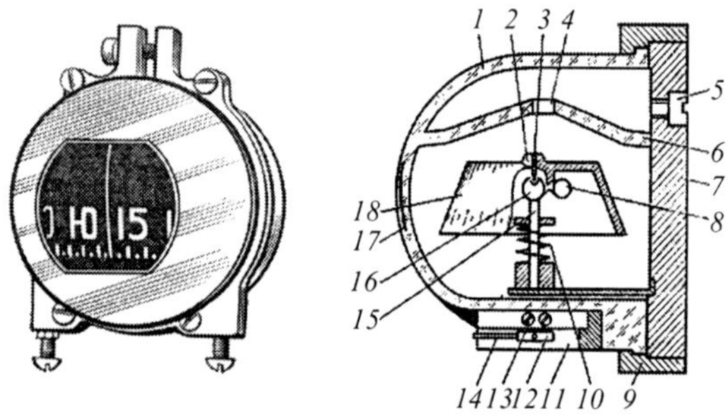


**Figure 7.17.** Appearance and a section of a small aircraft magnetic compass.

coil alone. As the current in the coil increases, more and more atoms are aligned along the field direction of the coil until all are aligned in the same direction. Further increments of current then increase the magnetic field only proportionally to the current and further strengthening of the magnetic field due the iron core does not occur. The iron core is then said to be saturated.

In magnetometers, the core is made of a ferrite such as Permalloy, in which saturation occurs abruptly and completely at a stably defined level. Figure 7.18 shows a simplified diagram of the sensor in which an exciter coil is wound around a center post. This coil carries an alternating current and the center post produces a magnetic field that has opposite directions in the top and bottom legs of the flux valve. Parameters of the system are selected so that when the amplitude of the current is maximal, magnetic saturation of the legs takes place. If alternating current of an appropriate magnitude is passed through the exciter coil, the magnetic polarity changes and saturation occurs symmetrically in each half of the cycle. If there is no external magnetic field the fluxes in legs A and B are equal, so that the total flux through a pick-off coil wound around both legs as shown is zero. Hence, there is no current induced in that pick-off coil (Fig. 7.19(a)). However, when such a magnetometer is located in an external magnetic field whose direction coincides with the axis of the ferrite core, the symmetry is upset. In that half of the cycle where the field of the coil is added to the external magnetization, saturation arrives a little earlier because it depends on the total magnetic intensity, that is, the external field plus that produced by the coil. In the other half of the cycle, where the magnetization due to the coil opposes that of the external field, saturation occurs a little later because the sum of the two is somewhat weaker than the field of the coil alone. So, in the presence of an external magnetic field such as that of the Earth, the total fluxes in legs A and B become different, the flux through the pick-off coil is not zero, and so the induced electric current in that pick-off coil becomes proportional to the external magnetic field (Fig. 7.19(b)).

Thus the sensor is capable of measuring the Earth's magnetic field, and such fluxgate compasses must employ two or three such sensors to measure its horizontal component.

Sensitive electronic magnetometers are used very widely. In space applications they are often used on board satellites as sensors of the Earth's magnetic field; and when searching for minerals, local anomalies in the Earth's magnetic field can be mapped by magnetometers installed on aircraft. There are also many magnetometer applications for military purposes, and for safety and security.

The fluxgate compass is one of the most widely used aviation devices because it is capable of providing remote directional indication. The compass sensors rotate together with an aircraft

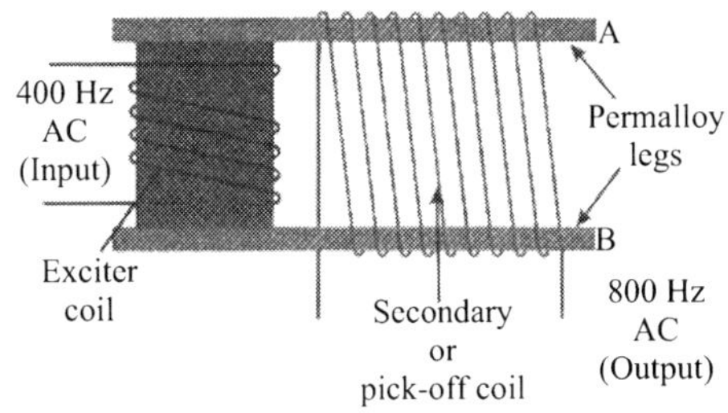


**Figure 7.18.** Simplified diagram of a flux valve.

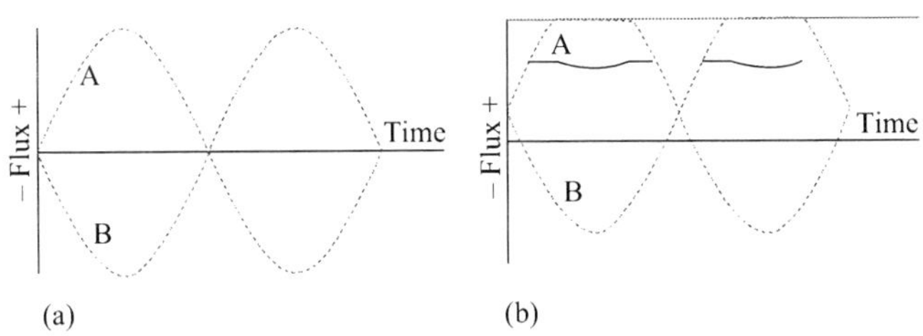


**Figure 7.19.** Magnetic field flows in a flux valve.

when its heading is changed, and the induced pick-off currents are capable of operating remote indicators after amplification.

However, despite essential basic and constructional differences, the fluxgate compass is still a version of the magnetic compass, and therefore most of the magnetic compass features, and in particular their errors listed above, concern fluxgate compasses too.

To overcome some negative features of fluxgate compasses, a new type of system has been designed. A fluxgate compass was combined with a heading gyro so that a gyro-magnetic compass was created. This new system has had a major impact on the development of aviation onboard equipment, having united the properties of both a magnetic compass and a heading gyro. Since the introduction of this gyro-magnetic compass, the magnetic compass has become an instrument of backup status for most large aircraft. Only if there is an interruption of electrical power to the gyro-magnetic compass, or a mechanical failure, does the magnetic compass returns to its primary status again.

In the first designs of the gyromagnetic compass, its magnetic system maintained the main axis of the gyro in the direction of a magnetic meridian with the help of a correction system, whereas in the latest designs the gyro provides steady indications of the pointer irrespective of compass card fluctuations.

### *7.3.2 EXAMPLES OF FLUXGATE AND GYRO-MAGNETIC STRUCTURES*

As was described above, the fluxgate compass (Figure 7.20) determines the direction of the horizontal component of the Earth's magnetic field $H$ relative to the longitudinal axis of an aircraft, that is, it defines a magnetic heading necessary for the azimuth correction of heading gyros.

A system of three electromagnetic probes (6) fixed on a pendant platform forming an equilateral triangle serves as the sensitive element of the fluxgate compass. Three pick-off coils on these probes are star connected and each probe consists of two Permalloy cores (8) in glass tubes (9). Each tube (9) is located in a Textolite skeleton (10) over which an exciter coil (11) and a pick-off coil (12) are wound. These skeletons are placed in pairs in parallel to each other and fixed in aluminum cases (14) to form complete electromagnetic probes (6). Cases (14) containing coils in each probe are located in pairs in the vertical plane to maximize compactness in the sensitive element.

Platform (7) along with its probes is suspended like a pendulum in gimbals (4) and the probe coils are connected directly to a lead-in socket by flexible spirals (15). This platform is grooved and is balanced horizontally by soldering small blobs into the grooves. The sensor case is filled with a viscous liquid to improve damping and any surplus liquid resulting from volume changes caused by temperature variations spills over into a compensatory chamber (13) located in the cover (1) of the sensor. A deviation compensator (3), located on top of this cover, is used for semicircular deviation compensation.

The fluxgate compass is connected to the heading system by cable (5) to a plug, and there are three rectangular apertures (2) in the base of the sensor for fixing purposes. The scale (16) marked on the flange of the compass assists in the elimination of installation errors. The final appearance of a typical fluxgate sensor is shown in Figure 7.21.

Fluxgate compasses are usually installed in wing tips or tail fins where any aircraft-generated magnetic disturbances are minimal.

Usually, the fluxgate sensor is a part of the complex heading system that forms a gyromagnetic compass (Figure 7.22). One of the main advantages of this system is its ability to transmit heading information to other sites, such information transfer being carried out by selsyns (self-synchronous systems). Selsyn stators and rotors are used to transmit the direction of the field generated at the fluxgate sensor when an aircraft makes turns within the Earth's magnetic field. Concurrently, a fluxgate compass measures the Earth's magnetic field and sends signals to the stator of the gyro control transformer CT. These signals vary in amplitude and sign according to the direction of the Earth's magnetic field relative to the longitudinal axis of the aircraft.

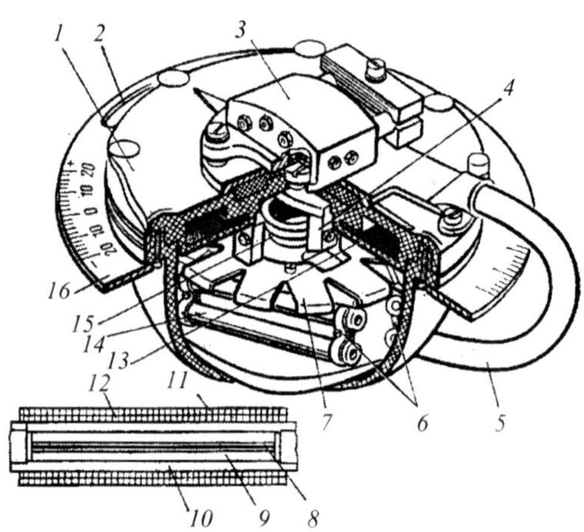


**Figure 7.20.** Design of a fluxgate sensor.

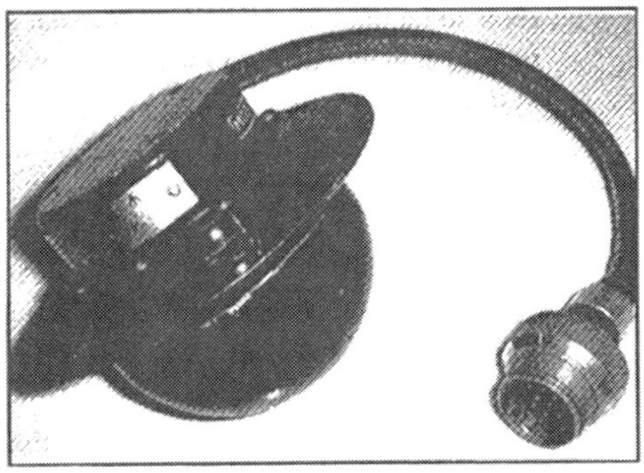

**Figure 7.21.** Example of a fluxgate sensor.

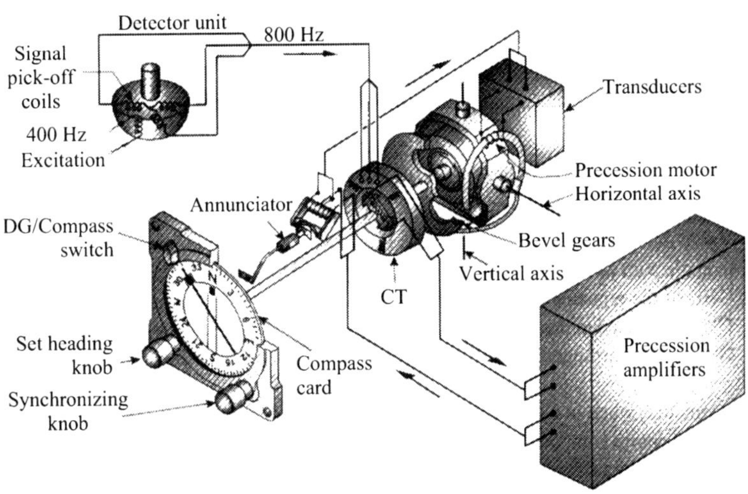


**Figure 7.22.** A gyro-magnetic compass layout.

The compass card is fixed on the same shaft as the rotor of the gyro control transformer and the system is adjusted so that if the rotor is located at right angles to the field produced by the stator, the compass card correctly indicates the aircraft's magnetic heading. If the rotor is not at right angles to the magnetic field set up in the stator, by the signals from the fluxgate compass a signal will be induced in the rotor proportional to the misalignment angle. This signal is fed to the precession amplifier and then through the annunciator unit to the transducer. The purpose of the annunciator unit is to show whether the compass is operating either in the "Gyro" or "Compass" mode and to indicate that the gyro is synchronized with the detector unit. The signal is then fed to the precession coil installed on the horizontal gimbals of the gyro and precession of the gyro begins around the azimuth axis. This movement is translated into a movement of the shaft on which is mounted the rotor of the control transformer and the compass card. As a result of this rotary movement the rotor of the control transformer comes to the "zero" position at which the precession signal is zero and the system is aligned with the direction of a magnetic meridian. When the aircraft makes a turn the gyro keeps its main axis fixed in space as the gyro case turns with the aircraft. The rotation of the gyro shaft that turns the compass card is then equal to that of the aircraft turn. Simultaneously, the signals from the fluxgate compass change according to the aircraft turns. The values of the two changes should be identical so that the compass remains synchronized.

Remote compasses were designed to overcome the main disadvantages of the combined compasses, which are as follows:

1. Installation had to be where the compass could be easily seen by the pilot, usually in the cockpit, and here they were very subject to major deviations due to aircraft magnetism.
2. They exhibited turning and acceleration errors.
3. They could not feed heading information to other equipment.

The first disadvantage may be overcome by using a flux valve that is a small sensor which can be installed as far from the main sources of aircraft generated magnetic fields as possible (wings or fin tip). Turning and acceleration errors can be almost completely eliminated by combining the directional stability of the gyro with the magnetic north-sensing property of the fluxgate compass. Finally, remote gyro-magnetic compasses can easily transfer digital information to other equipment such as indicators and navigation and flight management systems.

## 7.4 ELECTRONIC COMPASSES

The traditional design of fluxgate compasses includes two-axis gimbals to allow two degrees of freedom in pitch and roll, so providing a measurement of the horizontal component of the Earth's magnetic field. Advances in electronics have now allowed the design and production of solid-state magnetic sensors for aerospace applications, and such sensors are very small and light, lower in cost, and have very low power consumption. They are more reliable than fluxgate compasses. New generations of such devices (Figure 7.23) are strapdown types that can measure the direction and strength of the Earth's magnetic field. Usually, they resolve the three orthogonal projections of the magnetic field and are called *magnetometers*, of which there are several different types based on various physical principles. The four most widely used magnetometers are fluxgate sensors, sensors based on the Hall Effect, eddy current sensors, and magnetoresistive sensors. The strapdown design has no moving parts and provides consistent performance in spite of changing temperatures and magnetic distortions.

For obvious reasons, electronic devices are currently replacing the outdated technology of the "magnetized needle" whose indications are often erroneous because of external disturbances. Also, traditional magnetometers and compasses are difficult to adapt for digital readout or for the computer interfaces of modern aerospace systems. However, the new generation of sensors has no such problems and are well adapted for use in the digital environment.

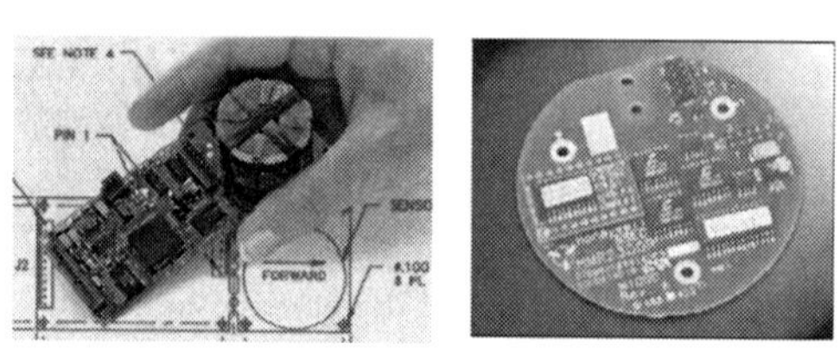

**Figure 7.23.** Examples of strapdown magnetometers.

## REFERENCES


Abanave, Swapnil T. 2005. *Navigation*. Oxford Aviation Services Ltd.

Aczel, Amir D. 2001. *The Riddle of the Compass: The Invention that Changed the World* (1st edition). New York: Harcourt.

The American Practical Navigator. 2002. *An Epitome of Navigation*. Originally by Nathanial Bowditch, LL.D. Bicentennial Edition. National Imaginary and Mapping Agency.

Backus, G.P., R. L. Parker, and C. Constable. 1996. *Foundations of Geomagnetism*. Cambridge, MA: Cambridge University Press.

Barton C. E. 1997. "International geomagnetic reference field: The seventh generation." *Journal of Geomagnetism and Geoelectricity* 49: 123–48. DOI: 10.5636/jgg.49.123.

Chapman, S., and J. Bartels. 1940. *Geomagnetism* (p. 1049). London: Oxford University Press.

Davis, Jeremy. 2004. "Mathematical modeling of Earth's magnetic field." Technical Note. Virginia Tech, Blacksburg, May 12.

Gurney, Alan. 2004. *Compass: A Story of Exploration and Innovation*, London: Norton.

McLean, S., S. Macmillan, S. Maus, V. Lesur, A. Thomson, and D. Dater. 2004. Tech. Rep. NOAA Technical Report NESDIS/NGDC-1, NOAA National Geophysical Data Center.

McLean, S., S. Macmillan, S. Maus, V. Lesur, D. Dater, and A. Thomson. 2005. The US/UK World Magnetic Model for 2005-2010. Retrieved from http://www.geomag.us/info/Smaus/Doc/WMM_2005.pdf

National Geophysical Data Center. International Geomagnetic Reference Field. 2005. http://www.ngdc.noaa.gov/IAGA/vmod/igrf.html

Wertz, J. R. 1978. Spacecraft Attitude Determination and Control. Boston, MA: D. Reidel Publishing Company.

Williams, J. E. D. 1992. *From Sails to Satellites: the origin and development of navigational science*. Oxford University Press.

http://pilotsweb.com/navigate/compass.htm

http://www.tghaviation.com

CHAPTER 8

# Propulsion Sensors

J. Paul Sims
*East Tennessee State University, USA*
Joe Watson
*Swansea University, United Kingdom (Retd)*

## 8.1 INTRODUCTION

There are six basic quantities that require measurement and instrumentation relevant to aerospace vehicle propulsion units, plus a host of secondary and tertiary parameters. In this chapter, only these six primary quantities will be considered because the plethora of others are specific to each type of vehicle, so are better treated in individual papers and similar publications.

These six basic parameters are fuel quantity, fuel flow, pressure, temperature, rotational speed, and vibration; fundamental methods for measuring each are presented below. Whether or not a sensor based on a given method can be applied to a particular vehicle depends entirely on whether that sensor is capable of withstanding the flight characteristics of the vehicle and continue to make valid measurements of the designated parameter—and the ranges of such parameters can themselves be extreme.

## 8.2 FUEL QUANTITY SENSORS

Fuel quantity sensors measure the amount of fuel in a tank and may be based on various different physical phenomena. For example, either the level (i.e., the height over a datum) of a column of the liquid, or its pressure, may first be determined. For level measurements, the volume of liquid in a tank can be computed if its density and details of the tank geometry and dimensions are known, under which circumstances the quantity may also be expressed in terms of mass provided that the density is either constant or measured simultaneously.

Level measurement can be fundamentally described in terms of the *head*, which is the height $h_l$ of a liquid column above a point where the pressure is measured. When the density (in terms of specific weight $w$) of the liquid is known, the level $h$ will be given by the pressure $p_l$ measured relative to the pressure above the liquid surface $p_s$:

$$h = \frac{p_l - p_s}{w}. \tag{8.1}$$

The actual measurement techniques include various mechanical and electromechanical systems, such as buoyancy methods and those using pressure sensors. Others include electronic systems for determining conductance or capacitance; or heat-transfer measurements and ultrasonic methods may be employed. Some of these are described below.

### 8.2.1 MECHANICAL AND ELECTROMECHANICAL METHODS OF LEVEL SENSING

These are the simplest methods of determining the head of a liquid, but because the specific weight is involved (as opposed to the specific mass) they are not easily adaptable for measurements in the more extreme aerospace environments.

#### 8.2.1.1 Buoyancy or Float Methods

Floats for the measurement of liquid in a tank are either hollow or made of material less dense than the measured fluid, as in Figure 8.1(a). The up/down motion of the float relative to the tank is converted into an output indicative of level by a transducer element mounted on the tank wall. This may be a simple potentiometer to produce a continuous measurement as in the figure. Fuel quantity transmitters based on this simple principle are found in the wing tanks of many light aircraft.

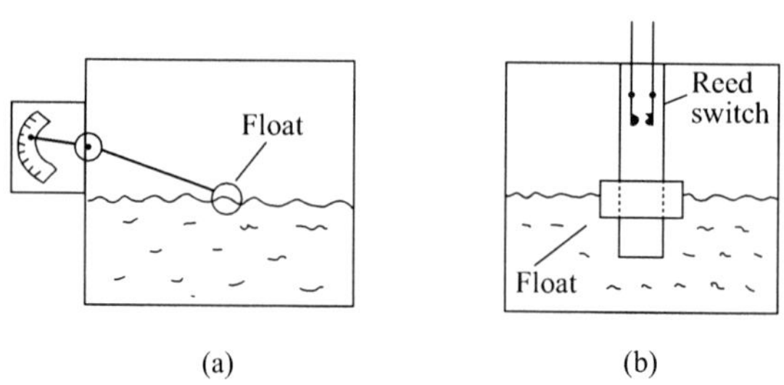


**Figure 8.1.** Basic buoyancy-type level gauges.

For a discrete measurement, a reed switch in a nonmagnetic tube, along with a permanent magnet embedded in an annular float, will result a point-level sensor as in Figure 8.1(b). This reed switch is actuated when the magnetic float rises to a position adjacent to it. For both types the case containing the actuation mechanism must be hermetically sealed.

In a related sensing method, the float does not actually move, the force acting on it due to buoyancy being sensed by an appropriate transduction element, typically of the strain gauge or force-balance type.

#### 8.2.1.2 Level Sensing Using Pressure Transducers

Figure 8.2 illustrates three basic methods for level-sensing using liquid head measurement via pressure sensors (Holman 1994; Norton 1992). The fundamental method is illustrated by the open tank of Figure 8.2(a), where a pressure transducer is flush-mounted close to the bottom of the tank. At the port *M*, the measured pressure $P_M$ is equal to *hw*, the head of the liquid multiplied its specific weight. Obviously, an open tank is not practical for any aerospace vehicle, and Figure 8.2(b) is relevant to a closed tank where a differential pressure $P_D$ is measured by a transducer having a measuring port *M* and a reference port *R*. Here, the specific weight of the liquid is assumed to be much greater than that of the gas above its surface, otherwise the gas head must be accounted for in the total measurement.

In Figure 8.2(b) the measuring port contains a differential transducer mounted externally to the tank as shown. This reference port system can be used only when there is no possibility of moisture entering the port even by condensation in the reference-pressure line, otherwise a condensation trap must be included as in Figure 8.2(c).

More sophisticated designs are required for more difficult circumstances, such as, for example, when the pressure transducer cannot be flush-mounted to the tank, or when the gas or liquid is corrosive. Various forms of pressure sensors suitable for aerospace applications will be described in Section 8.4.

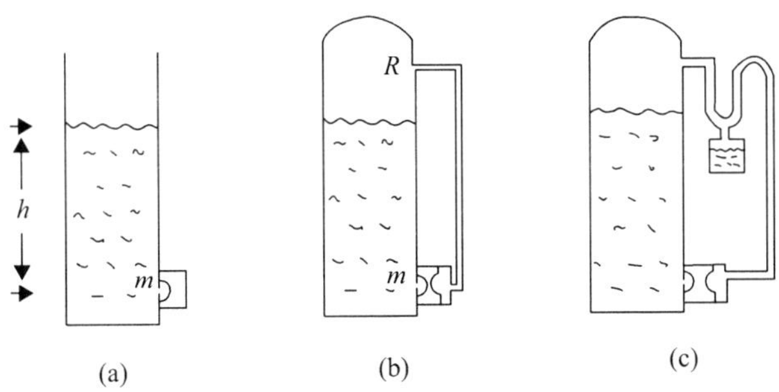


**Figure 8.2.** Level sensing using pressure sensors.

### 8.2.2 ELECTRONIC METHODS OF LEVEL SENSING

Basically, an electronic instrument will consist of a transducer that converts an appropriate physical parameter to an electrical signal, followed by various forms of signal-conditioning circuitry.

#### 8.2.2.1 Conductivity Level Sensing

The level of electrically conductive liquids can be sensed using two vertical electrodes in the liquid, between which the change in resistance is measured (Norton 1992). The tank wall, if metallic, can be used as one of the two electrodes. This method can be used for both continuous-level and discrete-level indications, but is obviously not useful for nonconductive liquids, such as aviation fuels.

#### 8.2.2.2 Capacitive Level Sensing

When one or more pairs of electrodes are vertically immersed in a tank containing a nonconductive liquid, any rise or fall in liquid levels will cause changes in capacitance between the electrode pairs. Again, the tank wall, if metallic, can be used as one electrode of such pairs (Bodner 1960; Kremliovskiy 2002).

Alternatively, a complete sensor can be configured as a pair of coaxial tubes, or as a coaxial group of four with alternate tubes ganged together to form the capacitive element. Such sensors are partly perforated or slotted to permit the free flow of liquid and vapor.

A four-arm AC bridge network may be used as the measuring circuit with the level-sensing capacitance element $C$ constituting one arm of that bridge. Also, the accuracy can be improved by placing a small submerged capacitive element below the level-sensing element to compensate for any changes in the liquid characteristics. Such a reference capacitance element $C_R$ sees a fixed head $h_R$, which constitutes a small fraction of the main head $h$ that corresponds to the sensed capacitance $C$. The tank level is then given by:

$$h = h_R \frac{c}{c_R}. \tag{8.2}$$

Sophisticated capacitive transmitters based on this principle are common in heavy aircraft and are distributed at many points in the wing and other tanks, and a computer system produces data relevant to both individual and average fuel quantities. As an example, the MD-11 aircraft employs 39 capacitive sensors. Each of these sensors has a unique dry capacitance (in picofarads) and the level indicated by each is monitored for every tank. In the case of the MD-11, a totalizer is also used to provide the pilots with information regarding the total fuel load and its gross weight. The reason for the large number of probes is to provide redundancy, accuracy, and reliable readout regardless of deck attitude.

For such aircraft, the medium surrounding the capacitive sensors is Jet-A fuel of dielectric constant 1.7 or air of dielectric constant 1.0. (Unity is the dielectric constant for dry air by definition.) For the number one capacitive fuel-level sensor on the MD-11, the dry capacitance is 187.38 picofarads. Therefore, if the sensor is surrounded by JET-A fuel, this value will be multiplied by 1.7 to give 318.5 picofarads. Therefore the measurement range for this sensor is 187.38 picofarads to 318.5 picofarads from empty to full and this relationship is linear (depending on interface electronics).

Probes used for point-level sensing are often covered with an antifouling coating for a portion of their length, and may be side-mounted so that the entire probe length becomes the active electrode, all of which senses the capacitance change when the probe becomes wetted by the liquid. Probe design, material, and finish should minimize any adherence of liquid when the level falls.

#### *8.2.2.3 Heat-Transfer Level Sensing*

From a heated element, the rate of heat transfer is usually higher in a liquid than in a gas, and this principle is applied in various discrete-level sensors. Resistive elements carrying an electrical current are frequently used for this purpose. When the liquid level rises so that it comes into contact with the warm element, the element will be cooled and the resultant step change in resistance is used for a point-level indication, and *vice versa*. The voltage drop across a hot-wire probe electrically heated by a constant-current power source will change as its resistance changes during transitions between the gas and the liquid environments. This voltage change may be used to indicate such a transition, and has been used in cryogenic applications. (The same principle is also relevant for a thermocouple attached to a wire-wound heater.)

Thermistors carrying currents sufficient to produce self-heating are also suitable for such applications, and again the resistance undergoes a step change when the liquid/gas environment changes. This resistance change can be converted into a voltage change via a simple voltage-divider bridge circuit (Klaassen 1996; Kremliovskiy 2002); or an operational-amplifier-type comparator may be employed. Thermistors are considerably more sensitive that hot wires, and can be obtained with either negative or positive resistance/temperature coefficients.

Various heat-balance designs may also used for point-level sensing. For example, a heated resistive element connected into a bridge circuit along with a similar but unheated resistive element connected into an adjacent bridge arm will form a differential-temperature sensing circuit. When a probe carrying the resistive elements is immersed in a liquid, the temperature difference between them will be significantly less than when the probe is in a gas.

Another heat-balance design uses the thermal coefficient of expansion of a heated metal rod provided with a good heat conduction path to the measured fluid. When transitioning from a gas to a liquid environment, the rod cools due to the increased heat transfer into that liquid, and its contraction mechanically operates a relay.

#### *8.2.2.4 Ultrasonic Methods*

Three methods of ultrasonic sensing are employed: *cavity-resonance sensing*, *sonic-path sensing*, and *damped-oscillation sensing*—the first two being the most common.

In the *cavity-resonance* method, ultrasonic or radio frequency oscillations are excited in the cavity bounded by the tank walls and the liquid surface within a tank by a coupling element at the top of that tank. As the liquid rises, the cavity volume shrinks and its resonant frequency changes accordingly. When the resonant frequency of the empty tank is known, the level of the liquid can be determined and hence its volume calculated using appropriate scaling factors. Variable-frequency feedback oscillators can be used to determine the resonant frequencies; and radio-frequency methods can be used with dielectric liquids.

*Sonic-path* methods can be used for both continuous and discrete level sensing, the reflectance mode being commonly used for the former and the transmittance mode for the latter. For continuous level sensing, either separate transmitting and receiving elements, or a single element operating alternately in the transmitting and receiving mode, can be employed.

Figure 8.3 represents a recent design for a pulsed sonic-path system using the single transmitter/receiver method. Here, pulsed ultrasonic energy is directed at the liquid/gas interface and the travel time of the pulse when reflected back from this interface is measured. When the velocity of sound in the vapor through which the pulse travels is known, the distance $L$ between the transmitting/receiving element and the liquid surface can be determined. Such transducers typically contain a temperature sensor to correct for changes in the speed of sound as a result of temperature changes.

As an example, let the time between transmission and reception be 5 ms. If the temperature is 0° the speed of sound in air is 331 $ms^{-1}$, so the relationship shown in the figure results in a distance from the transmitter to the liquid surface of 0.8275 m. Hence, if the tank is 1 m deep when empty the fluid level is 0.1725 m from the bottom.

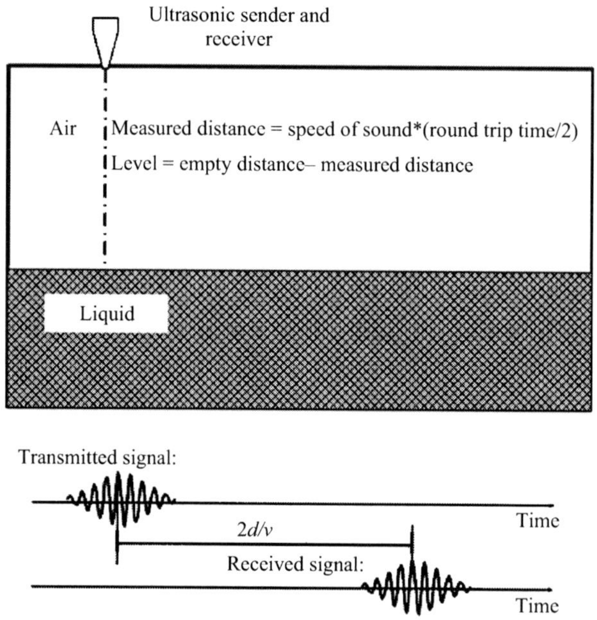


**Figure 8.3.** Typical pulsed ultrasonic level measurement system.

Such single transmitter/receiver transducers may also incorporate interference echo suppression to provide more accurate level measurements.

Continuous-level sensors usually see the surface of the liquid from the top of the tank; but they can also be mounted in the bottom, in which case the ultrasonic transmission characteristics of the system will be significantly different.

Discrete-level sensors normally use a separate transmitter and receiver. When liquid enters the sonic path between the two, the amount of sound energy at the receiver is attenuated significantly because of absorption by the liquid. With the exception of the gap type of point-level sensor, transmitting and receiving elements can be either of the wetted or externally mounted type, though the latter may not be feasible in some installations, or its use may be precluded by characteristics of the measured fluid. Point-level ultrasonic sensors can be installed at the top of a tank or at the side.

A sensor of this type can be realized as a single probe, the tip being separated from the probe body by a narrow metal rod. The body and tip contain the transmitter and receiver transducers respectively, facing each other across the gap. A variation on this design places the transmitting and receiving transducers facing each other across a horizontal gap. This sensor is usable for liquids whose temperature can be as low as 60 K, which is below the boiling point of nitrogen. The small gap used in both such above sensors permits very small crystals and very low power to be employed.

A generalized system block diagram is shown in Figure 8.4.

*Damped oscillation point-level sensors* can be of either the piezoelectric or magnetostrictive type.

Piezoelectric models employ a quartz or piezo-ceramic driver unit mounted at the tip of a hermetically sealed probe. This is driven by an oscillator circuit and resonates at a specific amplitude when immersed in gas and at a significantly reduced amplitude when immersed in

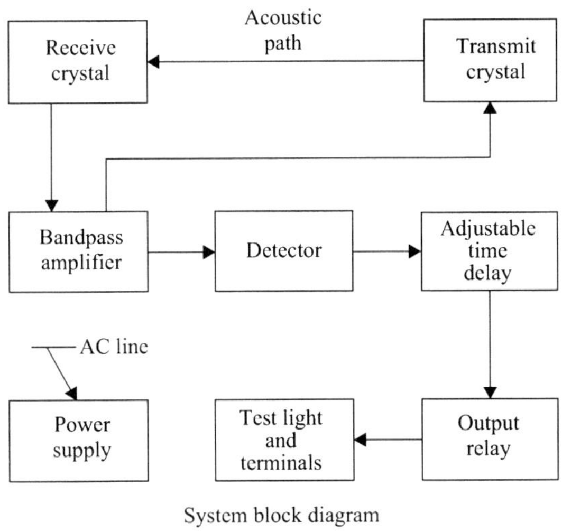


**Figure 8.4.** Block diagram of a generic ultrasonic sensor electronic system.

liquid. This amplitude change is measured and provides a discrete output signal via the relevant electronic module. Piezo-ceramic drivers are generally preferred over quartz because they exhibit much greater stability over substantial temperature ranges, and are more robust.

Magnetostrictive drivers use an assembly consisting of a drive coil and a feedback coil, both wound on the same ferromagnetic rod mounted at the inner-surface tip of a hermetically sealed probe. A feedback oscillator circuit incorporates these drive and feedback coils, and the output signal current level is so adjusted as to maintain vibration in the rod only when the probe tip is exposed to a compressible fluid such as air, froth, or foam. When the probe tip encounters a non-compressible fluid, the vibration is damped and the oscillation stops, which results in a discrete output signal. Such circuitry can be contained in a separate box with interconnecting cables kept short to minimize their capacitance, or it can be packaged into the probe head. The liquid sensed by such point-level sensors should not contain any material that may remain on the probe tip and dry and then harden, so disabling the whole sensing system.

## 8.3 FUEL CONSUMPTION SENSORS

### *8.3.1 INTRODUCTION*

Sensors used for measuring instantaneous or mean quantities of fluid consumption per unit time are usually termed *flow sensors* or *flow meters*, and many such devices also require pressure and temperature measurements in order to achieve overall accuracy. This chapter presents only the basic methods of operation of various flow measuring devices commonly used in aerospace applications, though there are several more, mostly experimental techniques.

### *8.3.2 FLOW-OBSTRUCTION METHODS*

Several types of flow meters fall in the category of obstruction devices (Helfric 1994; Holman 1994). They are also known as *differential pressure meters*, and all are based on the classic

Venturi tube depicted in Figure 8.5. Here, the flow is from left to right and application of the Bernoulli equation shows that whereas the flow velocity $U_2$ is greater to the right of the narrowed neck (the obstruction), the pressure $P_2$ is smaller. Measurement of these two pressures will therefore provide information on the fluid flow. This can be shown by considering the continuity equation, which gives the mass flow as:

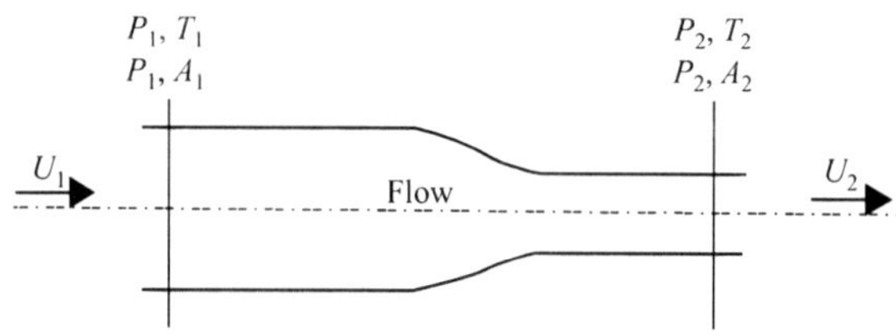


General one-dimensional flow system.

**Figure 8.5.** The basic Venturi tube.

$$\dot{m} = \rho_1 A_1 U_1 = \rho_2 A_2 U_2 , \tag{8.3}$$

where $U$ is the velocity, $\rho$ is the fluid density, and $A$ is the tube area.

If the flow is adiabatic and frictionless, the Bernoulli equation may be written as:

$$\frac{P_1}{\rho_1} + \frac{U_1^2}{2g_c} = \frac{P_2}{\rho_2} + \frac{U_2^2}{2g_c} . \tag{8.4}$$

For an incompressible fluid $\rho_1 = \rho_2 = \rho$ so combining the two equations gives:

$$P_1 - P_2 = \frac{U_2^2 \rho}{2g_c} \left[ 1 - \left( \frac{A_2}{A_1} \right)^2 \right] . \tag{8.5}$$

The volumetric flow rate may now be written:

$$Q = A_2 U_2 = \frac{A_2}{\sqrt{1 - \left( {A_2}/{A_1} \right)^2}} \sqrt{\frac{2g_c}{\rho} (P_1 - P_2)} . \tag{8.6}$$

However, no practical tube is frictionless and some losses always accrue in the flow. That is, the volumetric flow calculated using Equation (8.6) represents the ideal value, and it relates to the actual flow by an empirical *discharge coefficient C* where:

$$\frac{Q_{\text{act}}}{Q_{\text{ideal}}} = C . \tag{8.7}$$

The value of this discharge coefficient is actually dependent on the relevant Reynolds number and the channel geometry, too. Considerable further material on this and various other coefficients appear in the following works, which treat both incompressible and compressible fluids in detail: White (1979); Beck (1981); Holman (1994); Klaassen (1996); Kremliovskiy (2002); Norton (1992); Nuliten et al. (1983).

#### 8.3.2.1 Practical Considerations for Obstruction Meters

The geometry of Venturi tubes has been standardized by the American Society of Mechanical Engineers (Holman 1994; Norton 1982) and the recommended design is shown in Figure 8.6. Here, the pressure points are connected to annular manifolds at the upstream and throat portions of the tube. These manifolds receive pressure samples from all around the tube periphery at these points so that a good average value is obtained. The discharge coefficients for such tubes are required to be within specified tolerance limits.

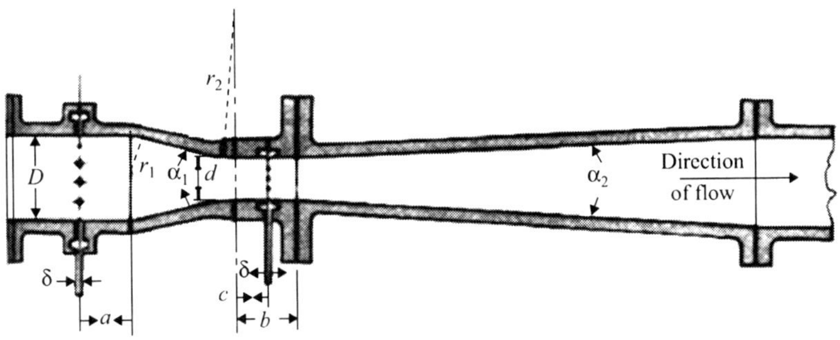


**Figure 8.6.** The standard Venturi tube.

### 8.3.3 THE TURBINE FLOW METER

A popular type of flow-measurement device is the turbine sensor depicted in Figure 8.7 (Benedict 1977; Bodner 1960; Helfric 1994; Holman 1994; Kremliovskiy 2002, 2004). Basically, it converts volumetric fluid flow to an equivalent angular speed in a rotor.

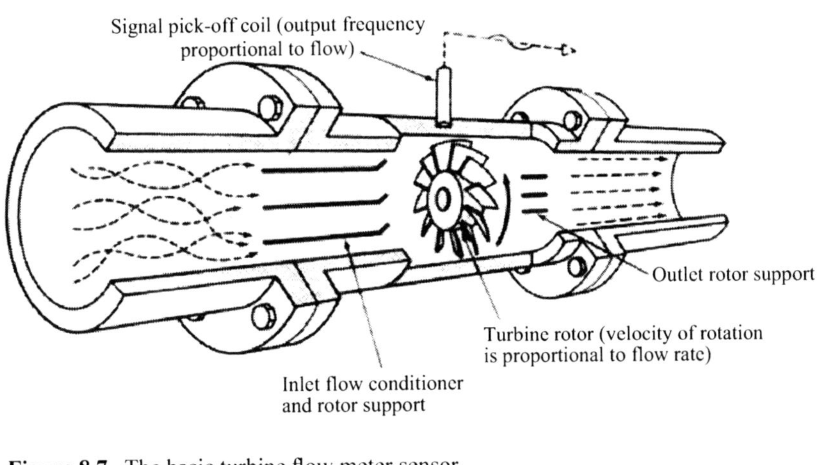


**Figure 8.7.** The basic turbine flow meter sensor.

As the fluid moves through the tube it causes the small turbine wheel to rotate. A permanent magnet enclosed within this wheel induces a pulse per revolution in a reluctance pick-up attached to the top of the sensor. The number of wheel revolutions (and hence pulses) per unit time is proportional to the volumetric flow rate $Q$ (Bodner 1960; Holman 1994; Kremliovskiy 2002), so that the sensitivity in terms of a $K$-factor may be defined. If the pulse rate is $f$ pulses per second (Hertz) and the flow rate is $Q$ cubic metres per second, then:

$$K = f/Q \quad \text{Hertz per cubic metre per second,} \tag{8.8}$$

which is obviously the same as pulses per cubic metre. (Actually, more practical units are normally used, such as pulses per gallon.)

The usefulness of this simple formula depends on $K$ remaining essentially constant over as wide a range of flow rates as possible. However, the rotor angular speed is dependent not only on the value of this flow rate, but also on factors such as the viscosity of the liquid and the friction in the rotor and its bearings. Hence, any changes in temperature or in the liquid specification will bring concomitant changes in the meter readings.

In a practical turbine sensor typified by that of Figure 8.8, the accuracy is good, and such turbine flow sensors can range in size from about 0.6 cm to over 25 cm. Some are available with more than one reluctance pick-up, which must obviously multiply the magnitude of $K$ by the number of pick-ups incorporated.

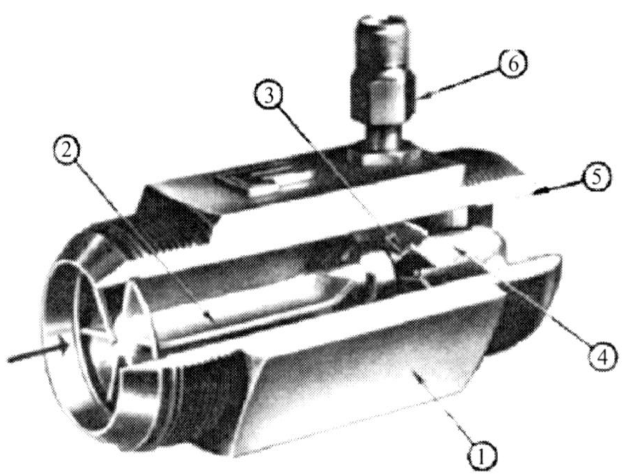


**Figure 8.8.** A practical turbine flow meter. (1) sensor case; (2,5) flow conditioners; (3) rotor; (4) bearing housing; (6) reluctance pick-off.

Calibration curves for a typical flow meter are given in Figure 8.9(b), which shows that the pulse rate varies linearly with the flow rate over a wide range. This graph also includes the $K$-factor, which is seen to remain essentially constant over most of the same range of frequencies, and hence flow rates. Figure 8.9(a) gives this in a different way and clearly shows the fall-off in $K$ at low flow rates.

The effect of viscosity changes may be measured by heating the fluid to a series of different temperatures and checking how the $K$-factor varies between each sample.

The use of the turbine flow meter in pulsating flow is discussed in Benedict (1977), Bochniak and Bisov (1968), Helfric (1994), Holman (1994), and Smith (1975).

Flow meter calibration
Model #FT6–8 AEXSRLEA-2022 serial #86XXX
Cal date: 05/26/98 call ref. Mil-C-7024D density 6.381 –40° Jet A

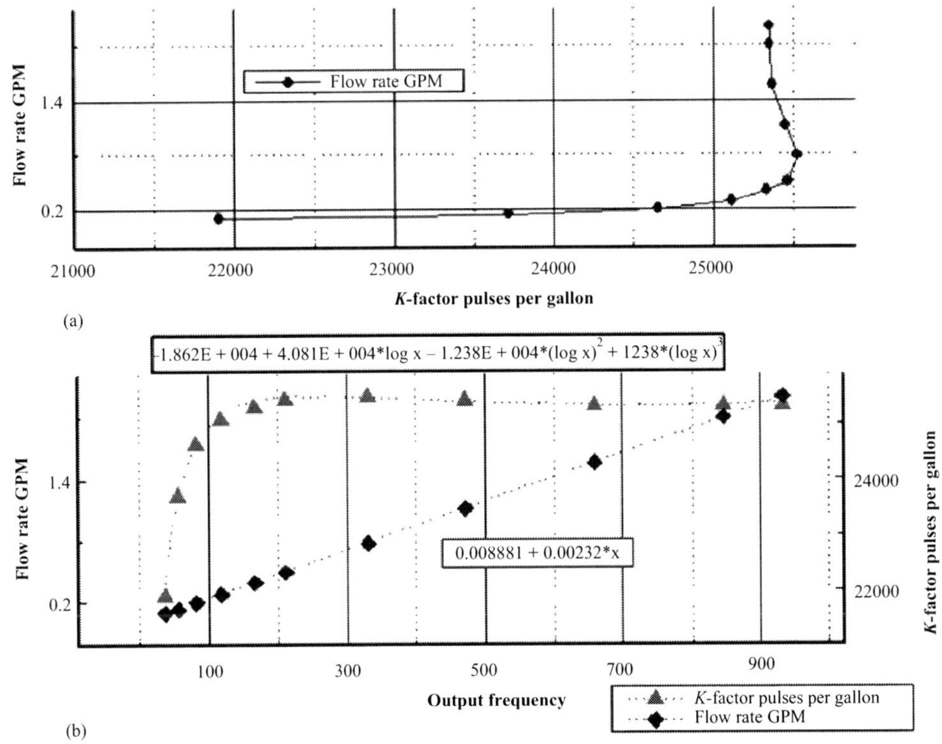


**Figure 8.9** Calibration curve for a typical flow meter.

## *8.3.4 THE VANE-TYPE FLOW METER*

This method of flow measurement is very simple and lends itself to direct measurement techniques from a pointer to a circular digitizer without difficulty. As seen in the (exaggerated) diagram of Figure 8.10, fluid enters a cylindrical chamber having a volute cross-section—that is, the radius of the cylinder increases around its inner periphery. A shaft bears a vane of almost the same length as the smallest radius of that inner cylinder, and is able to undergo a clockwise angular displacement from this position under the restoring force of a spiral spring. When a fluid enters the inlet port this movement is initiated by the flow and progresses until the restoring spring force equals that of the flow force. This point is defined by the increase in the cylinder radius, and hence gap area, that allows progressively more fluid to flow past the vane. Hence, the vane position is a function of the flow rate. Basically, it is a square-law device because the gap area is clearly the defining parameter and this is proportional to the square of the radius.

This simple mechanism has been manufactured to become a sophisticated sensor that will typically include a bypass valve that may open should the vane become jammed or to accommodate a greater fluid flow. An adjusting screw to calibrate the spring will be included, plus a counterweight attached to the shaft to balance the vane. The latter may be situated in a small damping chamber that uses the fluid itself as the damping medium.

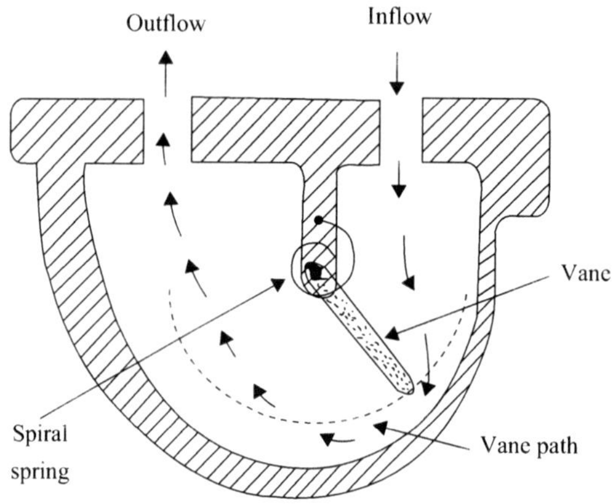


**Figure 8.10.** The basic vane-type flow meter principle (exaggerated for clarity).

## 8.4 PRESSURE SENSORS

### *8.4.1 BASIC CONCEPTS*

Pressure is defined as the force per unit area and the SI unit is the Pascal (Pa), which is defined as a Newton per square metre (N $m^{-2}$). However, this is rather a small quantity, so the kilopascal (kPa) and Megapascal (MPa) are therefore more common. Historical units such as the pounds per square inch (lbs $in.^{-2}$) of Imperial measure are also still in use.

In aeronautics, ambient air pressure decreases with altitude, so air pressure measurements are often used to determine altitude, airspeed, and rate of climb or descent (see Chapter 2). Furthermore, gauges for determining fuel, lubricant, and hydraulic pressures are also required.

Fluid pressure results from a momentum exchange between the molecules of the fluid and a containing wall. The total momentum exchange is dependent on the total number of molecules striking the wall per unit time and the average velocity of these molecules.

There are a number of definitions of pressure that may be applied in various practical situations, for example:

*Absolute pressure* is the pressure measured relative to zero
*Gauge pressure* is the pressure measured relative to the ambient pressure
*Differential pressure* is the pressure difference between two measuring points (for example, see Figure 8.2(b)).
*Static pressure* is the pressure in a fluid that is exerted normal to the surface along which the fluid flows
*Impact pressure* is the pressure in a moving fluid exerted parallel to the direction of flow due to flow velocity (see Chapter 2)
*Stagnation pressure* (also called total pressure) is the sum of the static pressure and the impact pressure (see Chapter 2).

Essentially, pressure is sensed by the deformation of a mechanical element, for example, an elastic member such as a plate, shell, or tube, which offers the pressure a surface area upon which to act, so producing a force. When this force is not balanced by an equal force acting on the opposite surface of the sensing element, the element is deformed, which may be sensed by either displacement or strain. Hence, pressure-sensing elements actually respond to the differential pressure across them, and so can be designed to measure either differential, gauge, or absolute pressure, depending on the reference pressure maintained in, or admitted to, the reference side of the element.

### 8.4.2 BASIC SENSING METHODS

The simplest pressure sensor is the flat diaphragm, and this may be used to illustrate the fundamental pressure measurement principle, as in Figure 8.11 (Benedict 1977; Norton 1982). Here, the reference side of an absolute-pressure sensing element is evacuated and sealed (a); gauge pressure is measured when the reference side is vented to ambient (b); and differential pressure sensing results when the diaphragm is open to two pressures (c), both of which may vary. In a special version of the differential-pressure configuration (d), a fixed pressure which is greater than zero is permanently maintained on the reference side.

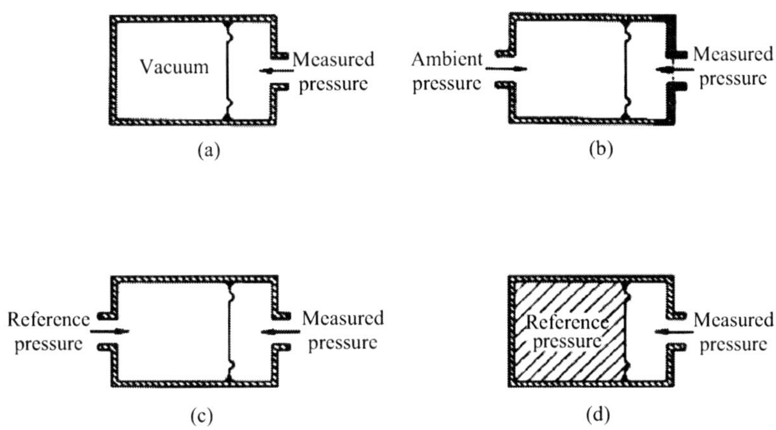


**Figure 8.11.** The basic pressure measurement principle.

#### 8.4.2.1 The Diaphragm

A practical diaphragm is essentially a thin circular plate supported continuously (Figure 8.11) around its edge. Two basic types of diaphragm are used as pressure sensors: the flat diaphragm and the corrugated diaphragm. Exact calculations are required for diaphragm design, and recent developments include the use of computer-assisted optimization methods. The basic flat diaphragm deflects in accordance with laws generally applicable to circular plates under conditions of symmetrical loading.

Diaphragms are either machined to include the edge support, or they are formed separately and then welded (sometimes brazed) to a separate support. Sometimes, the diaphragm is made slightly concave, and several other configurations have been used including central strengthening for robust connection to the deflection transducer.

Corrugated diaphragms contain a number of concentric corrugations that increase the stiffness as well as the effective area of the diaphragm, so providing a larger useful deflection than that of a flat diaphragm having the same diameter. The corrugations become progressively shallower from the periphery to the center because bending is maximal near the periphery and minimal at the center.

The deflection of a diaphragm varies inversely with the 1.2 to 1.6 power of its thickness and approximately with the fourth power of its diameter. Within a specific band of pressure magnitudes, the deflection changes linearly with pressure, and this band is defined by the design of the corrugations (if any), the material and its preparation and treatment, the manner of attachment to its peripheral supporting wall (and its fillet radius at the wall interface, if machined), and the diameter of the central reinforcement (if any).

Materials used for diaphragms are elastic metal alloys such as brass, bronze, phosphor-bronze, beryllium-copper, stainless steel, and proprietary alloys such as Monel, Inconel-X, and Ni-Span-C (a ferrous nickel alloy with good thermal properties). Choice of diaphragm material is strongly influenced by the chemical properties of the measured fluid that comes in contact with the diaphragm. Heat-treating and pressure cycling help to reduce elastic after-effects (drift) and hysteresis.

#### *8.4.2.2 Capsules*

Corrugations are typically found in hollow sensing elements such as the *aneroid capsule* detailed in Chapter 2 with reference to air pressure. In that chapter, the measured pressure is shown as being applied to the inside of the sensing element or the outside, or both, depending on the application.

The simple capsule consists of two annular corrugated diaphragms formed into shells of opposite curvature and sealed together at their peripheries. Usually, one diaphragm is provided with a pressure port and the other with a boss connected to a mechanical displacement transducer. Alternatively, one diaphragm may be provided with an internal boss attached to a pushrod passing through a port in the opposite diaphragm through which a reference pressure is admitted into the capsule. The use of two diaphragms in the form of a capsule nearly doubles the deflection obtained from a single diaphragm. Additional multiplication of deflection can be obtained by ganging two or more capsules together to form a *bellows*, as shown in Figures 2.15 and 2.16.

#### *8.4.2.3 The Bourdon Tube*

The basic Bourdon tube is roughly elliptical in cross-section and sealed at one end (the tip). A simple Bourdon tube is C-shaped and this will tend to straighten as pressure increases inside it, so providing a measurable displacement. The C-shaped Bourdon tube usually has an angle of curvature between 180 and 270 degrees. The theory behind the straightening process is not simple, but depends initially on the fact that an elliptical tube bore contains less fluid than if it were circular, so that as the pressure increases within it, this bore tends to become less elliptical, which leads eventually to the straightening process. More sensitive types include both spiral and helical types, but unfortunately, all are sensitive to vibration, which precludes their usage for most aerospace applications.

### *8.4.3 SIGNAL ACQUISITION*

The pressure sensors described above all respond in the form of deflections, and these must be converted to electrical (or sometimes mechanical) signals. There are several common methods of achieving this including capacitive, inductive, and potentiometric methods, and the associated null-balance servo techniques. Whether any of these, or less common methods involving piezosensors or strain gauges, should be used depends, like the sensor itself, on the required range, accuracy, response sensitivity and speed, the environmental conditions, and the nature of the measured fluid.

### 8.4.3.1 Capacitive Deflection Transducers

The capacitive transduction principle utilized in pressure sensors appears in either of the following two designs (Bodner 1960; Fribance 1962; Holman 1994; Norton 1982) or in modifications thereof. In one, pressure is applied to a diaphragm that moves with respect to a single stationary electrode; in the other, the pressure is applied to a diaphragm supported between two stationary electrodes.

In the former, the diaphragm can be either the grounded or the ungrounded electrode. In the design shown in Figure 8.12, it is integrally machined with its support member and moves with respect to the other electrode, which is mounted on an insulating substrate. The full-scale diaphragm deflection is only about 0.1 mm. A lead connects the stator to an external terminal, and the case acts as the alternate terminal. The internal cavity of the sensor shown is evacuated and then sealed, so producing an absolute-pressure sensor unit wherein the final signal is in the form of a small capacitance change, so necessitating a high-frequency applied voltage in order to produce a measurable response.

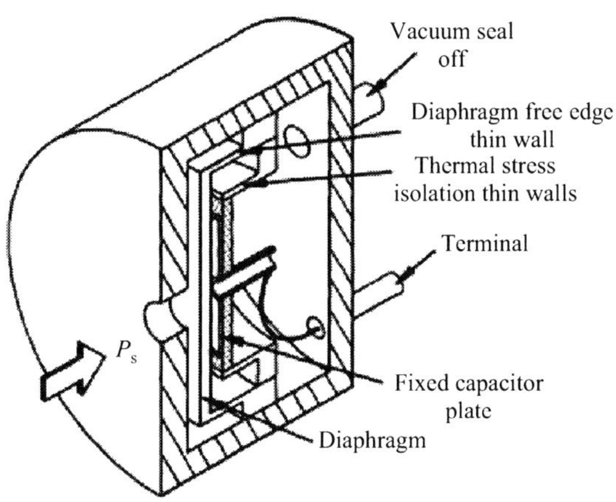


**Figure 8.12.** A capacitive pressure sensor.

Insulated stators, or insulated diaphragms supported between stators, are now frequently made of quartz or ceramic with the electrode vacuum-deposited or sputtered onto the substrate. Dual-stator designs offer greater capacitance changes because as the diaphragm deflects, its capacitance to one stator increases whilst simultaneously decreasing to the other stator.

### 8.4.3.2 Inductive Deflection Transducers

In an inductive deflection transducer the self-inductance of a single coil is varied by pressure-induced changes in the displacement of a metallic diaphragm in close proximity to that coil. Past designs have used a diaphragm of magnetic material moving with respect to a ferric core around which a coil is wound, or conversely, the displacement of a movable ferromagnetic core within a coil, both leading to small inductance changes. Some more recent designs have used a metallic diaphragm and a coil excited by a radio-frequency current to produce eddy current changes in that diaphragm leading to the required self-inductance changes.

#### *8.4.3.3 Potentiometric Deflection Transducers*

This type of transducer is a device having a high sensitivity and hence needing no amplification for most applications (Benedict 1977; Norton 1982). Single or multiple pressure-sensing capsules can be used for relatively low pressure ranges up to about 3.5 MPa. (For nonaerospace applications special types of Bourdon tubes are useful for high-pressure ranges.)

#### *8.4.3.4 Null-Balance Servo Pressure Transducers*

These designs are generally more complex than other transducer types but provide considerably better accuracy. Simplified block diagrams illustrating the three basic types are shown in Figure 8.13. All depend on a negative-feedback loop to return or maintain a sensor in a null-balance condition.

In Figure 8.13(a), the sensing element, usually a capsule or bellows, is allowed to deflect freely, and any displacement moves a transduction element such as a push-rod mounted ferric slug (as shown) and is detected by the coils of a differential transformer. The resulting electrical output becomes an error signal that is amplified and applied to a servomotor that returns the coil system to a null-output position whilst simultaneously driving a display device such as a potentiometer (shown here), or a synchro, or a digital shaft-angle encoder.

A motor-driven output device may also be used in the force-balance system illustrated in Figure 8.13(b). Here, the sensing element is not allowed to deflect freely but is restrained by a force generator. When the sensing element tries to deflect in response to applied pressure, this is again detected by a transduction element that produces an error signal. This transduction element can be inductive (as implied by the sensing coil shown), capacitive, or any one of several different displacement-sensing mechanisms. The error signal is amplified and applied to a motor that drives a display device whilst simultaneously causing the force generator to apply sufficient force to the sensing element to restore a balanced condition.

The operation of the force-balance transducer illustrated in Figure 8.13(c) is less complex and therefore used more frequently in production designs. The error signal produced by the transduction element again forms the input signal to an amplifier, but in this case the amplifier output drives a forcing coil that operates electromagnetically to maintain the sensor in a condition of zero displacement. Hence, the current through the forcing coil is proportional to applied pressure and may be measured to provide a pressure reading.

As an example of the force-balance approach, the Sundstrand Data Control Company manufactures a pressure sensor of this type, but using a capacitive displacement transducer. The output from this is detected and amplified, eventually feeding a force-balance coil that maintains the bellows/beam assembly in its null-balance position. The current through the force balance coil also produces the transducer output signal. This design has a measuring range of 0–200 kPa in its absolute-pressure version and a range of ±1 bar in its differential-pressure version shown. Repeatability is reported as within 0.02% of the pressure excursion, and the threshold (smallest detectable pressure change) is reported as 0.1 Pa.

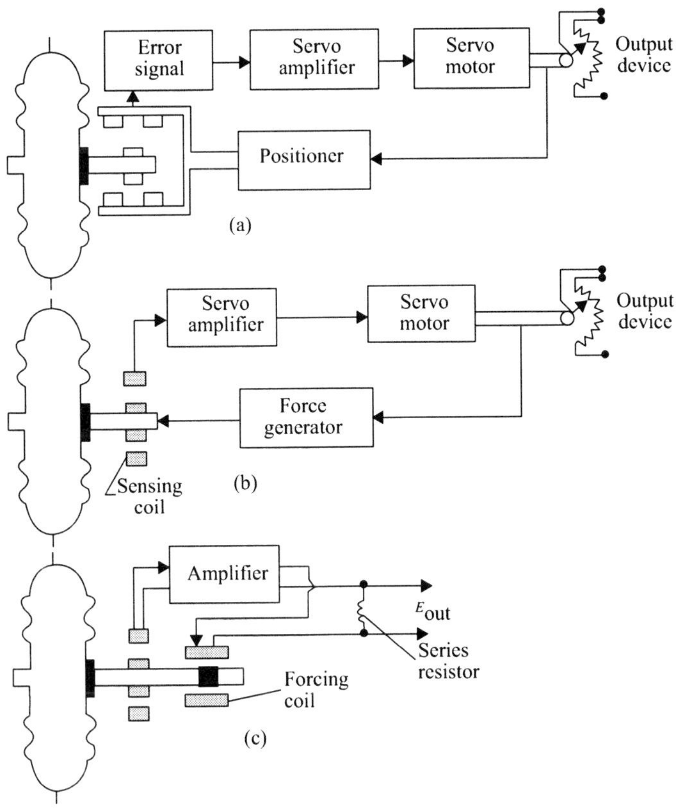


**Figure 8.13.** Principles of null-balance pressure sensors.

### 8.4.4 OPERATIONAL REQUIREMENTS

In aircraft, the pressures of fuel, oil, and hydraulic fluids are of major importance and require transducers that can provide signals for both indicators and warnings of pressures that are too high or too low. For example, a low fuel pressure condition may presage a failing pump or a leak, whilst a high pressure condition may suggest a blockage or partially opened control valve. Each manufacturer will have a different set of operational limits, and as an example, for the SAAB 2000 turboprop the normal engine oil pressure limits are 35 psig to 110 psig.

Another important application of pressure transducers is for the indication of *engine pressure ratio* (EPR), which may be defined as the differential pressure between the turbine discharge total pressure and the compressor inlet total pressure. These two pressures and other fixed engine parameters provide indications of thrust. The EPR is used to select the minimum thrust required for cruise and other operational conditions. The minimum required thrust will also be the thrust for minimum fuel usage. As another example, the Boeing 727-100 Performance Handbook defines cruise performance in a table entitled "Maximum Cruise EPR" which shows that at a pressure altitude of 40,000 ft and an outside air temperature (OAT) of –40°C, the maximum cruise EPR is 2.17 for engines 1 and 3 and 2.23 for engine 2. However, with the addition of more computer control systems, EPR numbers are becoming less important to crews. In the Boeing Quick Reference Handbook (QRH) for the 757, performance settings are still provided in relation to EPR indications, whilst in the 787 QRH these numbers are provided as

Turbofan Pressure Ratios (TPRs). In the 787 and newer aircraft, pilots are provided with markings on the relevant indicator to show maximum performance selections for minimum fuel consumption for a given flight condition. Furthermore, the advent of the LCD display has allowed aircraft instruments to be reconfigured as a function of operational mode.

Though only indirectly connected to propulsion, cabin pressurization sensors are critical for the provision of a safe operational envelope for the crew and passengers. A typical pressurization system will have pressure sensors for cabin pressure level, differential pressures, bleed pressures, and other inputs on more complex systems. These systems, like most aircraft systems, contain redundancy and are designed to regulate cabin pressure automatically. Typical cabin pressures are maintained at levels similar to pressure altitudes between 7,000 and 8,000 ft.

Certain pressure indicators are required by regulation, and in the United States, examples include oil pressure indication according to the Code of Federal Regulation (CFR) Title 14 Section 25.1305 and fuel pressure indication according to Section 25.1337. When sensors are required under the CFR they must meet certain performance specifications. These are typically satisfied when a pressure sensor meets the FAA Technical Standard Order (TSO) for that sensor. For example, TSO C47a provides specifications for the fuel, oil, and hydraulic sensors required by regulation; and also includes environmental, hardware, and software qualification requirements. Aircraft sensors cannot be selected without these qualifications unless an independent qualification test is established and approved by the national regulatory body.

## 8.5 ENGINE TEMPERATURES

The control of the internal temperatures of a turbine engine is critical to the life of the engine. Hence, temperature sensors play a key rôle in the operation of such engines and provide the crew with information about their operational state. As Figure 8.14 shows, the temperature and pressure extremes inside the engine are at the limits of most physical temperature sensors. Most aircraft engines use thermocouples to provide temperature indication, but some new designs also use optical pyrometers to indicate burner temperatures. However, this type of sensor is more commonly found in the test environment rather than the operational environment.

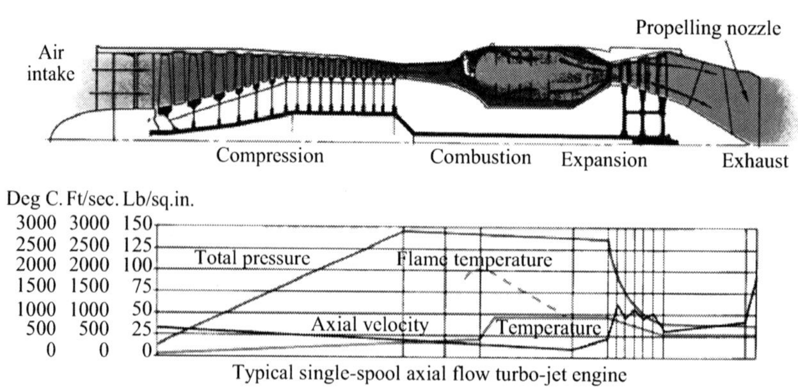


**Figure 8.14.** Environmental extremes for the turbine engine (Courtesy of Rolls Royce plc.).

For most new aircraft designs, Full Authority Digital Engine Controller (FADEC) units maintain engine temperatures and rotational speeds within prescribed limits. They automatically limit such temperatures by controlling fuel flow, this being the primary means of controlling the turbine engine itself. The FADEC also looks at a multitude of sensors to keep the engine within operational limits whilst providing the thrust output selected by the pilot. In older engine designs that are still in operation in a great number of aircraft, the crew takes the place of the FADEC and monitors temperatures to keep the engine within operational limits, the single engine controller being the thrust lever that actually directly controls fuel flow.

In addition to thermocouples, other types of sensor may be employed for external engine measurements such as nacelle temperature, fuel temperature, and oil temperature, for example.

The list below is a sample of those engine indications that depend on temperature measurement.

### *8.5.1 INTERMEDIATE TURBINE TEMPERATURE (ITT)*

This is the temperature measured between the high and low pressure turbine systems of a turbojet or turbofan engine. On the turboprop or turboshaft engine it is the temperature taken between the gas production turbine and the free power turbine. The temperature limits for the ITT will vary from engine to engine but controlling this through fuel flow is critical to the health and performance of any engine. There are different limits for the ITT during start and operation as shown in Figure 8.15, and pushing an engine beyond these limits can result in its damage or destruction.

In most cases the ITT is measured using a thermocouple or bank of thermocouples. Such thermocouples are typically encased in a highly temperature-resistant alloy such as Inconel.

Another type of sensor in use for these high temperature measurements is the noncontact infrared temperature sensor, which can provide accurate results in the 800–1500°C (1500–3000°F) range, and can operate in high background temperature ranges up to about 180°C.

The basic design of the infrared temperature sensor involves a lens to focus the infrared energy on to a detector which converts some of this energy into an electrical signal that can be processed by signal conditioning electronics and displayed in units of temperature after compensation for ambient temperature variation. The infrared thermocouple is a special case of the infrared sensor, being self-powered and providing the same output as would a normal thermocouple. These sensors can operate in the high ambient conditions of the engine nacelle. When installing them, however, calculations must be made to ensure that the optics are correctly aligned for accommodating the measurement area it is required to examine—this is called the "spot size" calculation. Another parameter of concern for such sensors is the emissivity, which is the relative ability (compared with the standard "black body") of a surface to emit energy by radiation. These IR probes come in a wide range of emissivity values, and some manufacturers such as Exergen define their emissivity units in two ranges Hi E (nonmetal) and Lo E (metal). Care must be taken to install them only when spot size and emissivity are known.

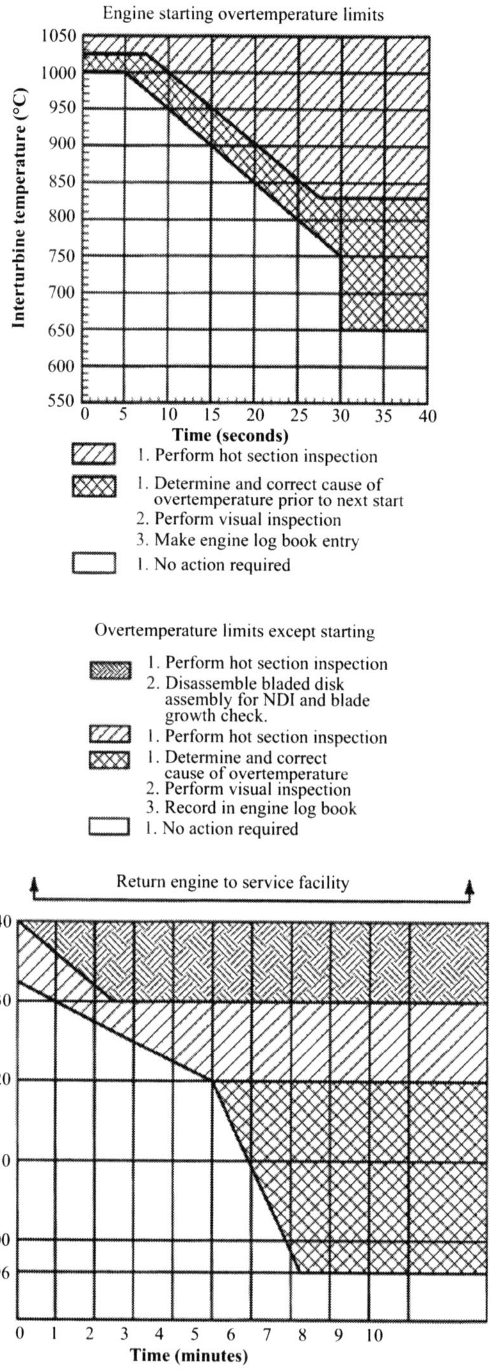


**Figure 8.15.** ITT Limits for the Williams FJ44-1A turbofan "Graphic Courtesy of Williams International".

### *8.5.2 OIL TEMPERATURE / FUEL TEMPERATURE*

The sensors used in these applications are typically thermistors or RTD (Resistance Temperature Device) elements that are appropriate over the range –50 to 200°C. Oil temperature warning levels vary from engine to engine but are typically 90°C and above. An example is the Allison AE2100A engine, for which the limits are 86°–93°C. Between these temperatures the FADEC will turn on an amber caution message to the crew.

The need for fuel temperature measurement is critical for the high altitude jet because of the low temperature at altitude and the fact that JET A (the typical turbine fuel in use by civilian aircraft) gels at temperatures below –40°C. The sensor used is typically an RTD to provide a readout typically from –50°C to 150°C, and is located past the fuel/oil heat exchanger used on high altitude aircraft. The FADEC will normally energize a caution light preset to operate at some specific temperature within this range. The critical nature of fuel temperature measurement cannot be overstated: in one case on January 17, 2008, a Boeing 777-236ER suffered a dual engine failure resulting in loss of thrust command by the pilots during landing. This was found to be due to ice formation in the tanks and the resultant occlusion of fuel flow by this ice at the fuel/oil heat exchanger.

### *8.5.3 FIRE SENSORS*

There are many fire detection systems used in today's aircraft, and all turbine aircraft must include these in some form. Such sensors may be self-generating or passive types, the former being typified by thermocouples and the latter, thermistors. A less common type is the capacitive temperature sensor, which consists of a tube containing a dielectric material with a conductor running through the center. As the temperature increases, the dielectric constant changes accordingly, and the corresponding capacitance change is detected and a fire warning is signaled. A major advantage of this type of fire loop is in the comparatively large area of this rather unusual sensor element. Another unusual example is the sealed-tube sensor wherein trapped gas in a tube is purged when in the presence of fire to result in an appropriate warning signal. Finally, there are new types of fire detection devices using light sensors that respond to the light radiation frequencies present in a kerosene fire.

The measurement of temperature is critical to engine health and successfully reaching overhaul time limits. In the case of large engines FADEC controllers have made the start process safe and error-free in most cases. However, the majority of small engines depend on pilot throttle input to control engine temperatures. Furthermore, the temperature acquisition and display must be sufficiently rapid for the pilot to properly control the start procedure. Figure 8.16 shows two engine starts for a small 675 hp turboprop engine. Both of these are actually restarts after an engine shutdown, as indicated by the initial temperatures. This is the most critical initial condition, and can cause an over-temperature situation. In the case of Figure 8.16 one start is performed by the pilot in full control of engine fuel flow the higher of the two temperatures. In the second case a simple temperature control (rate limiter) is used to trim fuel flow to avoid an over-temperature situation this is the lower of the two temperatures. An outstanding feature of the graphs is the rapid changes in temperature that take place over only a few seconds. In a normal start cycle the pilot will spool the engine up to a given rpm, activate the igniters and initiate the fuel flow by bringing the throttle control forward. In the manual control case he/she must watch the temperature closely to avoid an over temperature situation, and this is accomplished by carefully controlling the throttle position.

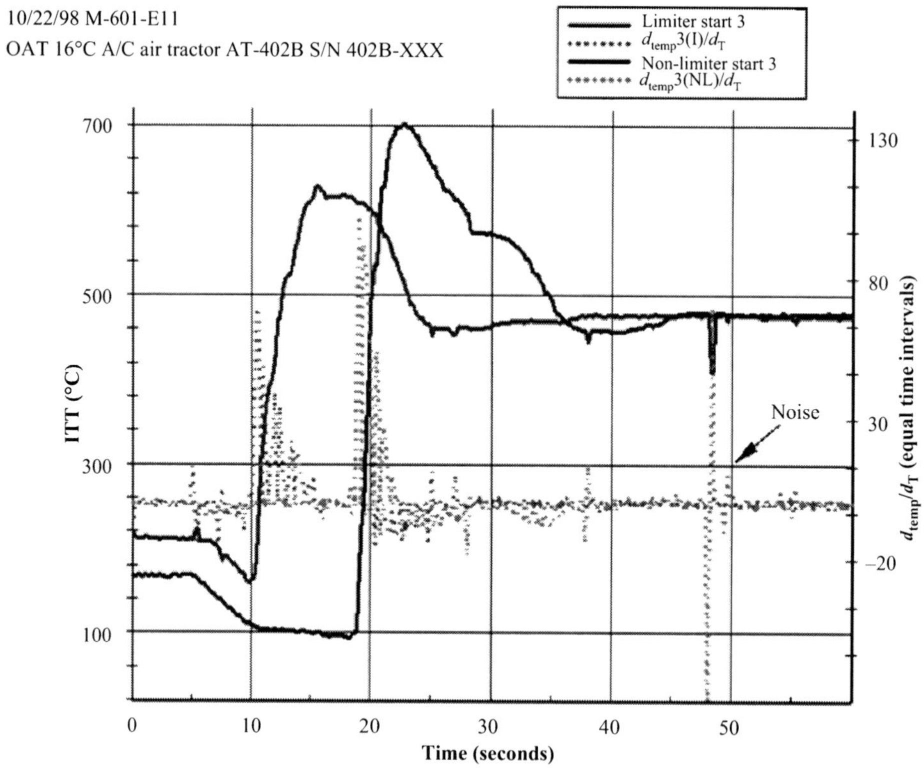


**Figure 8.16.** Turboprop engine start temperatures showing manual and fuel limiter assisted starts.

### *8.5.4 EXHAUST GAS TEMPERATURE (EGT)*

The EGT is the total temperature measured at the turbine exit, usually by means of a thermocouple-based system. The total temperature accounts for both the ram heat (friction) and the static heat from the exhaust gases. In the big turbines common in larger aircraft this is normally the principal indication of engine operational temperature. In newer aircraft IR methods may also be used to provide this readout. As with the ITT discussed above, the primary control of engine temperature is the set point for fuel flow selected by the throttle. As an example, in the case of the DC-10, 11 thermocouples are used as sources for the cockpit EGT display.

### *8.5.5 NACELLE TEMPERATURE*

This temperature is typically measured using a resistive device such as a thermistor, or RTD, usually mounted at one location inside the nacelle. In the case of the thermistor, which is (usually) a negative temperature coefficient device, it may form one of the legs of a Wheatstone Bridge whose output is fed directly into a readout or as an input to a FADEC. Typical display values for the readout are 0 to 300°C. However, there are also examples of thermocouples being used for this task, as in the case of the SAAB 2000, where engine overheat warnings are generated by the FADEC at or above 232°C.

## 8.6 TACHOMETRY

The rotational speeds of engines, either piston or turbine, are of major importance to the operating conditions of those engines. In particular, engine rotational speeds are indicative of the power being produced, and they are also subject to both upper and lower limits that must not be exceeded. Hence, the tachometer is also used to provide appropriate safety indications and warnings. There are several forms of sensors that can be used, and the major ones are introduced below.

### *8.6.1 THE EDDY CURRENT TACHOMETER*

The primary application of this form of tachometer was on small general aviation aircraft, and though simple and reliable it is no longer in common use; but it well illustrates the basic principle of the method. It consists of a permanent magnet fitted to an axle and surrounded by a nonferrous (aluminum) *drag cup* that has its own axle, as shown in Figure 8.17. The magnet axle may be coupled to the engine drive shaft so that the magnet rotates at the same speed, so inducing eddy currents in the nonferrous cup. These eddy currents produce their own magnetic fields that interact with the permanent magnet to produce rotation. However, the cup axle is restrained by a spiral hairspring that limits its movement to an angle where the two rotational forces reach equilibrium, and this angle will therefore represent the rotational speed of the engine shaft. In practice, a carefully calibrated hairspring is used, and the drag cup is mounted in jeweled bearings to minimize mechanical friction.

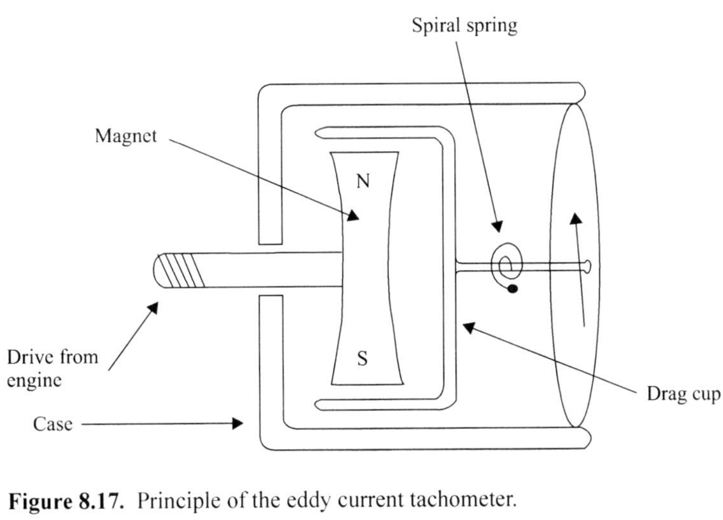


**Figure 8.17.** Principle of the eddy current tachometer.

The magnet axle may be coupled to the engine shaft via a flexible drive cable and located elsewhere, usually on the instrument panel. Also, it may be driven via a gear train at an accessory drive point to achieve a more appropriate speed.

### *8.6.2 THE AC GENERATOR TACHOMETER*

From the above, it will have been noted that the eddy current tachometer consists basically of two parts, the magnet drive and the drag cup, and that the drive from the engine is transmitted mechanically to that assembly if there is a requirement for a remote indicator. A flexible drive cable does have reliability problems however, so a reasonable solution is to substitute an electrical coupling. For example, the engine may drive a DC generator, the output of which could be arranged to operate a DC motor for driving the rotating magnet in the tachometer. However, this would introduce a further problem in that the resistance of the wiring between the two would produce a voltage drop that would alter the desired speed of the magnet, and this could also change with temperature. Furthermore, the need for a commutator would introduce some mechanical unreliability plus some transient radio interference that might disturb electronic devices in the vicinity. Although such units have been built, a better solution is to use an AC generator operating a synchronous motor driving the tachometer magnet. This has the result of using frequency rather than voltage as the operational parameter so that the wiring resistance becomes much less important, which allows longer wiring runs from the engine to the rpm display.

Drag cup display units are self-generating and require no external power, so offering an increased margin of safety. They are now primarily used in small general aviation aircraft where the propeller is directly coupled to the engine so that the engine rpm is also the propeller rpm.

Usually, the AC generator consists of a star-connected three-phase machine driven by the engine and electrically connected to a synchronous motor in the tachometer proper. Typically, a tachometer generator will produce a sinusoidal, 26V, 400Hz output. The basic principle of the generator/display combination is illustrated in Figure 8.18 from which it will be appreciated that the whole system is actually still an eddy current tachometer, but with the magnet drive mechanically isolated from the engine. The transmitter unit can be small and driven directly from the engine or an accessory drive via a simple splined shaft.

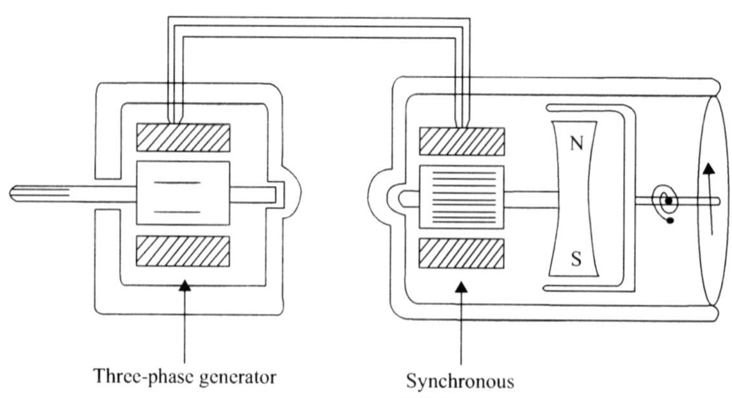


**Figure 8.18.** Principle of the three-phase AC generator tachometer.

There are, of course, many improvements that can be made to such a system. One is the inclusion of longitudinal copper bars in the synchronous motor stator to form a collocated squirrel cage motor, which has much better starting characteristics than does a purely synchronous machine. Another is that a set of gears may be included on the drag cup axle so that both a coarse and a fine readout can be displayed. Also, as progressively more aircraft are being retrofitted with digital monitoring systems for engine parameters, it has become necessary to convert the AC signal from the generator to a digital equivalent. An example of such a device is the ST26 converter made by Sandia Aerospace.

### *8.6.3 THE VARIABLE RELUCTANCE TACHOMETER*

This instrument uses a simple pick-up sensor consisting basically of a permanent magnet along with a coil wound on a ferrite core, as in Figure 8.19. When a ferrous object moves across its field, the reluctance changes and this can be detected as a current pulse in the coil. Hence, if such an assembly were positioned close to the teeth of a gear wheel or the outer edges of turbine blades, their passage could be counted. Thus, unlike the tachometers described earlier, the device is essentially a digital rather than an analog sensor. This leads to great accuracy because, whereas the rotational speed of a gas turbine spool is already high, there will also be a number of pulses per revolution depending on the number of gear wheel teeth or the number of turbine blades. Actually, modern engines include dedicated pick-off rings to provide the pulses—for example, the Pratt and Whitney PW4460 engine uses a notched ring that provides 60 pulses per revolution. The electronic system that follows the sensor is based on well-known pulse-counting and frequency measurement technology.

Modern turbine engines may also incorporate up to three concentric shafts (spools): one driving the fan (N1), one driving the compressor (N2), and the fastest one being the turbine driven by the hot gases (N3). As an example, the maximum allowed speeds of the three spools in one version of the Rolls Royce Trent engine are 3,500 rpm for the fan spool, 7,700 rpm for the compressor spool, and 10,000 rpm for the turbine spool.

Manufacturers specify optimal rotational speeds for various operational conditions and altitudes because such rpm settings at altitude will vary from those at sea level; and because temperature also affects some rpm limits such as those for idle speeds. Therefore, using the

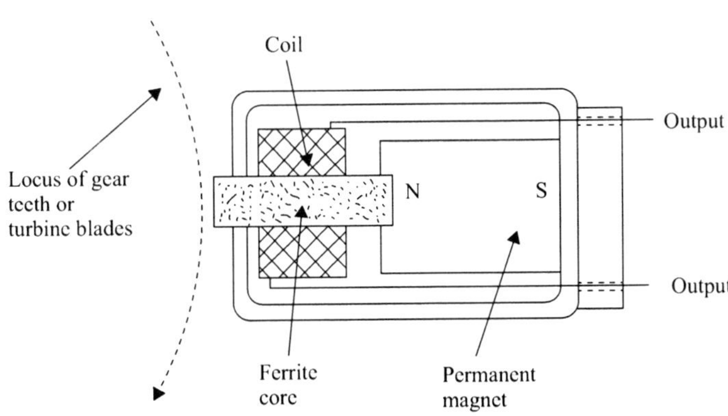


**Figure 8.19.** Principle of the variable reluctance pick-up.

modern reconfigurable displays found in "glass cockpits," an instrument may indicate the actual rpm as a percentage of the maximum rpm and show a target rpm marking for a specific operational condition. As another example, for the Boeing 757 at a cruise condition of Mach 0.78 and 25,000 ft, the flight manual recommends 76.3% N1, whilst at 40,000 ft it is 82.8% N1.

### *8.6.4 THE HALL EFFECT TACHOMETER*

Another transducer used for rpm measurement is the Hall Effect sensor. The active element of this consists of a small thin chip of a semiconducting material such as indium antimonide, with contacts on both sets of opposing edges and the faces, as shown in Figure 8.20. A drive current is passed through this chip using one pair of edge contacts. When a magnetic field is passed through the chip perpendicular to the faces, a voltage appears across the remaining edge contacts that is proportional to the magnetic field strength (Ramsden 2006). However, this output voltage is only at the microvolt ($\mu$V) level and therefore requires electronic amplification to produce a usable output signal. When this Hall element is combined with integrated electronics to provide the drive current, amplification, and appropriate signal conditioning, it becomes a complete Hall Effect sensor. Of course, one or more magnetic elements must also be included with the gear wheel or turbine ring to enable pulse detection to take place at all.

These sensors offer many advantages over variable reluctance devices in terms of size, sensor area, zero position (stationary) indication, and long life. Also, they may be used not only as engine sensors, but also as fuel flow pick-offs; and because of their ability to operate in the stationary mode, they may also be used to indicate the positions of selectors or other moving parts in the aircraft environment.

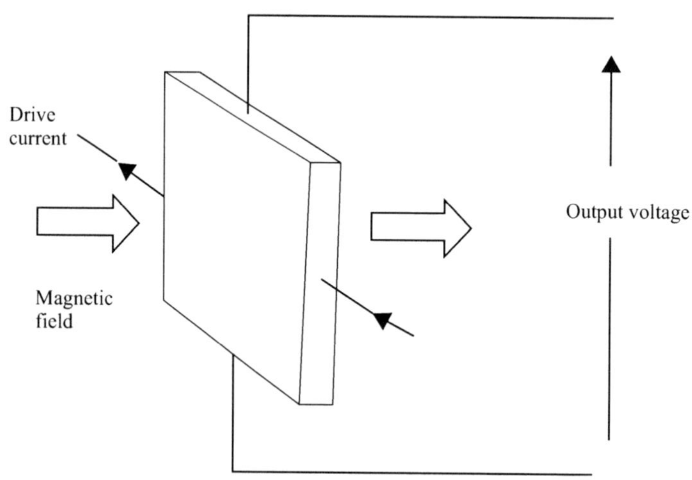


**Figure 8.20.** Principle of the Hall Effect sensor.

## 8.7 VIBRATION SENSORS—ENGINE AND NACELLE

Many aircraft from helicopters to large transport category types employ vibration sensors as a means of indicating unbalanced conditions and/or to provide component vibration data for preventive maintenance. In early applications they were used to indicate off-balance conditions to the crew, but have now transitioned to include unbalance warnings as inputs to predictive maintenance systems. As an example, the DC-10 was equipped with a vibration indicator using sensors mounted on the fan and turbine sections of each engine. This data was interfaced to an indicator system showing the relative level of vibration for each engine. Thresholds were also used to set amber cockpit warning lights that indicated unduly high engine vibration. In the MD-11, this same sensor arrangement was employed but some added interfaces were used to provide improved engine vibration monitoring. For the MD-11 the first stage filter after the vibration sensors was set in a range from 15 to 187Hz depending on the aircraft and engine.

In more advanced applications such as those employed in newer helicopters, the vibration sensor is an integral part of both in-flight monitoring systems and post-flight data analysis systems. Such sensors help to identify aircraft components that are at risk of failure. In flight, they can provide predictive warnings to crews of system anomalies sooner than could the old-style CHIP sensor. In early aircraft the primary means of indicating engine mechanical issues was a magnetic probe placed in the oil flow system and designed to form an electrical circuit when metal chips from the engine built up around that probe, so forming a current path between the probe itself and its metal casing. This activated a red warning light typically marked "CHIP." The production of such metal chips is obviously an excellent indicator of an incipient engine malfunction. Engine CHIP sensors are actually still in use today in all types of engines but are particularly important in helicopter operations.

On the ground, data from the sensor system is extracted and time histories are converted to frequency components in a search for unusual vibration frequencies. Long term measurements are compiled and employ probability density functions to provide indications of failing components based on spectral measurements moving toward, or outside of, normal limitations. These systems are often called HUMS—Health and Usage Monitoring Systems. HUMS can also use a mixture of sensor systems including those for sound and temperature, but vibration is one of the key components. The following is a quotation from an article in "Aviation Today" dated February 1, 2006, that discusses the importance of HUMS to aircraft safety: "The widespread use of HUMS is expected to be a major contributor to the helicopter industry's goal of reducing the accident rate by 80 percent within 10 years."

For vibration measurement, the usual sensor is the piezoelectric accelerometer, which measures only dynamic changes in vehicle conditions. These types of accelerometer require a charge amplifier (Figure 8.21) to produce an output that can be used as an input to a data acquisition system. Basically, this is an operational amplifier that accepts a charge at its summing point produced by the piezo accelerometer and converts it to a voltage pulse at the output.

A typical material used in piezoelectric accelerometers is quartz, and for such a sensor using the transverse effect, the charge measured is given by:

$$Q = Fd\frac{b}{a}, \tag{8.9}$$

where $F$ is the force in Newtons applied to the crystal, $d$ is the piezoelectric coefficient (Coulombs/Newton) and $b/a$ is the ratio of crystal height to width in metres. A typical quartz element has a piezoelectric coefficient in Coulombs per Newton of $2.3 \times 10^{-12}$; whereas for some piezoceramic materials this value can be $390 \times 10^{-12}$, resulting in much higher outputs. As already mentioned, the output signal is produced only during dynamic motion and no signal is produced under static conditions.

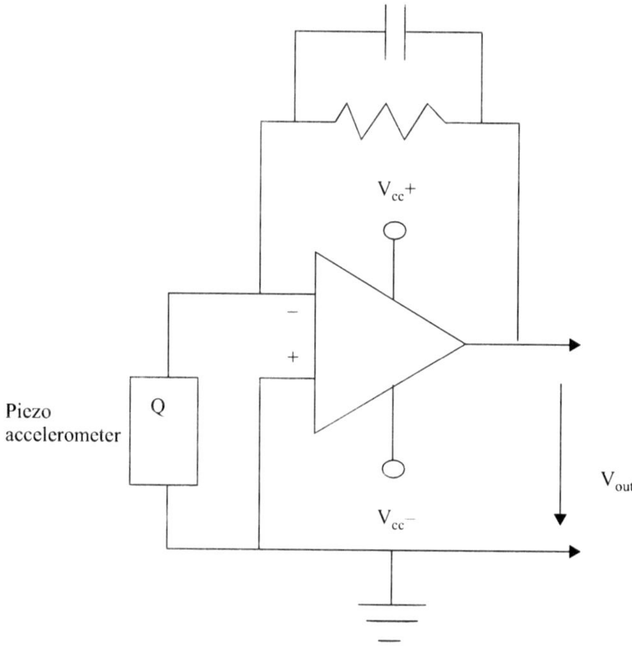


**Figure 8.21.** Basic charge amplifier circuit.

Such sensors have very wide frequency ranges and can withstand large accelerations without failure or loss of calibration. Figure 8.22 shows accelerometers integrated into stainless steel housings specifically designed for application in HUMS installations. These sensors, by PCB Inc., are designed to mount via a bolt or stud on an engine or airframe component, and each has a very high frequency range of 1 to 5,000Hz, will withstand a peak acceleration of ±500 g, and will produce an output of 10 mV $g^{-1}$ when properly amplified. Some dynamic accelerometers are designed to withstand ±2500 g, but these units typically have lower resolutions of about 2mV $g^{-1}$.

**Figure 8.22.** Specially designed accelerometers for use in HUMS measurements. (Courtesy PCB Inc.).

Vibration sensors are also used to help in the determination of the natural frequencies and response characteristics of nacelle installations. When mating an engine to an airframe, care must be taken to make sure that the design does not excite any undamped natural frequencies. Typically, a modal exciter such as a Bruel & Kjaer model 4824 is used to test the response of the engine mounting frame to dynamic inputs. These units inject a pulsed load into the engine frame and the resulting responses are measured by acceleration pickups. Figure 8.23 shows the data from a test where a step input was used to excite the frame, the resulting accelerations being shown as a function of the input.

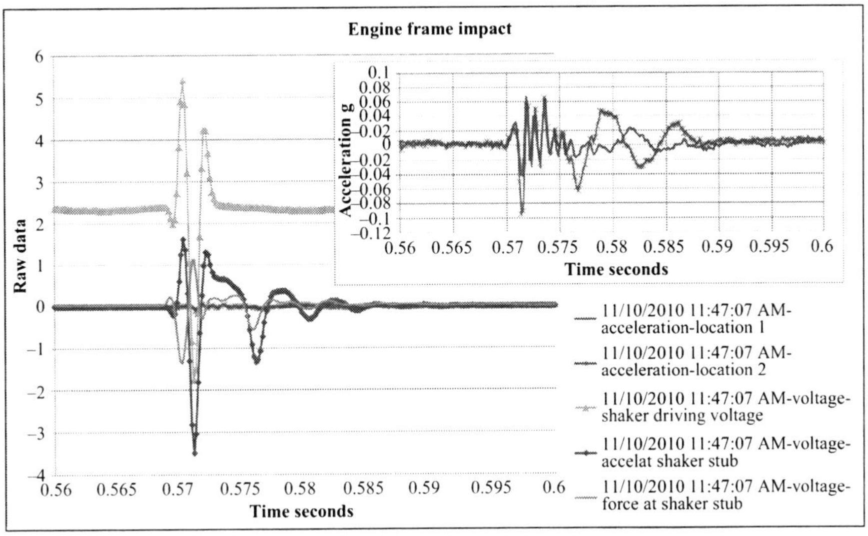


**Figure 8.23.** Engine nacelle responses to modal exciter input. The lines in inset graph are the responses to impacts. The effective data rate is 1000 samples/sec per channel simultaneously.

## 8.8 REGULATORY ISSUES

As mentioned in Section 8.4 there are certain government regulatory issues related to the adoption of components in aerospace applications for each country. Today, in the United States, even the military is beginning to use Commercial Off The Shelf (COTS) components in its aircraft. In fact, every sensor so far noted has some form of regulatory guidance that establishes design and environmental requirements, and the stringency of such requirements has contributed immeasurably to the excellent safety record of modern aircraft.

Table 8.1 lists the sensor types mentioned in this chapter and gives examples of the relevant American FAA Technical Standard Orders (TSO). The examples given are therefore only samples of the regulatory documents related to each sensor type. Every component in an aircraft requires some form of regulatory engineering approval in order to be installed on that aircraft, from agricultural crop dusters to the largest commercial jet. Even the lavatory smoke detector used in commercial aircraft must meet TSO-C1d to be eligible for installation on such aircraft! For the United States, these regulations can be found at the Federal Aviation Administration's Regulatory and Guidance Library (RGL) at http://rgl.faa.gov/. The EASA regulations are very similar and can be found at http://www.easa.europa.eu/agency-measures/certification-specifications.php.

**Table 8.1.** A sampling of the FAA Technical Standard Orders (TSO)

| Sensor type | FAA regulatory compliance reference |
|---|---|
| Fuel Quantity | TSO-C55a, TSO-C55, |
| Fuel Flow | TSO-C44c |
| Pressure Sensors | TSO-C47a, TSOC45a |
| Temperature Sensors | TSO-C43c |
| Engine Fire Sensor | TSO-C79. TSO-C11e |
| Tachometer Sensors | TSO-C49b |
| Vibration Sensors | TSO-C153 |

## REFERENCES

McShea, R. 2010. *Test and Evaluation of Aircraft Avionics and Weapons Systems* (1st edition). SciTech Publishers.

Moir, I., and Seabridge, A. 2008. *Aircraft Systems: Mechanical, Electrical, and Avionics Subsystems Integration* (3rd edition). John Wiley & Sons.

Pallett, E. H. J. 1992. *Aircraft Instruments and Integrated Systems*. Longman Sc & Tech.

Walter, P. 2002. *The Handbook of Dynamic Force, Pressure and Acceleration Measurement* (1st edition). Endevco Corporation.

The authors would also like to acknowledge material initially provided by the late Professor M. S. Katkov, formerly of the State University of Aerospace Instrumentation, St. Petersburg, Russia.

## BIBLIOGRAPHY

Beck, M. S. 1981. "Correlation and instruments: Cross-correlation flowmeter." *Journal of Physics E.: Scientific Instruments* 14 (1): 7–19. DOI: 10.1088/0022-3735/14/1/001.

Benedict, R. P. 1977. *Fundamentals of Temperature, Pressure and Flow Measurements* (p. 465). New York: John Wiley and Sons, Inc.

Bodner, V. A., G. O. Fredlender, and N. I. Chistiakov. 1960. *Aviation Instruments* (p. 512). Moscow: Defence State Publishing house. (In Russian.)

Bochniak, L. L., and L. N. Bisov. 1968. *Tachometers Flow Meters* (p. 212). Moscow: Engine. (In Russian.)

Fribance, A. E. 1962. *Industrial Instrumentation Fundamentals* (p. 328). New York: McGraw-Hill Book Comp.

Helfric, A. D., and W. D. Cooper. 1994. *Modern Electronic Instrumentation and Measurement Techniques* (p. 446). Upper Saddle River: Prentice-Hall International, Inc.

Holman, J. P. 1994. *Experimental Methods for Engineers* (p. 616). New York: McGraw-Hill Inc.

Klaassen, K. B. 1996. *Electronic Measurement and Instruments* (p. 352). Cambridge: Cambridge University Press.

Kremliovskiy, P. P. 2002. *Flowmeters and Meter of a Quantity of Matter: Book 1* (p. 409). Saint-Petersburg, Russia: Politechnica. (In Russian.)

Lee, W. F., M. J. Kiric, and J. A. Bonner. 1975. "Gas turbine flow-meters measurement of pulsating flow." *Journal of Engineering for Gas Turbines and Power* 97: 531.

Norton, H. N. 1992. *Sensors and Analyzer Handbook* (p. 562). Englewood Cliffs, NY: Prentice-Hall Inc.

Nuliten, J. A., R. P. Keech, and J. Coulthard. 1983. Industrial mass flow measurement trials using an ultrasonic cross-correlation flowmeter/Flow measurement Proceedings, pp. 113–23.

Ramsden, E. 2006. *Hall-Effect Sensors: Theory and Applications*. Elsevier.

Smith, R. V., and J. T. Leang. 1975. "Evaluation of correlations for two-phase flow-meters, three current-one new." *Journal of Engineering for Gas Turbines and Power* 97: 579.

CHAPTER 9

# Principles and Examples of Sensor Integration

Alexander V. Nebylov
*State University of Aerospace Instrumentation*
*St. Petersburg, Russia*

## 9.1 SENSOR SYSTEMS

### *9.1.1 THE SENSOR SYSTEM CONCEPT*

If there is only one sensor for obtaining information on the value of any measured parameter onboard a vehicle, opportunities for determining the accuracy and reliability of this measurement are essentially limited. For example, measurement of noise can be partially suppressed only along with some of the spectral components of the useful signal, and optimal one-dimensional filtration of a signal mixed with noise defines the limit of measurement accuracy. Also, the failure of a single sensor obviously leads to the complete loss of the desired measurement. Finally, the possible range of the relevant physical values is often difficult to cover by any single sensor. Therefore, only the integration of several sensors within one system makes possible a measurement of the desired quality.

The problem of integrating two or more sensors is one of the major difficulties in designing navigation and motion control systems for aerospace vehicles. However, new integration algorithms along with improvements in the characteristics of primary sensors permit the realization of ever-increasing requirements for measurement accuracy and reliability.

It is clear that structural redundancy in a measuring system, plus multiple sensors, expands opportunities for improvement in measurement quality. Such redundancy may include the following:

- Some identical sensors
- Some similar sensors with different ranges and measurement accuracies
- Some sensors working on the basis of different physical principles, but measuring the same or functionally connected parameters.

The actual structures of integrated measuring systems and the relevant connections between the sensors can therefore take various forms. For example, it is possible that the output signal of one sensor might be injected into another sensor to facilitate its operation in some way. Such an example implies that the optimization of the mutual influences of sensors is possible. The well-known tightly coupled GPS/INS integration system may be considered a good example of such an approach.

An extremely generalized configuration of an integrated measuring system is shown in Figure 9.1 where $S_i$, for $i = \overline{1,l}$, are the available onboard primary sensors whose output signals, $x_i(t)$, are applied to a computer that produces appropriate values for the measured signals $\hat{g}_j(t)$, $\eta > 1$. If necessary, it is possible to augment the scheme with direct connections between sensors, but this is not of real importance for the potential facilities of an integrated system.

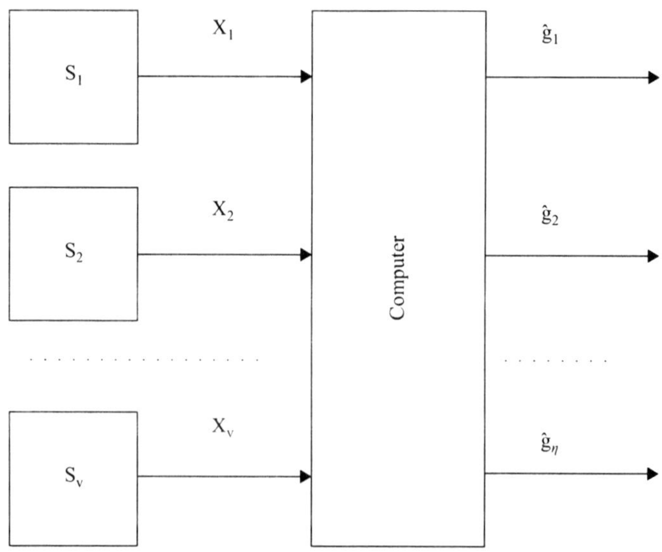


**Figure 9.1.** Generalized integrated measuring system scheme.

The relevant algorithm depends on many factors and can be described by a system of differential/difference equations of a rather generalized form. Normally, all the problems to be solved by the computer can be divided into three groups. The first group is concerned with the reduction of all sensor outputs to conform to a uniform system of measurement units, that is, within a single frame of reference. Also, taking into consideration that the sensors may be placed in different regions with different geometrical proportions, different scale factors will be needed for sensors responding to kinematical values (coordinates, speeds, or accelerations) and other factors affected by differences in sensor placement. Algorithms for solving similar problems are often nonlinear, for example, they may be based on spherical trigonometrical formulae, as in the case of navigation and orientation parameter measurement for aerospace vehicles. As a result, all sensor outputs must be transformed in such a way that the calculated estimate of one physical value can be compared with another and should be completely identical, so resulting in an ideal situation that resembles sensors without measuring errors.

The second group of problems is concerned with maintaining the best weighting of each sensor output to make the final calculated estimate valid. Clearly, the most accurate sensor in comparison with other available sensors over a specific range of measurements should provide the greatest contribution to the final result. For example, if two sensors have identical error spectral structures, but the dispersions of errors $D_{n1}$, $D_{n2}$ are different and are in the ratio $D_{n2}/D_{n1} = r$, and these errors are mutually independent, then it is expedient to calculate the final value as a sum in which the output of the first sensor is weighted with the coefficient $k_1 = r/(r+1)$ and the second with the coefficient $k_2 = 1/(r+1)$. Thus the dispersion of the error in a final value will reach the theoretical minimum

$$D_e = D_{n1} r/(r+1). \tag{9.1}$$

At $r = 1$ when the sensors have equal accuracies, the dispersion for an integrated system will be a half of the initial dispersion, $D_e = D_{n1}/2$. If the sensors are widely different in accuracy, only the best sensor will have a significant influence on the resultant accuracy. (It should also be noted that different sensor reliability indices can also affect the choices of weighting coefficients.)

If *a priori* information on distinctions in the spectral structure of sensor errors is available, the parity of weight coefficients should depend on frequency, that is, the computer should filter the sensor signals before their summation. The theory of optimal linear filtration may be broadly applied in the relevant processing algorithm design if the errors follow a normal or unknown distribution law. If the error distribution is not Gaussian, optimal nonlinear filtration may be applied.

The third group of problems is concerned with the automatic technical diagnosis of sensor condition. The automatic detection of any single failure in a measuring system is often needed, and the failure of any primary sensor is more likely than a failure in the computer itself. For solving such problems, some functions of the individual sensor outputs and final weighted outputs are formed (usually their weighted sums). These functions are normally close to zero, but increase sharply when a sensor fails. In the simplest case of two identical sensors the difference of their outputs may provide such a test function. In the case of three identical sensors, the differences of each output and final weighted output may be considered as three test functions.

Such a function (called an *invariant*) is compared with a threshold value, and any sharp increase implies sensor failure. This threshold value is chosen from an acceptable false alarm level, and the calculated maximal error value of the final estimation is stored from the moment of failure detection. After any sensor failure is detected, the measuring system is reconfigured, but continues operating using inputs from the remaining sensors, though the quality of the final estimation decreases.

The separation of algorithm design problems into the above three groups is somewhat arbitrary because the boundaries between these groups are blurred. However, in principle, they could be stated and solved simultaneously on the basis of the mathematical theory of optimal filtration. It is quite clear from Figure 9.2, where "the scheme of compensation" (as it was called in Bendat and Piersol (1966)) is shown, that for the optimal estimation of a measured parameter it is necessary to extract the best filtration error from a mixture of both errors. Numerous filtration methods are well developed for all possible variants of the problem, the most popular being the algorithms of Kalman's linear filtration and its many modifications.

Appearing in the early sixties, algorithms for the Kalman–Bjusi evaluation immediately found numerous applications in many practical problems, including those for data fusion synthesis. Nowadays, it is difficult to visualize a real integrated navigational system in which a Kalman–Bjusi algorithm or one of its modifications would not be used. An essential feature of these algorithms from the practical point of view is that it is not necessary to remember any prior information—the future state of a measuring system is determined only

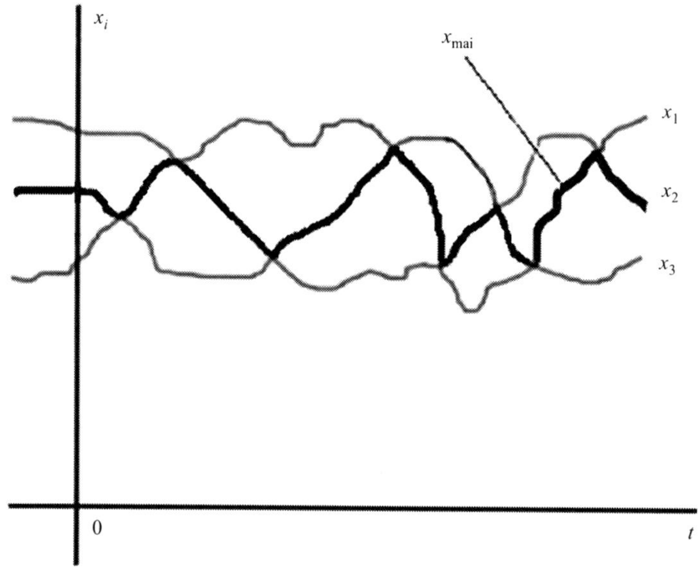


**Figure 9.2.** Majority processing of sensor outputs.

via currently-obtained data and up-to-date estimation. However, the problem of any *a priori* uncertainties in the error properties of sensors and the properties of the measured parameter, the dynamical error of which may be not equal to zero compared with an invariant meter, usually requires the use not of canonical Kalman filters, but modifications with robust properties (Besekerskiy and Nebylov 1993; Kassam and Poor 1985; Magni, Bennani, and Terlouw 1997; Nebylov 2004).

### *9.1.2 JOINT PROCESSING OF READINGS FROM IDENTICAL SENSORS*

If there are two identical sensors aboard a vehicle, it is possible, in principle, to activate the second sensor only after the detection of a failure in the first (or basic) sensor. This approach is economical in the use of the second sensor. However, in addition to problems associated with the fast detection of sensor failure, other disadvantages of this option include the nonuse of potential opportunities of improving measurement accuracy, and this does not match modern concepts of complex measuring system construction. Therefore it is expedient to consider only a variant of the parallel operation of two or several identical sensors.

Let the output signals of several sensors be $x_i(t)$, where $i = \overline{1,l}$ for $l > 1$, and assume that they conform with a uniform frame of reference in which each output signal contains the same useful component $g(t)$ corresponding to the true value of the measured parameter, but exhibit different values of noise $n_i(t)$. That is, $x_i(t) = g(t) + n_i(t)$. Noise associated with such measurements is usually considered to be Gaussian and due to mutually independent random processes but with identical dispersions, $D_n$, for all serviceable sensors (basically, it is possible to consider a nonstationary case where this dispersion depends on time).

If the basic requirement for a multiple sensor signal synthesis algorithm is measurement error dispersion minimization for all serviceable sensors, then the weighted summation of the sensor signals when the measurement estimation is calculated will be optimized as in the following equation:

$$\hat{g}(t) = \frac{1}{l}\sum_{i=1}^{l} x_i(t). \tag{9.2}$$

This implies that the resulting error dispersion will be in $l$-times smaller than the dispersion of a single sensor error, that is,

$$D_e = D_n/l. \tag{9.3}$$

However, if any sensor fails, its false output will be included in the estimation $\hat{g}(t)$ with a multiplier $1/l$, which can entail a significant error until the moment of failure detection.

If the basic requirement is high-fault tolerance, then for an odd number of sensors $l$ it is expedient to use majority processing:

$$g(t) = \text{maj}\{x_1(t), x_2(t), \ldots, x_l(t)\}. \tag{9.4}$$

Majority processing consists of selecting those sensor outputs that occupy intermediate values between other sensors outputs. The current outputs from all the sensors are arranged depending on their values, and the average value is taken from a row above and below for which an identical number of output values appear. This is also explained by the graph of Figure 9.2, which is calculated for $l = 3$.

Such processing of $l$ serviceable sensors with Gaussian error distribution laws will give a resulting error dispersion of:

$$D_e \cong \frac{\pi}{2(l-1)+\pi} D_n, \tag{9.5}$$

This is greater than for the weighted summation—for example, at $l = 3$, Equations (9.3) and (9.5) will give $0.33D_n$ and $0.44D_n$. However, failure of one of the sensors will have hardly any effect on the resulting error, showing that failure detection does not demand excessive promptness and can be carried out with only a small mistake probability.

Except for the specified cases of weighted summation and majority processing, the use of various combinations with intermediate indices of accuracy and fault tolerance is possible.

### *9.1.3 JOINT PROCESSING OF READINGS FROM COGNATE SENSORS WITH DIFFERENT MEASUREMENT RANGES*

Consider the following cases:

1. *Sensors having adjacent or partially overlapping measurement ranges*: One example is a low altitude frequency radioaltimeter (up to 1500 m) and a pulse radioaltimeter for higher altitudes (above 1500 m). Together, these two sensors can cover all possible altitudes for the vehicle in question.

2. *One sensor with a wide measurement range but low accuracy, and another sensor to provide high accuracy but having a narrow measurement range*: In this case the working zone of the second sensor can be defined via indications from the first sensor, so making it possible to take advantage of the precise indications of the second sensor to give an accurate value of the measured parameter in that zone. In this narrow range, the second sensor can provide unequivocal measurements and the function of the first sensor then consists solely of the elimination of ambiguous indications from this second sensor. For example, the second sensor (for a phase channel) precisely distinguishes the values of any periodic processes only in a range ±180 degrees, whereas the first sensor is able to define the number of periods. For example, if the second sensor output is 39°, then in reality the measured value may be $39° + n360°$, $n = 0,1,2....$ Hence, the value of $n$ can come only from the first sensor. A typical case might be the code channel of a GPS receiver with a coordinate measurement error of a few meters (the first sensor); whereas the phase channel (the second sensor), having a measurement error of a few centimeters, can correct any ambiguity. In all such cases, the sensor signal processing algorithm is constructed so that the rough sensor can deduce that the accurate sensor is in an operating condition and provide nonambiguity for its indications.
3. One sensor has a continuous output signal and another (usually the more accurate one) has a pulse output. The second sensor can then periodically correct the indications of the first.

It should be noted that in many practical cases a set of different sensors, among which are some similar ones, unite in the general system to allow joint processing of their output signals using the filtration methods described subsequently.

### 9.1.4 JOINT PROCESSING OF DIVERSE SENSORS READINGS

Usually, such sensors have different error spectral properties that make the dynamic processing of their signals expedient. So, from each of the sensors the spectral components of the useful signals least deformed by measurement noise are taken. Usually, only the lowest-frequency components are taken from the sensor with a broadband error, and the highest-frequency components are taken from the sensor with the most low-frequency errors. Other sensors (given that there are more than two) provide components in the average frequency domain.

Dynamic signal processing is also necessary when sensors of different types are used and sensor output reduction to one dimension is required if some sensors measure the first or second derivatives instead of the signal itself. In all the three cases listed earlier, the processing is multidimensional, mostly via the linear filtration of primary sensor signals. The principles of algorithm optimization for such filtration are described below.

It must be noted that in many cases the main mathematical apparatus dealing with the pithy problems of sensor integration is best served using the theory of optimal filtration. After the formalization of a problem, the whole matter is usually reduced to linear multidimensional or single-dimensional statistical filtration using complete *a priori* information on the characteristics of errors in sensors. It can be solved using Wiener filters (Wiener 1949) or, in nonstationary cases, by Kalman filters. The imposition of a condition of invariance eliminates the necessity of accepting a similar complete model for the measured values. However, the inadequacy of complete spectral and correlation models for navigational system signals compared with real *a priori* information about their properties, creates a problem in distinguishing between real and theoretical measurement accuracy and can depreciate the results of theoretical synthesis.

For overcoming this problem in an integrated measuring system, it is possible to use a robust approach for synthesis that is now being actively developed in filtration and identification theory. This approach involves the acceptance of rather rough but authentic models of signals that correspond to the bounding of allowable classes of their spectral-correlation and stochastic characteristics (Besekerskiy and Nebylov 1993; Greenlee and Leondes 1977; Huber 1984; Kassam and Lim 1977; Kassam and Poor 1985; Looze and Poor 1983; Magni, Bennani, and Terlouw 1997; Nebylov 2004; Tsipkin and Pozniak 1981).

### 9.1.5 LINEAR AND NONLINEAR SENSOR INTEGRATION ALGORITHMS

The linearity of the filtration algorithms is justified not only by simplicity in their realization but also because, if Gaussian input excitation models and square-law (or close to it) criteria of accuracy are accepted, the optimum processing algorithm is linear. The hypothesis about the Gaussian distribution of useful and interference components in the processes $x_i(t)$, $i = \overline{1,l}$ is most persuasive at ordinates of these processes close to the expected values.

In particular, sensor errors usually have rare but large surges relevant to certain types of failure where real distribution curves appear above Gaussian for large (but not infinite) values of the argument. These circumstances generally result in nonlinear filtration algorithms that may be suboptimal and only reject sensor indications essentially distinguishable from those expected, and hence correct such failures. Because the probability of failure in any primary sensor is small, and for serviceable sensors the best resulting accuracy is provided by linear filtration, it is expedient to include linear filters in the main channels of an integrated measuring system. The logic for automatically finding a failure in a primary sensor and reconstructing the evaluation algorithms after such detection should only supplement the main linear structure of the system. Details of the synthesis of such logic and the efficiency of its operation are not considered here because they cannot be resolved before definitions of the main linear structure of systems have been presented.

The generalized configuration of an integrated measuring system shown in Figure 9.1 can be associated with the equivalent block-diagram presented in Figure 9.3, where $W_{si}(s)$ are transfer functions of sensors; $H_{ci}(s)$ are transfer functions of calculator channels; and $v_i(t)$ are measurement noises referred to sensor inputs, $i = \overline{1,l}$. The relationships between the various signals in this scheme can be written in matrix form. The matrices ($\eta \times 1$) of the measured coordinate images $\boldsymbol{G}(s)$, their estimations $\hat{G}(s)$, and the measurement errors $\boldsymbol{E}(s)$, and also the matrix ($l \times 1$) of the primary sensor output signal images $\boldsymbol{X}(s)$ and measurement noise $V_x$ are as follows:

$$\hat{G}(s) = H_c(s)X(s), \tag{9.6}$$

$$E(s) = [I - W_s(s)H_c(s)]G(s) - W_s(s)H_c(s)V_x(s), \tag{9.7}$$

where $H_c(s)$ and $W_A(s)$ are the nonsquare matrices ($\eta \times l$) and ($1 \times \eta$) of the transfer functions of the calculator and primary sensor channels, and $\boldsymbol{I}$ is a unit matrix ($\eta \times \eta$).

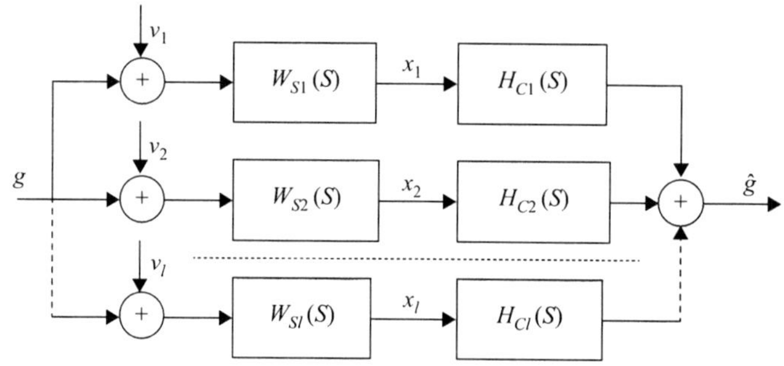


**Figure 9.3.** Block diagram for an integrated measuring system.

Notice that for matrix equality,

$$W_s(s)H_c(s) = I \tag{9.8}$$

so it follows from Equation (9.7) that the measurement errors do not depend on the measured coordinates. This means that the system acquires a property of invariance for the measured coordinates. For analog (continuous) implementation of such a system, the equality (9.8) can be achieved because among the primary sensors there are some that measure coordinate derivatives, and in the transfer functions of such sensors the number of zeros exceeds the number of poles. Otherwise, the fulfilment of Equation (9.8) would require the generation of some calculator channels having transfer functions wherein the number of zeros exceeds the number of poles, which is physically impossible. A digital implementation of the calculator for the exact fulfilment of a condition of invariance is impossible, even in principle, if not all the sensors are time-lag-free dynamic units.

With reference to the decomposed block diagram (Figure 9.3) of an integrated measuring system containing $l$ independent isolated channels, the general expressions (9.6) and (9.7) for the images of the measured coordinate estimations and measurement errors may be transformed into the following:

$$\hat{G}(s) = \sum_{i=1}^{l} H_{ci}(s)X_i(s) = \sum_{i=1}^{l} W_{si}(s)H_{ci}(s)[G(s)+V_{xi}(s)],$$

$$E(s) = G(s) - \hat{G}(s) = \left[1 - \sum_{i=1}^{l} W_{si}(s)H_{ci}(s)\right]G(s) - \sum_{i=1}^{l} W_{si}(s)H_{ci}(s)V_{xi}(s). \tag{9.9}$$

Thus the condition of meter invariance for the coordinate $g(t)$ means it is possible to treat it as a component of matrix equality (9.8):

$$\sum_{i=1}^{l} W_{si}(s)H_{ci}(s) = \sum_{i=1}^{l} H_i(s) = 1, \tag{9.10}$$

where the symbol $H_i(s) = W_{si}(s)H_{ci}(s)$ is introduced.

# 9.2 FUNDAMENTALS OF INTEGRATED MEASURING SYSTEM SYNTHESIS

## 9.2.1 SYNTHESIS PROBLEM STATEMENT

The dynamic (statistical) synthesis of an integrated measuring system having a known primary sensor structure with transfer functions $W_{si}(s)$, $i=\overline{1,l}$, consists of choosing the transfer functions $H_{ci}(s)$ of the calculator channels or (which is the same thing) the transfer functions of complete channels within the system $H_i(s)=W_{si}(s)H_{ci}(s)$. Given such a choice, it must be ensured that the required—or even the extreme—value of an acceptable measure of system accuracy (e.g., the root-mean-square, r.m.s., or maximum error) is provided. Also, certain restrictions must be fulfilled, as detailed subsequently.

Dynamic synthesis must precede the solving of the problems involved in the structural design of a calculator with $l$ inputs and one output. Here, a requirement for simplicity in the realization of the calculator (either analog or digital) implies giving preference to fractional rational transfer functions $H_i(s)$ for the calculator channels in which the denominators are identical, the numerators being the distinguishing features. This defines one of the restrictions imposed on functions $H_i(s)$ for dynamic synthesis.

## 9.2.2 CLASSES OF DYNAMIC SYSTEM REALIZATION

The simplest task in measuring system synthesis involves the consideration of a class of linear filters with constant parameters that enable the block diagram of Figure 9.3 to be used. Such a limitation essentially simplifies system realization, but in many practical cases cannot appreciably improve the potential measurement accuracy. The problem is that in the absence of valid models for the distribution laws of excitation, there is no basis for the use of nonlinear filters. Moreover, as the case in point is a synthesis made on the assumption of sensors in working order, the hypothesis of normal or restricted-normal laws of error distribution over all sensors and measured coordinates is much more justifiable than any other hypothesis. As already noted, for normal excitations the filtration problem may be solved most successfully by using linear methods.

Regarding the acceptance of time-invariant system parameters, it is necessary to notice that the first steps in this direction consist of discounting consideration of all sets of possible sensor error magnitudes for various modes of aerospace vehicle motion, and taking account only of maximally significant values—in other words, the synthesis of filters for the worst case. This does not result in a synthesis problem for a strictly stationary situation because the process of measurement lasts indefinitely long and begins at the moment of meter switch-on under the presence of some initial mismatch. Therefore, an integrated meter with constant parameters cannot ensure accuracy for any moment in time after measuring system switch-on. In this context, the advantages that nonstationary Kalman filtration has over Wiener filtration are known, given complete *a priori* information. However, taking into account that an initial mismatch should be detected rather quickly, and the meter accuracy during such a transient mode is ignored, it would be wrong to reject an opportunity to seek a satisfactorily operating meter structure using a class of filters with constant parameters, at least within a first investigation phase. The advantages that alteration of the measuring system parameters may confer require special study in particular cases.

The condition of invariance, Equation (9.10), may be among the essential restrictions for the selection of functions $H_i(s)$, but though it can be imposed in many cases it is not always expedient to do so. However, the imposition of the condition of invariance is absolutely necessary in the complete absence of information about the characteristics of the measured coordinate $g(t)$, because otherwise the dynamic component of measurement error does not yield to restriction and theoretically can be indefinitely large. However, in such a situation, complete *a priori* definition is not available in practice, though some partial information on the properties of excitation $g(t)$ can usually be used. It is also noteworthy that the condition of invariance, accepted in many known works on sensor integration, sometimes does not reflect any objective requirements about meter quality but a desire to reduce synthesis to a classical filtration problem using complete *a priori* information (as in the case of known sensor errors characteristics).

It is obvious that there are only two circumstances that could really form a basis for the acceptance of a condition of invariance, Equation (9.10), without the preliminary analysis of properties of excitations and errors in sensors $v_i(t)$. The first is the assumption of complete noninertia in the invariant measuring system, which is promoted as providing good dynamic properties in a vehicle control loop using a coordinate $g(t)$ in which the integrated system is considered as a measuring unit. However, even for partial invariance of the meter, this is an inexact fulfilment of the (9.10) condition, though the dynamic properties of integrated meters usually appear as rather good. Second is the maximization of simplicity in the realization of the invariant integrated meter. Notice that the fulfilment of a condition of invariance also guarantees an acceptance of transfer functions $H_i(s)$ (distinguished only by their numerators) and removes some restrictions on their choice. However, it may be shown (Nebylov 2004; Nebylov and Wilson 2002) that it is possible to design rather simple structures for the calculator without fulfilment of a condition of invariance.

Summarizing the above, it may be concluded that the fulfilment of Equation (9.10) should not be an aim in itself, but is useful only when the highest measurement accuracy is required. Such a high-accuracy condition should be obtained as the result of optimizing the transfer functions $H_i(s)$, or from the comparative analysis of properties of effects, rather than simply accepting the fulfilment of Equation (9.10) as an initial synthesis restriction. The method of synthesis described subsequently permits this approach because it is based on using partial prior information about excitation properties.

### 9.2.3 MEASUREMENT ACCURACY INDICES

Because the main measure of the accuracy of measurements depends on features of their usage, it is expedient to accept the maximum or "practically maximum" error, or the root-mean-square (r.m.s.) values of an asymptotically unbiased error (Nebylov 2004). The maximum value of an error well-characterizes a measurement quality such as is needed for the altitude of a low-flying vehicle for which unacceptably large deviations of this coordinate from a nominal value can result in an emergency situation and possibly a fatal accident. By contrast, meters for pitch angle and ground speed indicating even major error extremes cannot usually directly create emergency situations (provided they are rare). In such a case the loss function* can be

* In statistical decision theory, the loss function describes the loss incurred by an incorrect decision based on the indicated data. In solving the problem of extracting the parameter estimation signal from the noise background, the loss function results from the discrepancy between the true value of the parameter and the estimated value.

considered not as rectangular but as continuously increasing and close to a square-law; and as a main measure of accuracy an r.m.s. error can be accepted. However, it is useful to consider an alternate variant having an analyzed dependence of synthesis results on the chosen accuracy criteria.

It should be noted that the universal use of Wiener and Kalman methods of synthesis for the complete spectral–correlation description of excitations has resulted in the maximum error as a criterion of accuracy being superseded by the r.m.s. error. If the Gaussian law of distribution is applied to all multidimensional linear dynamic system excitations, then the normal law of distribution of system output signals and errors, the r.m.s. value of centered error, completely characterizes the filtration accuracy and unequivocally defines the stochastic properties of this error. The concept of a "practically maximum error" is connected with the rule of "three sigma" ($3\sigma$, or $2\sigma$, or $5\sigma$, etc.) and in this sense does not have independent significance, but only emphasizes the importance of knowledge of the r.m.s. error value as the main measure of accuracy.

However, being different from the normal, or an unknown, distribution law for parts of excitations applied to systems, the maximum error should be specially examined. It is possible to talk with confidence about only the Gaussian components of those integrated meter errors caused by errors in radar sensors, positioning sensors (including GPS and other SNS), and vehicle speed components. It is not only that errors in listed sensors are usually normal, but it is also essential to have effective smoothing of these errors in the appropriate calculator channels of the integrated meter. This is usually accomplished by narrow-band low-frequency filters. Therefore, by virtue of the central limiting theorem, normalization of the output signal and measurement errors is achieved irrespective of the distribution law of the input signal.

Another situation arises because of error components in inertial and gyro sensors. In the integrated meter channels appropriate to these sensors there are usually high-frequency filters that do not ensure signal normalization. The indicated signals (or errors) cannot be attributed to normal stochastic processes with confidence. It has been shown (Nebylov 2004) that, for errors in inertial and gyro sensors, it is possible to indicate the maximum possible values of their derivatives that permit evaluation of maximum values of corresponding error components in integrated meters. Similarly, the aforesaid concerns regarding meter error components caused by invariance condition defaults that are influenced by the changing measured parameters of the aerospace vehicle motion, are not normal and have limited values of their derivatives.

Thus, in a structure having resulting measurement errors for some of its output components, the distribution densities of which are finite functions, it is possible to define maximal probable values; and for other components with normal distribution laws, only "practically maximum" values, for example, at a $5\sigma$ level. Thus, in a structure with such resulting measurement errors, it is also possible to define a practically maximum value, but this will in no way be relevant to the r.m.s. value of a resulting error defined by a $5\sigma$-type elementary formula, and hence it acquires an independent value representing an important measure of measurement accuracy.

### 9.2.4 EXCITATION PROPERTIES

If measurement accuracy is accepted as being represented by an r.m.s. error, and complete spectral-correlation models of the excitation are available, an examination of accuracy can be executed within the framework of correlation theory. Meter synthesis is then reduced to a classical problem of multidimensional MIMU, or of single-dimensional MISO linear filtration (at $l = 2$ and imposing the condition of Equation (9.9)). Such an approach was used mainly in sensor integration for about four decades, and is in many respects also used for the synthesis of integrated algorithms on the basis of Kalman filters.

An alternative development direction for linear filtration theory is characterized in particular by a tendency to employ coarser excitation models, and by using nonparametric random process classes of excitation model construction. This approach imposes the essential features of "statement" and "decision" for integration problems.

In the statement of a problem, there is a large variety of forms in which *a priori* information about excitation properties may be given. For errors in some radar sensors, a uniform model of equal spectral density within the passband of the appropriate meter channel remains acceptable. However, in the majority of cases the presentation of a specific graph or an analytical expression of spectral density would not correspond to real *a priori* knowledge. It is possible to allocate authentically only some allowable area in the functional space of spectral densities. Some examples of the allocation of such areas are the assignment of bounds from above and below in the form of known frequency functions (the band model), the fixing of several points on the spectral density curve (the dot model), and under the restriction of several generalized spectral density moments (in an important specific case, power moments having the sense of variances in excitation derivatives).

In some cases, presenting nonstatistical numerical excitation characteristics is expedient, including maximum values for derivatives, finite differences, the highest frequency in a spectrum, and so forth. It is mainly justified when describing such low-frequency excitations as errors in inertial and gyro sensors and the changing of measured motion parameters. It is possible to increase the information involved in similar models at the expense of dividing the excitation into some additive components, each of which has various sets of values of the numerical characteristics. For example, an excitation can be considered as the sum of two components with different limitations for the first or second derivatives of these components. For radar systems the first component may be generated by the target motion and the second component by the carrier vehicle motion (Kassam and Lim 1977; Kassam and Poor 1985; Nebylov 2004).

Uncertainty about excitation properties makes it impossible to find a specific value for a measurement accuracy index for which only evaluations from above and below are available. Using the robust mini-max approach to synthesis, minimization of the upper accuracy index evaluation can be achieved, and the relationship between the upper and lower evaluations characterizes the maximum loss at the expense of incompleteness in *a priori* information. Sometimes it is expedient to ignore the mini-max approach and not minimize the upper evaluation, but only to fix it at an allowable level. This provides some freedom of choice in meter properties, so allowing for improvement in other quality indices in addition to the accuracy index.

The problems of integration considered in this chapter are formulated within the framework of the above-stated robust approach to dynamic filter synthesis. Thus, from the variety of possible forms in which nonparametric classes of excitation may be presented, and according to the real availability of *a priori* information, only the restrictions of maximum or r.m.s. values of sensor error derivatives and of measured motion parameters are used. Error spectral densities are assumed to be known only for some sensors.

### 9.2.5 OBJECTIVE FUNCTIONS FOR ROBUST SYSTEM OPTIMISATION

It is possible to write the required transfer functions of complete channels of meter $\{H_i(s)\}_{i=1}^{l}$ in the following general form:

$$H_i(s) = \frac{b_{i0} + b_{i1}s + \cdots + b_{in}s^n}{1 + a_1 s + \cdots + a_n s^n}. \tag{9.11}$$

Here $\{a_j\}_1^n, \{b_{ij}\}_{j=o}^n \in [0, \infty)$ are the factors to be determined.

The order $n$ should be taken as rather large so that actually the structural, instead of the parametric meter, synthesis would take place. If the optimum meter structure corresponds with the transfer functions—for example, of $(n-1)$ power, instead of $n$ power—then under the optimization of factors (9.11), it should be found that $a_n = 0$ and $b_{in} = 0$. In practice, the value of $n$ should be no less than a power of the forming filter that transforms the white noise into the excitation (or error) of the sensor with a known fractional-rational spectral density, and no less than a power of the oldest limited derivatives of excitation (or error) of the sensor in the case of an unknown spectral density.

As is shown by Nebylov (2004), the case of known spectral densities is characteristic for errors in radar sensors and the case of the restriction of derivatives is characteristic of errors in inertial and gyroscopic sensors. It should be noted that the values $n = 3$ or $n = 2$ correspond most frequently to the above-mentioned rule for errors in inertial and gyroscopic sensors.

Extending the set of $l$ filters with transfer functions (9.11) by an additional filter $(l + 1)$, the numerator factors of this filter are defined by the following formula:

$$b_{l+1\,j} = a_j - \sum_{i=1}^{l} b_{ij}, \tag{9.12}$$

where $a_0 = 1$. It follows from Equations (9.11) and (9.12) that for the transfer function of such a filter the following formula is valid:

$$H_{l+1}(s) = 1 - \sum_{i=1}^{l} H_i(S). \tag{9.13}$$

Using Equation (9.13), the expression (9.9) may be written for a total measurement error as follows:

$$E(s) = -\sum_{i=1}^{l} H_i(s)V_i(s) + H_{l+1}(s)G(s). \tag{9.14}$$

The equivalent schematic for investigating integrated meter accuracy, shown in Figure 9.4, corresponds to expression (9.14). Here, $v_i(t)$ are sensor errors referred to their inputs, $g(t)$ is a measured coordinate, $e_i(t)$ are error components from the measurement noise in each sensor channel, $i = \overline{1,l}$, and $e_g(t)$ is the dynamic error. If condition (9.10) is imposed, $H_{l+1}(s) = 0$ and $e_g(t) = 0$.

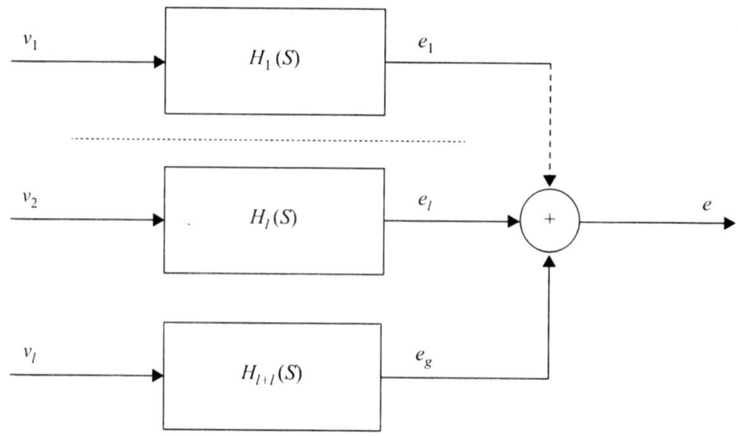


**Figure 9.4.** Equivalent schematic for investigating integrated meter accuracy.

All the output excitations shown in Figure 9.4 can be considered as mutually independent and mutually uncorrelated. Therefore, for the maximum value and for the variance of the resulting integrated meter error, the formulae

$$e_{\max} = \sum_{i=1}^{\ell} e_{i\max} + e_{g\max},\ D_e = \sum_{i=1}^{\ell} D_{ei} + D_{eg}, \tag{9.15}$$

are valid.

Methods for determining the maximum or r.m.s. evaluations of output signals from linear dynamic filters receiving maximum or r.m.s. source signals are considered by Nebylov (2004). Their use permits reception of the upper values of $e_{\max}$ and $D_e$ that depend upon the partial or complete given characteristics of sensor errors $v_i(t)$ and measured coordinates $g(t)$. Designating a set of such characteristics as $S$, then if a set of meter parameters $P = \{a_j, b_{ij} / i = \overline{1,\ell}, j = \overline{0,n}\}$ are known, it is possible to make the following definition:

$$e_{\max} = e_{\max}(S,P), \overline{D_e} = \overline{D_e}(S,P).$$

Then, a criterion function for integrated meter optimization based on the best accuracy may be written down in the following form:

$$e_{\max}(S,P) \rightarrow \min_P \text{ (or } \overline{D}_e(S,P) \rightarrow \min_P) \text{ at } B(P) \leq B_0, \tag{9.16}$$

where $B(P)$ is the time of initial mismatch processing in the meter, this being a function of all its parameters, and $B_0$ is an allowable maximum value for this time.

It is also possible to use other optimization variants, for example, using the criterion of maximum response speed for a given accuracy with the criterion function B($P$) $\rightarrow$ min at $e_{\max}(S,P) < e^0{}_{\max}$. However, criterion (9.16) can be considered as the most common version.

Notice that for the problems of meter optimization for motion parameters considered here, the restriction of time $B$ usually appears immaterial and does not influence the optimization criterion for best accuracy. Furthermore, it is important that the potential measurement accuracy is not usually so high that it would be impossible to replace the criterion of best accuracy by accuracy in the category of restrictions.

### 9.2.6 METHODS OF DYNAMIC SYSTEM ACCURACY INDEX ANALYSIS UNDER EXCITATION WITH GIVEN NUMERICAL CHARACTERISTICS OF DERIVATIVES

A compulsory phase of an investigation of the accuracy that is integrated into a measuring system (in the general case, any dynamic system) is the definition of maximum or r.m.s. values of each of the total error components included in expression (9.16). If the excitation causes the considered error component, and only the numerical characteristics of the derivatives are known (or limited), but the spectrum is unknown, then solving this problem requires some special explanation. Here, only the main principles involved in investigating the accuracy of excitation filtering with limited maximum or r.m.s. values of derivatives will be presented. Detailed descriptions of such methods, and other approaches to the analysis and synthesis of robust dynamic filters, will be found in the book (Nebylov 2004).

#### 9.2.6.1 Estimation of Error Variance

Let the input of a linear dynamic system with an amplitude/frequency characteristic $A(\omega)$ be a random excitation with unknown spectral density $S(\omega)$ but known variances of stationary centered derivatives, as follows:

$$D_i = \frac{1}{\pi}\int_0^\infty \omega^{2i} S(\omega)d\omega, \quad i = \overline{K,N}, \quad K \geq 0; \tag{9.17}$$

The variance of the signal at the system output (below) does not yield to exact calculation.

$$I = \frac{1}{\pi}\int_0^\infty A^2(\omega)S(\omega)d\omega; \tag{9.18}$$

However, it is possible to estimate its upper and lower bounds, $\overline{I} = \sup I$ and $\underline{I} = \inf I$, if $A^2(\omega)$ corresponds to a transfer function connecting the Laplace images of a system error component and of the excitation inducing this error component, in which case the value of $\overline{I}$ will characterize the accuracy of the system; and the ratio of $\overline{I}$ and $\underline{I}$ will be a measure of indefiniteness in accuracy evaluation at the expense of incompleteness in the *a priori* information.

This raises the problem of extracting extreme values of integral (9.18) from the product of a known function $A^2(\omega) \in R_1^+$ and a function $S(\omega)$ given by its power (in particular, even-numbered) moments (9.17). This is actually an important one within a classical mathematical group forming the overall problem of moments (Ahiezer 1961; Karlin and Studden 1966; Krein and Nudelman 1973).

Amongst problems in this group, particular attention should be paid to one, namely under what conditions can the sequence of numbers $\{D_i\}_{i=K}^N$ be submitted in the form of Equation (9.17)? The resulting solution of this problem is reduced to an inequality $D_{i+1}^2 \leq D_i D_{i+2}$, $i = \overline{K, N-2}$.

Notice that in the more general case, any integration limits covering an area of nonzero values of function $A^2(\omega)$ in Equation (9.18) are allowable. The function $S(\omega)$ can be set by generalized moments concerning some system of basic functions. For example, generalized moments of spectral density are easily found experimentally by passing the excitation through filters with certain amplitude/frequency characteristics and measuring the output signal powers.

The problem of moments was defined and investigated by the famous Russian mathematicians P. L. Tchebysheff and A. A. Markov. They proved that the limits of integral (9.18) can be achieved for functions $S(\omega)$ in the form of δ-image pulses having nonzero values at a finite number of points on a half-axis $\omega \geq 0$. Such functions, being the elementary solutions of the system of equation (9.17), have been named as the initial (canonical) representations of a sequence of moments $\{D_i\}_K^N$. However, for this purpose the function $A^2(\omega)$ should have a continuous convex $(N-1)$-number derivative for $\omega^2$.

Such methods of canonical representations are important in problems of moments and provide opportunities for finding functions $S(\omega)$, supplying extreme values $I_{max}$ and $I_{min}$ to integral (9.18), and also for locating well-defined formulations for the conditions of solution existence. In a slightly modified form, canonical representations can also be used in the $L$-problem of moments, as has been shown by Krasovsky (1968).

Unfortunately, in many applied tasks the conditions for applicability of the method of canonical representations are not possible. However, solutions can be found by using other methods that do not require any particular definition of a function $S(\omega)$ and so are free from restrictions on $A^2(\omega)$. These are based on representations of $A^2(\omega)$ functions by polynomials, which in this connection must be taken as approximate. In the general case, such methods do not permit the finding of exact bounds for $I_{max}$ and $I_{min}$ as are achievable using exact $S(\omega)$ functions, but only guarantee a fitting of value $I$ to an interval $I \in [\overline{I}, \underline{I}]$, where $\overline{I} \geq I_{max}$ and $\underline{I} \leq I_{min}$.

To explain the kernel of the approximate method, let real coefficients $\{c_i\}_K^N, \{q_i\}_K^N$ of polynomials be defined:

$$C_{2N}(\omega) = \sum_{i=K}^{N} c_i \omega^{2i}, \; Q_{2N}(\omega) = \sum_{i=K}^{N} q_i \omega^{2i}, \tag{9.19}$$

which at $\omega \geq 0$ are found to satisfy the conditions

$$C_{2N}(\omega) \geq A^2(\omega), \; Q_{2N}(\omega) \leq A^2(\omega). \tag{9.20}$$

After premultiplication of the right and left-hand parts of inequalities (9.20) by function $S(\omega) \geq 0$ and after integration along the interval $\omega \in [0, \infty)$ the following is obtained:

$$\frac{1}{\pi}\int_0^\infty C_{2N}(\omega)S(\omega)d\omega \geq I \geq \frac{1}{\pi}\int_0^\infty Q_{2N}(\omega)S(\omega)d\omega.$$

After allowing for Equations (9.17) and (9.18), this gives the required upper and lower bounds:

$$\overline{I} = \sum_{i=K}^{N} c_i D_i, \; \underline{I} = \sum_{i=K}^{N} q_i D_i. \tag{9.21}$$

The interval $[\overline{I}, \underline{I}]$ will be narrowest if the choice of factors $\{c_i\}_K^N, \{q_i\}_K^N$ is optimally made using the following criteria:

$$\overline{I} \to \min_C \overline{I}, \; \underline{I} \to \max_Q \underline{I}, \tag{9.22}$$

with restrictions as in Equation (9.20). This is a problem in linear programming in which the restrictions are functions, and in some simple cases analytical solutions are possible (Nebylov 2004).

It is possible to determine by an approximate method whether values for the bounds of $I$ are exact, that is when $\overline{I} = I_{max}$ and $\underline{I} = I_{min}$. For this purpose, the following interpretation of results derived in Krein and Nudelman (1973) may be used.

*Theorem 1*. If some polynomials $C_{2N}(\omega)$ $(Q_{2N}(\omega))$ convert the inequalities (9.20) into equalities at $\omega = \omega_j$, $j = \overline{1,v}$, then there is a set of factors $\rho_j > 0$ producing the function

$$S(\omega) = \sum_{j=1}^{v} \rho_j \delta(\omega - \omega_j), \tag{9.23}$$

which is suitable as the solution of a system of equations (9.17). Then, as determined by formula (9.21) the bounds are exact and are acquired using the function $S(\omega)$ in the form of Equation (9.23).

It will be noticed that abscissas $\{\omega_j\}_1^v$ correspond to points of contact of curves $C_{2N}(\omega)$ $(Q_{2N}(\omega))$ and $A^2(\omega)$, but also, probably, the bounds of an actual interval of integration in Equation (9.18).

*Theorem 2*. If some function $S(\omega)$ of a form (9.23) satisfies the system of equations (9.17), and it is possible to select factors of polynomial $C_{2N}(\omega)$ $(Q_{2N}(\omega))$ having values at the points $\omega = \omega_j$ that coincide with the values of function $A^2(\omega)$ and fulfill the condition (9.20), then the exact bounds of integral (9.18) are reached for the function $S(\omega)$ expressed by the formula

$$I_{max(min)} = \sum_{j=1}^{v} \rho_j A^2(\omega_j). \tag{9.24}$$

If the finding of the exact bounds of integral (9.18) is treated as a dual to the (9.22) infinite-dimensional problem of linear programming, then theorems 1 and 2 can be considered as analogues of duality and balance theorems for the infinite-dimensional case.

Essentially, approximate methods using the criterion of Equation (9.22) permit the best evaluation of the variance $I$ (narrowest interval $[\overline{I}, \underline{I}]$), and this cannot be improved upon by any other method without needing additional information about excitation properties. During the accuracy evaluation, this method permits the taking into account of such additional information as the width restriction of an excitation spectrum and of the monotonic nature of the curve $S(\omega)$ at $\omega > 0$ (Nebylov 2004).

### 9.2.6.2 *Example of Error Variance Analysis*

As an elementary example, consider the variance of a dynamic error in a system having an integrator with a transfer factor $K_1$ and closed by unit negative feedback. The transfer function of such a closed loop system is $H(s) = K_1/(K_1 + s)$, and the transfer function for the error is $H_e(s) = s/(K_1 + s)$. Hence, $A^2(\omega) = \omega^2/(K_1^2 + \omega^2)$. Such a property could describe a GPS receiver output signal filter for smoothing the added noise.

Both the canonical representation and the approximative methods permit the easy obtaining of formulae when the excitation variances and two of its derivatives are known:

$$\overline{I} = I_{max} = \frac{D_1}{K_1^2 + D_1/D_0}, \underline{I} = I_{min} = \frac{D_1}{K_1^2 + D_2/D_1}. \tag{9.25}$$

Notice that the upper and lower bounds of the dynamic error variance are found by processing the harmonic excitations using variances $D_0$, $D_1$ and $D_1$, $D_2$. For example, if the values $D_0 = 10^7$ m$^2$, $D_1 = 4 \times 10^2$ m$^2$ s$^{-2}$, D$_2$ = 10 m$^2$ s$^{-4}$ and $K_1$ = 3 c$^{-1}$ can be accepted for an aircraft during an approach to landing, formulae (9.25) give for the dynamic error in its positioning $I_{max} = 4.44$ m$^2$, $I_{min} = 4.43$ m$^2$. That is, the r.m.s. dynamic error $\sigma_{eg} = \sqrt{I}$ is within an interval $2.10\text{ m} < \sigma_{eg} < 2.11\text{ m}$. Two excitation harmonics that give $I_{max}$ have the frequencies $\omega_1 = 6.55\,10^{-3}\,s^{-1}$ and $\omega_2 = 0.155\,s^{-1}$.

If only the variance of the first excitation derivative is known, that is $D_0 \to \infty$ and $D_2 \to \infty$, then formulae (9.25) give the trivial result $I_{max} = D_1 / K_1^2$, $I_{min} = 0$.

Because it is very convenient to use harmonic excitation in the analysis of accuracy and the simulation of various dynamic systems, consider the conditions for its applicability in the general case.

### *9.2.6.3 Use of Equivalent Harmonic Excitation*

Given the variances of two excitation derivatives when $N - K = 1$, the system of equations (9.17) is satisfied by a function $S(\omega) = \rho_1 \delta(\omega - \omega_0)$, where $\rho_1 = D_{N-1}^N / D_N^{N-1}$, $\omega_0 = \sqrt{D_N / D_{N-1}}$. Such a spectral density corresponds to a harmonic excitation with an amplitude $\sqrt{2\rho_1}$, frequency $\omega_0$, and a uniformly distributed random initial phase. This is frequently used as typical in the analysis of dynamic system accuracy. Similar equivalent harmonic excitation is easy to find by providing the maximum values of two further excitation derivatives.

The question of in which case this excitation will lead to an extreme of value $J$ (for example, a variance in the integrated measuring system error) may be answered using Theorem 2.

Let $K = 0$, $N = 1$ and let the function $A^2(\omega)$ have the form shown in Figure 9.5, where the abscissa is the square of the frequency. Then according to Theorem 2, formula (9.24) at v = 1 will give an exact upper bound $I_{max}$ provided it is possible to find factors $c_0$ and $c_1$ that ensure that curves $C_2(\omega) = c_0 + c_1\omega^2$ and $A^2(\omega)$ contact at a point $\omega = \omega_{10}$. A condition $\omega_{10} \in [a, \gamma]$ should be applied for this purpose. The location of the section $[a, \gamma]$ is explained by Figure 9.5: the exact low bound $I_{min}$ will not be achieved at any value $\omega_0 > 0$. If $\omega_0 < a$, the exact upper

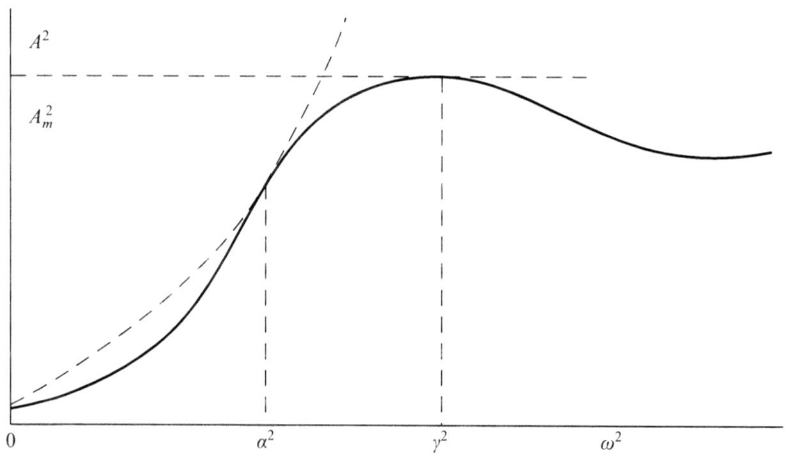


**Figure 9.5.** Determination of the interval $\omega \in [a, \gamma]$.

bound can be found using formula (9.24) at $\nu = 2, \omega_1 = 0, \omega_2 = a$, $\rho_2 = D_1 / a^2, \rho_1 = D_0 - \rho_2$. If $\omega_3 > \gamma$, the upper bound will not be exact.

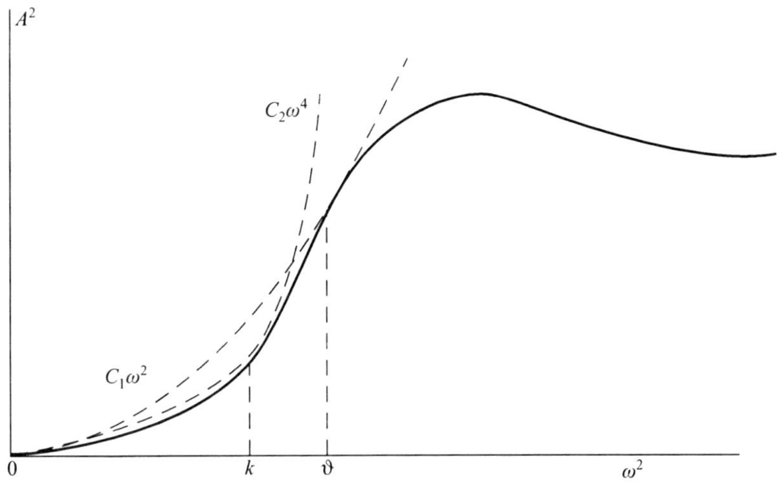


**Figure 9.6.** Determination of the interval $\omega \in [\kappa, \upsilon]$.

If $K = 1$ and $N = 2$ for a function $A^2(\omega)$ shown in Figure 9.6, then formula (9.24) at $\nu = 1$ will give an exact upper bound at $\omega_{\dot{y}} \in [\kappa, \vartheta]$, where $\kappa$ and $\vartheta$ are the abscissas of curves $A^2(\omega)$ at points $C_4(\omega) = c_1\omega^4$ and $C_2(\omega) = c_2\omega^2$where they come into contact (Figure 9.6). The exact lower bound in a mode of operation with the equivalent harmonic excitation can be achieved only in systems of the first order because in higher order systems it is impossible to fulfill the condition $Q_4(\omega) = q_1\omega^2 + q_2\omega^4 \le A^2(\omega)$, for $q_1 > 0$, $q_2 < 0$ at low frequencies.

Thus, depending on the form of function $A^2(\omega)$, the equivalent harmonic excitation found on variances $D_N$ and $D_{N-1}$ can cause either a maximum or a minimum, or a nonextreme value $I$. It follows that this must be taken into account when finding functions simulating sensor errors and changes in measured parameters during computer simulation of integrated measuring systems.

### *9.2.6.4 Estimation of Error Maximal Value*

Strict analysis of the maximum value of an error component in a dynamic system produced by an input excitation with bounded maximum values of derivatives,

$$\left|g^{(i)}(t)\right| \le g_M^{(i)},\ i = \overline{K, N},\ 0 \le K \le N, \tag{9.26}$$

is connected to the definition of the most unfavorable form of excitation $g_{mu}(t)$. The basis of this problem can be explained for the case when $K = N = 0$ by considering the convolution formula,

$$e(t) = \int_0^t w_e(\tau) g(t - \tau) d\tau, \tag{9.27}$$

where $g(t)$ is an input excitation, $e(t)$ is the error produced by this excitation, and $w_e(t)$ is an appropriate system weighting function.

It is clear that for error maximization for a skew symmetric weighting function, the excitation should be identical in sign with $w_e(\tau)$ and have a maximal absolute value. That is, $g_{mu}(t-\tau)=g_M \operatorname{sign} w_e(\tau)$. This concept was first introduced by Bulgakov (1946), the maximal output value of a dynamic system having been investigated by the so-called method of deviation accumulation (Andronov, Witt, and Hikin 1969).

If $K = N > 0$, instead of formula (9.27) it becomes necessary to use the convolution formula of function $g^{(k)}(t)$ and a weighting function appropriate to the system under consideration, along with a $K$-order input integrating unit. It then becomes obvious that the final value of the maximum error can be guaranteed only in a system containing an astatism of the $K$-th or higher order. In such a system, the excitation $g(t)$ in the form of a $K$-order polynomial will cause a limited error. Even for a $(K+1)$-order of input polynomial, the error would become unlimited and its maximum value could not be estimated.

If $K < N$, determination of the maximum error is complicated by the impossibility of a multistep change in the function $g^{(k)}(t)$. Some general methods for defining a maximum error have been developed, and are detailed by Nebylov (2004). Amongst them, consider a simple frequency domain method having much in common with the above approximative method for the analysis of an error variance.

It is necessary to note that an exact definition of the maximal value of an original $e(t)$ using the Laplace image $E(s)$ directly, and without considering the presence of the original, is impossible in essence. However, a developed frequency domain method along with a choice of the most adverse excitation in the form of the sum of several harmonic functions, gives an acceptable evaluation of maximum error. An underestimate in such an evaluation cannot surpass a value in the range $k=1-1.27$, dependent upon the oscillation properties of the system, and indices $K$ and $N$. With an increase in the number of bounded excitation derivatives, the evaluation accuracy is increased when the most adverse excitation becomes smoother.

Transition from the above-mentioned problem of finding a lined excitation spectrum that maximizes the error to a dual problem of linear programming (in other words, transition from the frequency to the time domain), permits obtaining the formula

$$e_M \le \kappa(c_K g_M^{(K)} + c_{K+1} g_M^{(K+1)} + \cdots + c_N g_M^{(N)}), \tag{9.28}$$

where $\{c_i\}_{i=K}^{N}$ are factors of the power polynomial $C_N(\omega)=c_K\omega^K+c_{K+1}\omega^{K+1}+\cdots+c_N\omega^N$ satisfying the condition $C_N(\omega)\ge A(\omega)$, $\omega\ge 0$, and optimal at criterion $\sum_{i=K}^{N} c_i g_M^{(i)} \to \min_C$.

Regarding the value of factor $\kappa$ when investigating problems of integrated measuring system accuracy, it is possible to initially accept $\kappa = 1.27$, and this could be improved using a special analysis (Nebylov 2004), if necessary.

### 9.2.7 SYSTEM OPTIMIZATION UNDER MAXIMUM ACCURACY CRITERIA

The problem of synthesizing a measuring system with a minimum upper bound of the error variance or the minimal maximum value of the error, for known numerical characteristics of the excitation derivatives, can be stated in a similar manner to the problem of synthesizing a Wiener filter for minimizing the variance of the total error at a known excitation spectral density. In both these cases it is possible to speak about robust system optimization on the criterion of best accuracy.

If the uniform spectral density of an error for one sensor $S_\upsilon$, and the variance or maximum value of only the $N$-th derivative of an error for another sensor, are both known, then the problem of invariant two-channel integrated measuring system optimization has an analytical solution (Nebylov 2004). Finding it is enough to optimize the parameters of the transfer function $W(s)$, appropriate to a system of the $N$-th order with astatism of the $N$-th order. The physical treatment of such a result involves the fact that such transfer functions provide a minimum passband for a channel with a transfer function $H_1(s) = W(s)/[1+W(s)]$ at a certain level and fulfilling specific stability margin requirements.

At $N = 1$ it is necessary to accept a transfer function

$$W(s) = K_1/s. \tag{9.29}$$

For the known variance $D_1$ the criterion function will be

$$\bar{D}_e(K_1) = D_1 / K_1^2 + S_\vartheta K_1 / 2 \rightarrow \min,$$

and its study on extremum gives an optimum value for the open loop gain factor and the minimum upper bound of the total error in the form

$$K_1^0 = 2^{2/3} D_1^{1/3} S_\vartheta^{-1/3} = 1.59(D_1 / S_\vartheta)^{1/3}, \tag{9.30}$$

$$\bar{D}_{e\,\min} = 3 \cdot 2^{-4/3} D_1^{1/3} S_\vartheta^{2/3} = 1.19 D_1^{1/3} S_\vartheta^{2/3}. \tag{9.31}$$

Similarly, if the maximum value $g_M^{(1)}$ is known, it is possible to obtain

$$e_M(K_1) = \vartheta_M^{(1)} / K_1 + 3\sqrt{K_1 S_\vartheta / 2} \rightarrow \min,$$

$$K_1^0 = 0.962(\vartheta_M^{(1)})^{2/3} S_\vartheta^{-1/3}, \tag{9.32}$$

$$e_{M\,\min} = 3.12(\vartheta_M^{(1)} S_\vartheta)^{1/3}. \tag{9.33}$$

At $N = 2$, for the transfer function

$$W(s) = K_2(1 + \tau s)/s^2, \tag{9.34}$$

in the case of a known variance $D_2$, it is possible to obtain (at $k_2\tau^2 \le 2$))

$$\bar{D}_e(K_2, \tau) = D_2\left[k_2^3\tau^2(1 - k_2\tau^2/4\right]^{-1} + S_\vartheta(1 + k_2\tau^2)/(2\tau) \rightarrow \min,$$

$$K_2^0 = 1.77(D_2 / S_\vartheta)^{2/5},\ \tau^0 = \sqrt{3/(2K_2^0)}, \tag{9.35}$$

$$\bar{D}_{e\min} = 1.70 D_2^{1/5} S_\vartheta^{4/5}, \tag{9.36}$$

and in the case of a known maximum value $g_m^{(2)}$,

$$K_2^0 = 1.28(\kappa\vartheta_M^{(2)})^{4/5} S_\vartheta^{-2/5}, \tau^0 = \sqrt{3/(2K_2^0)}, \tag{9.37}$$

$$e_{M\min} = 4.03(\kappa\vartheta_M^{(2)})^{1/5} S_\vartheta^{2/5}. \tag{9.38}$$

At $N = 3$, for the transfer function

$$W(s) = K_3\left[1+\tau_1 s+(\tau_2 s)^2\right]/s^3, \tag{9.39}$$

in the case of a known variance $D_3$,

$$K_3^0 = 1.73(D_3 / S_\vartheta)^{3/7}, \tau_1^0 = (8/(K_3^0)^{1/3}, \tau_2^0 = \tau_1^0/\sqrt{2}, \tag{9.40}$$

$$\bar{D}_{e\min} = 2.33 D_3^{1/7} S_\vartheta^{6/7}, \tag{9.41}$$

and in the case of a known maximum value $g_m^{(2)}$

$$K_3^0 = 1.46(\kappa\vartheta_m^{(3)} / \sqrt{S_\vartheta})^{6/7}, \tau_1^0 = (8/(K_3^0)^{1/3}, \tau_2^0 = \tau_1^0/\sqrt{2}, \tag{9.42}$$

$$e_{M\min} = 4.81(\kappa\vartheta_m^{(3)})^{1/7} S_\upsilon^{3/7}. \tag{9.43}$$

### 9.2.8 PROCEDURES FOR THE DIMENSIONAL REDUCTION OF A MEASURING SYSTEM

#### 9.2.8.1 Determination of an Optimal Set of Sensors

For writing a criterion function (9.15), the number of sensors $l$ that are included in an integrated meter system is assumed to be known. However, it is possible that one or more of these sensors exhibit errors with such adverse properties that the use of their signals does not improve the resulting measurement accuracy. Consequently, the transfer functions of the relevant meter channels should become zero. It is therefore useful to have a rule that enables the results of some simple comparative analyses of sensor error properties to immediately reject the useless sensors, because their inclusion would only complicate the optimization procedure. For cases of known error spectral density, such a rule would consist of comparing the plots of spectral densities constructed on logarithmic scales. A sensor would be recognized as useless if the plot of its error spectral density curve did not pass below other plots in any range of frequencies. Such a rule could also be applied in cases where, for each $i$-th sensor or for a selection of sensors, only the maximum (or r.m.s.) values of some error derivatives $\upsilon_{iM}^{(K)}, \upsilon_{iM}^{(K+1)}, \ldots, \upsilon_{iM}^{(N)}$ are given. Using a frequency domain method for investigating maximal errors as described by Nebylov (2004), it is possible to use the expression for the maximum possible amplitude $A_i$ of a harmonic function $\upsilon_i(t)$ with given restrictions on the derivatives, and so construct plots of piecewise-hyperbolic functions (on logarithmic scales):

$$A_i(\omega) = \min\left\{\frac{\upsilon_{i\max}^{(K)}}{\omega^K}, \frac{\upsilon_{i\max}^{(K+1)}}{\omega^{K+1}}, \ldots, \frac{\upsilon_{i\max}^{(N)}}{\omega^N}, \right\}, i = \overline{1, l}. \tag{9.44}$$

If one of the plots is such that at all frequencies there are other plots below it, then the sensor exhibiting the error corresponding to this plot can be excluded from consideration during the integrated meter synthesis.

#### 9.2.8.2 *Analysis of the Advantages of Invariant System Construction*

Together with the plots of the functions $A_l(\omega)$ constructed for sensor errors, it is expedient to consider the plot of the function

$$A_g(\omega) = \min\left\{\frac{g_{\max}^{(K)}}{\omega^K}, \frac{g_{\max}^{(K+1)}}{\omega^{K+1}}, \ldots, \frac{g_{\max}^{(N)}}{\omega^N},\right\}, \tag{9.45}$$

constructed for measured coordinates with known maximum (or r.m.s.) values of derivatives $g_{\max}^{(K)}, g_{\max}^{(K+1)}, \ldots, g_{\max}^{(N)}$. If at any frequency this plot does not pass below plots of functions $A_l(\omega)$, the availability of a "filter" with transfer function $H_{l+1}(s)$ in Figure 9.4 does not result in any increase in accuracy, then it is reasonable to put $H_{l+1}(s) = 0$ so that, allowing for Equation (9.13), it becomes equivalent to the condition of invariance (9.10).

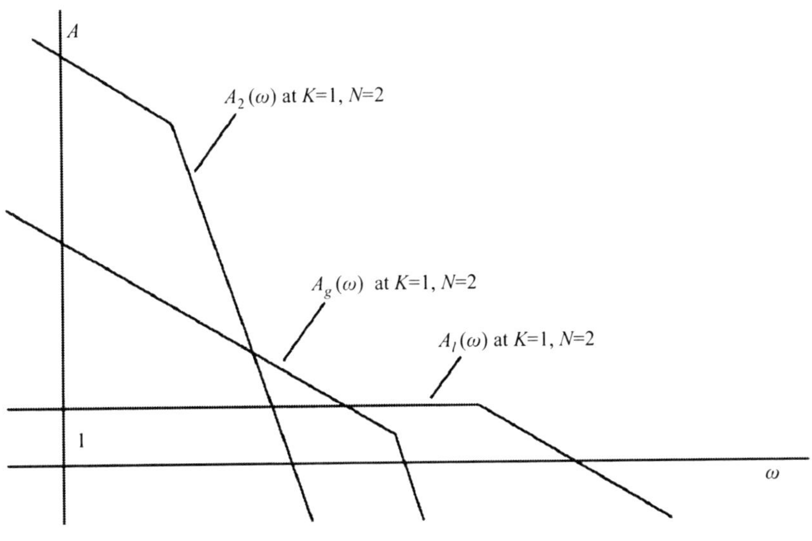


**Figure 9.7.** Plots for investigating the advantages of invariant systems.

Thus, for a synthesis based on the criterion of best accuracy, a rule can be formulated to ensure the condition of invariance. An example of its use is illustrated in Figure 9.7, where asymptotic amplitude/frequency plots of functions expressed by Equations (9.44) and (9.45) appear using logarithmic scales at $K = 0$, $N = 1$ and at $K = 1$, $N = 2$. The line $A_g(\omega)$ passes above that of $A_1(\omega)$ at low frequencies and that of $A_2(\omega)$ at high frequencies, so that it is the best choice for processing all the signals in the measuring system. If the function $A_g(\omega)$ corresponds to measured coordinate $g(t)$ and is found by Equation (9.45), it testifies to the validity of imposing the condition of invariance (9.10). However, if it corresponds to an error in one of sensors and is found by Equation (9.44), it testifies that such sensor is useless in an integrated meter.

### 9.2.8.3 Advantages of the Zeroing of Several System Parameters

In addition to the exclusion of useless sensors or the imposition of a condition of invariance, other methods are also possible for reducing the number of optimized parameters in transfer functions (9.11). Firstly, if the lowest of the bounded derivatives of the $i$-th sensor error is of the order $K$, then for the restriction of an error in the appropriate meter channel, the factors of the numerator of transfer function (9.11) should be a subject to the condition $b_{ij} = 0$ at $j = \overline{0, K-1}$. Otherwise, the error could become indefinitely large and the problem of optimization would lose meaning.

Secondly, if the highest of the bounded derivatives of the $i$-th sensor error is of the order $N$, and $N < n$, then in the numerator of transfer function $H_i(s)$ part of some parameters are fixed by the equality $b_{ij} = 0$ at $j = \overline{0, K-1}$ and $j = \overline{N+1, n}$.

A similar equality is valid for factors in the numerator of transfer function $H_{\ell+1}(s) = 1 - \sum_{i=1}^{\ell} H_i(s)$, connecting the images of the dynamic error and the measured coordinate.

Thirdly, the channel of a sensor with a broadband error (usually of the radar sensor) should have a transfer function, for example, $H_1(s)$, with a numerator of degree $n - 1$, whence $b_{1n} = 0$.

## 9.2.9 REALIZATION AND SIMULATION OF INTEGRATION ALGORITHMS

The considered methods of dynamic (statistical) synthesis of robust integrated algorithms permit the finding of transfer functions for the linear filters of Figure 9.3 that comprise the filtration circuit. Basically, each filter, having a linear rational-fractional transfer function $H_i(s)$, $i = \overline{1, \ell}$, can be realized irrespective of the other filters, the totality comprising a calculator unit circuit consisting of a parallel connection of $l$ detached analog filters. However, taking into account the uniformity of the denominators of the transfer functions $H_i(s)$, it is expedient to realize the calculator unit as a uniform filter circuit with $l$ inputs and not divided into detached channels, so containing a smaller number of elements. The actual circuit design of the calculator unit of such an integrated meter may be conducted using either an analog or a digital approach.

The best-known solution variants of the circuit synthesis problem for the special case when $l = 2$, and under a condition of invariance (9.10), include a compensation circuit (Figure 9.8(b)) and closed loop circuits (Figure 9.8(d),(f)), where $W(s) = H_1(s)[1 - H_1(s)]^{-1}$. This circuit, shown in Figure 9.8(b), provides part compensation of the measurement error $v_2$ by estimation of this error.

The equivalence of the dynamic properties of these two circuits and the filtration circuit (Figure 9.8(a)) can be confirmed by the structural transformations shown in Figure 9.8(a)–(f).

Notice that the compensation circuit conveniently reveals the physical basis of any gain in accuracy being at the expense of integration and the potential quality of the filter with transfer function $H_1(s)$. It is clear that this filter should allocate an error of one sensor $\upsilon_2(t)$ from its additive mix to an error of another sensor $\upsilon_1(t)$ in the most optimum way. Furthermore, the practical realization of the two-channel invariant meter circuits in closed loop form has additional advantages (Nebylov and Wilson 2002).

Now consider the general case of an $l$-channel noninvariant integrated meter (multiple input-single output MISO system), the channel transfer functions $\{H_i(s)\}_{i=1}^{l}$ having the form of Equation (9.11). It can be shown that the calculator unit of such a meter can be realized as an

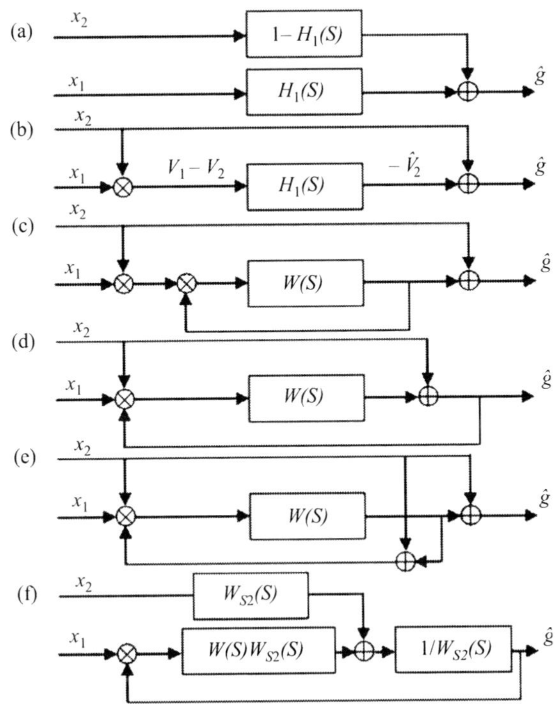


**Figure 9.8.** Structural transformations for an integrated measuring system scheme.

electronic multiparameter circuit containing $n$ integrating elements along with a certain number of scaling and summing elements.

For such an $l$-channel integrated meter, and allowing for Equation (9.9), the dependence of the output signal $\hat{g}(t)$ (using measured coordinate values) on source signals $x_j(t) = g(t) + \upsilon_j(t),\ j = \overline{1, \ell}$, is described by a linear differential equation of the $n$-th order,

$$\sum_{i=0}^{n} a_i \hat{g}^{(n-1)}(t) = \sum_{j=1}^{l} \sum_{i=1}^{n} b_{ji} x_j^{(n-i)}(t), \tag{9.46}$$

where $a_0 = 1, \{a_i\}_1^n, \{b_{ji}\}_{i=0}^n \in [0, \infty)$.

Any ordinary linear differential equation of the $n$-th order with constant factors can be related to an electronic model having the form of $n$ consecutive integrators, and including feedback with various points in the circuit connected through scaling units to an input signal source. The feedback transfer factors are determined by the factors in the left-hand part of the differential equation; and for direct communication with the input signal source, the transfer factors are determined by the right-hand part of the equation. Naturally, if there are several input signal sources (MISO system), derivatives of the input signals $x_i(t)$ will be entered into the right-hand part of the differential equation, and as in Equation (9.46), some direct communications with various input signal sources should be entered into various points in the model in parallel. However, the integrators and the feedback factors will stay the same. It is therefore possible to construct an analog calculator circuit for an $l$-channel integrated meter on the basis of an ordinary circuit for a linear control system model, and adding direct communication with each input signal source. Consider two variants of such circuits, distinguished by the structure of the feedback loops, and assume unity sensor transfer factors.

The first circuit variant for the calculator is shown in Figure 9.9 where the scaling factors for both direct and feedback circuits coincide with the factors in the numerators and denominators of the transfer functions (9.11).

The second variant of the calculator unit circuit is shown in Figure 9.10 where the factors $\beta_{ij}$ do not coincide with factors $b_{ij}$ but are defined using the formulae:

$$\beta_{j,n-i} = \begin{cases} \left( b_{j,n-i} - \sum_{v=0}^{i-1} a_{n-i+v} \beta_{j,n-v} \right) a_n^{-1} & at\ 0 \le i < n, \\ b_{j,0} - \sum_{v=0}^{n-1} a_v \beta_{j,n-v} & at\ i = n \end{cases} \tag{9.47}$$

Almost all known integration networks described in the literature may be reduced to variants of the general circuit of an $l$-channel integrated meter based on the use of $n$ integrating operational amplifiers, as represented in Figures 9.9 and 9.10. Equivalent structural transformations in these circuits can result in modifications of its separate parts, for example with the purpose of excluding excessively large values for the scaling and integrating amplifier transfer factors. However, the main aim of the integrating circuit design in the form of the structural models of Figures 9.9 and 9.10 is to permit an approach to the problem of circuit synthesis via uniform methodical items and not to reduce its realization to stochastic constructions.

## 9.3 EXAMPLES OF TWO-COMPONENT INTEGRATED NAVIGATION SYSTEMS

### *9.3.1 NONINVARIANT ROBUST INTEGRATED SPEED METER*

Consider the optimization of an integrated ground speed meter containing a Doppler sensor and an accelerometer with a longitudinal axis of sensitivity. Assume that for the Doppler sensor error referred to the input, the spectral density $S_{v1}(\omega) = S_{v1}$ is known. Assume too that for the accelerometer error (also referred to the input and having the dimension of speed), the maximum variance value $D_{\dot{v}2}$ of the first derivative is known, but the spectral density of that error is not known.

The measured parameter $g(t) = V(t)$ has variance, bounded from above by some given value $D_g$, at some mathematical expectation $\overline{g(t)} = M_g$ accepted for the conditional zero of a scale of speed, and not influencing the measurement accuracy. The criterion of optimality is the minimum upper bound of the centered error of measurement variance. The given characteristics of the measured parameter and the sensor errors do not permit reaching a conclusion about the expediency of excluding one of the sensors from the meter structure, or about the *a priori* imposition of a condition of invariance. Because the highest order of the bounded excitation derivatives is equal to unity, it is advisable to take the denominator order of the transfer functions $n = 1$. This defines not only a form of transfer function for the channel of the radar sensor $H_1(s)$, but

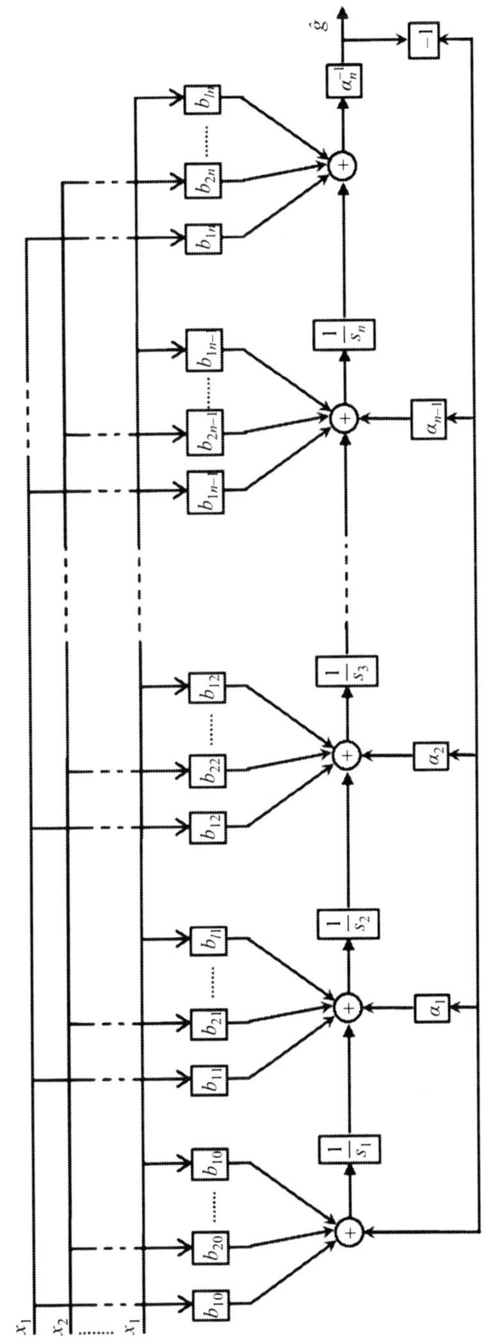


**Figure 9.9.** First variant of a calculator for an *l*-component integrated meter.

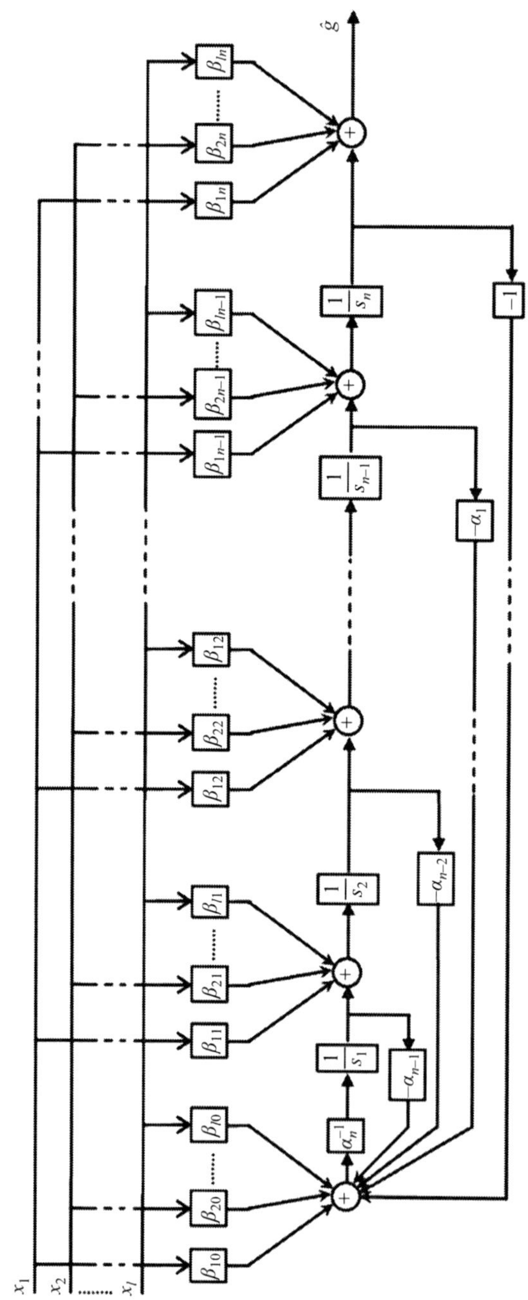


**Figure 9.10.** Second variant of a calculator for an $l$-component integrated meter.

also, taking into account the restriction of only the first derivative of the accelerometer error, a form of transfer function for the channel of accelerometer $H_2(s)$:

$$H_1(s) = b_{10} / (1 + a_1 s),\ H_2(s) = b_{21} s / (1 + a_1 s). \tag{9.48}$$

Then, the transfer function connecting the images of the measured parameter and of the dynamic error, will be:

$$H_3(s) = 1 - H_1(s) - H_2(s) = \left[1 - b_{10} + (a_1 - b_{21})s\right] / (1 + a_1 s). \tag{9.49}$$

Accepting that the optimization criterion function in the form

$$\bar{D}_e = D_{e1} + \bar{D}_{e2} + \bar{D}_{eg} \to \min,$$

where

$$D_{e1} = S_{\upsilon 1} \cdot \frac{1}{2\pi} \cdot \int_{-\infty}^{\infty} \left|H_1(j\omega)\right|^2 d\omega = \frac{b_{10}^2 \cdot S_{\upsilon_1}}{2a_1} \tag{9.50}$$

is an error variance in the radar sensor channel, $\overline{D_{e2}}$ is the upper bound of error variance in the accelerometer channel, and $\overline{D_{eg}}$ is the upper bound of the dynamic error variance, then from Nebylov (2004) the following formula can be written:

$$\overline{D}_{e2} = b_{21}^2 D_{\upsilon 2}, \overline{D}_{eg} = D_g \left[\max\left\{1 - b_{10}, 1 - b_{21} / a_1\right\}\right]^2. \tag{9.51}$$

Having subjected analytical research on a minimum obtained function of three variable $D_e(a_1, b_{10}, b_{21})$, one can find from (9.48)–(9.51) the expressions for optimum values of meter parameters,

$$a_1^0 = \left(\frac{S_{\upsilon 1}}{4D_{\upsilon 2}}\right)^{1/3}, b_{10}^0 = \left(\frac{3D_{\upsilon 2}^{1/3} S_{\upsilon 1}^{2/3}}{4^{2/3} D_g} + 1\right)^{-1}, b_{21}^0 = a_1^0 b_{10}.$$

For example, at $S_{\upsilon 1} = 1\,\mathrm{m^2 s^{-1}}$, $D_{\upsilon 2} = 10^{-3}\ \mathrm{m^2 s^{-4}}$, $D_g = 0.5\,\mathrm{m^2 s^{-2}}$ it is easy to obtain $a_1^0 = 6.30s$, $b_{10}^0 = 0.808$, $b_{21}^0 = 5.09\,s$, $\bar{D}_e(a_1^0, b_{10}^0, b_{21}^0) = 0.096\mathrm{m}^2 s^{-2}$.

For comparison, consider the minimum possible value $\overline{D_e}$ in the case of an *a priori* imposition of condition of invariance (9.10),

$$\bar{D}_e(a_1^0, 1, a_1^0) = 1.19 D_{\upsilon 2}^{1/3} S_{\upsilon 1}^{2/3} = 0.119\,\mathrm{m^2 s^{-2}}.$$

Here, it is seen that acceptance of a condition of invariance results in a loss in potential accuracy.

During meter realization, the following should be taken into account, that the accelerometer used as the sensor for speed information has a transfer function $W_{s2}(s) = k_a$ s because the transfer function of the appropriate channel of the calculator unit has a form $H_{e2}(s) = H_2(s) / W_{s2}(s) = b_{21} k_a^{-1} / (1 + a_1 s)$.

Notice that the optimal values of the parameters resulting in the transfer function $H_3(s)$ in a form $H_3(s) = 1 - b_{10}^0$ ($0 < b_{10} \le 1$ always) is appropriate to a noninertial unit. Physically, this is explained by the absence of *a priori* information on the frequency structure of the causal process $g(t)$. It is therefore expedient to make all its spectral components "equal in rights" during the formation of a dynamic error of measurement. If $D_g \to \infty$, this means that there is not enough information about the measured coordinate, $b_{10}^0 \to 1$, $H_3(s) \to 0$ and a condition of invariance is executed. For unlimited high measurement accuracy, such a result is obtained as $S_{\upsilon 1} \to 0$, when the radar sensor error is absent (in this case only the radar sensor is used), and at $D_{\upsilon 2} \to 0$, when the accelerometer error is quasi-constant. In the latter case the optimum parameters provide a zero steady-state error only over an indefinitely large observational time.

### 9.3.2 INTEGRATED RADIO-INERTIAL MEASUREMENT

An integrated system for measuring the altitude $h$ of a vehicle in flight involves a radio altimeter and an accelerometer with a vertical axis of sensitivity.

A radio altimeter has a broad-band error $v_{ra}$ that may be considered as white noise with a spectral density $S_{ra}(\omega)$. An accelerometer has an error $v_{ain}$ (referred to the input) with a limited maximum value for its second derivative $v_{ainM}^{(2)}$ (having the physical sense of zero drift). The output signal of the accelerometer could then be written as $x_a = k_a s^2 (h + v_{ain})$, where $s$ is an operator of derivation.

During synthesis of the integrated measuring system it is necessary to provide low pass filtration for the radio altimeter and high pass filtration for the accelerometer.

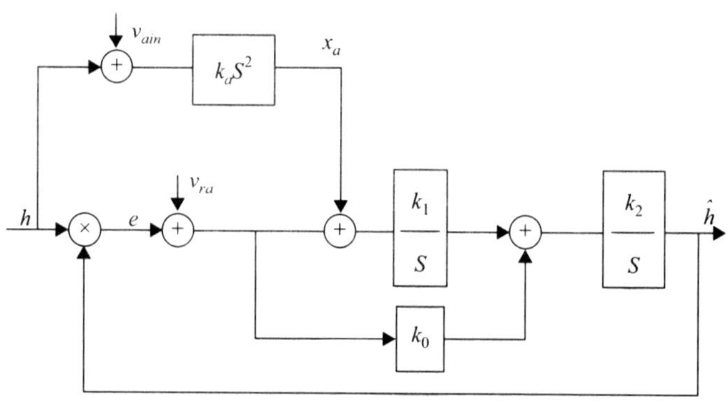


**Figure 9.11.** Block diagram for integrated radio-inertial measurement system.

Figure 9.11 shows a block diagram for an integrated radio-inertial measurement system that provides an estimation of vehicle flight altitude $\hat{h}$ with an error $e = h - \hat{h}$. Because the order of the limited derivative of disturbance $v_{ain}$ is 2, the system must include two integrators. Three factors $k_1$, $k_2$, and $k_3$ can be optimized, and the following expression for the output signal may be written:

$$\hat{h} = \frac{k_2}{s}\left[k_0(e + v_{ra}) + \frac{k_1}{s}(x_a + e + v_{ra})\right]. \tag{9.52}$$

This gives the following expression for the measurement error:

$$e = \frac{(1 - k_a k_1 k_2)s^2 h - k_2(k_1 + k_0 s)v_{rv} - k_1 k_2 s^2 v_{ain}}{s^2 + k_0 k_2 s + k_1 k_2}. \tag{9.53}$$

At $k_a k_1 k_2 = 1$ the condition of invariance (9.10) will be met, and the dynamic component of the error will be zero—that is, the error $e$ will not depend on the altitude $h$ to be measured. Hence, the total error will have only two components, caused by the mutually independent disturbances $v_{rv}$ and $v_{ain}$.

For such an error image it is correct to use the following expression:

$$E(s) = -H(s)V_{rv}(s) - [1 - H(s)]V_{ain}(s), \tag{9.54}$$

where

$$H(s) = \frac{K_2(1+\tau s)}{s^2 + K_2 \tau s + K_2}. \tag{9.55}$$

Here $K_2 = k_1 k_2$, $\tau = k_0 / k_1$, $V_{rv}(s)$, and $V_{ain}(s)$ are the images of $v_{ra}(t)$ and $v_{ain}(t)$.

The maximum error will be the sum of the maximum values of these two components:

$$e_M = e_{V_{ra}M} + e_{V_{ain}M}.$$

Transfer function (9.55) corresponds to the transfer function (9.34) of the equivalent open loop system. Hence, it is possible to apply formulae (9.37) and (9.38) when calculating the optimal values of $K_2$ and $\tau$:

$$K_2^0 = 1.28(\kappa v_{ainM}^{(2)})^{4/5} S_{ra}^{-2/5}, \tau^0 = \sqrt{3/(2K_2^0)}, \tag{9.56}$$

Here the factor $\kappa$ has the value 1.27 (see Section 9.2.7).

According to formula (9.38), at optimal values of $K_2$ and $\tau$ the maximal error will have the minimal value:

$$e_{M\min} = 4.03(\kappa v_{ainM}^{(2)})^{1/5} S_{vra}^{2/5}. \tag{9.57}$$

The 3D plot of values $e_{M\min}$ at different values of $v_{ainM}^{(2)}$ and $S_{vra}$ is shown in Figure 9.12.

For example, at $S_{vra}^{2/5}$ = 5 m$^2$ s$^{-1}$ and $v_{ainM}^{(2)}$ = $10^{-2}$ m s$^{-2}$ formulae (9.56) and (9.57) give $e_{M\min}$ = 3.21 m, $K_2^0 = 3.24\ 10^{-3}\,\mathrm{s}^{-2}$, $\tau^0 = 21.5$ s. Then, accepting $k_a = 1$, $k_0 = 1$, two other factors for the measurement calculator will be $k_1 = 1/\tau^0 = 0.04654\ \mathrm{s}^{-1}$, $k_2 = K_2^0\tau^0 = 0.0697\mathrm{s}^{-1}$.

### *9.3.3 AIRBORNE GRAVIMETER INTEGRATION*

The estimation of anomalies in the acceleration due to gravity (AGA) using airborne gravimeters (aviation gravimetry) has wide applications. AGA maps are very useful in the search for mineral resources, and in special navigational and guidance tasks. Any system of AGA mapping should encompass various areas of the Earth "seen" from different altitudes. Hence, the process of constructing such maps is labor-intensive, time-consuming, and requires the combining of data obtained by means of land, airborne, and spaceborne gravimeters. The velocity of motion of the vehicle in which the gravimeter is installed influences both the accuracy and efficiency of the measurements and can be optimized. In the case of aviation gravimetry the optimal

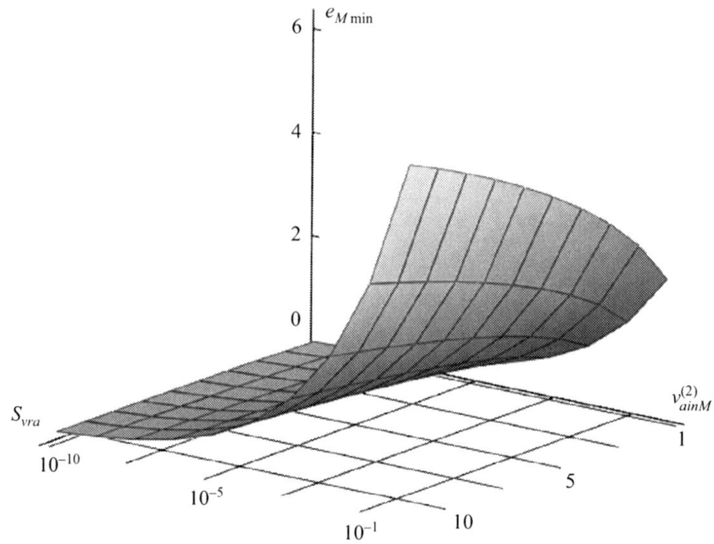


**Figure 9.12.** Plot of $e_{M\,\min}$ for different values of $v^{(2)}_{ainM}$ and $S_{vra}$.

velocity is rather insignificant and lies in the middle of the subsonic range. Airborne gravimetry is a highly productive and comparatively inexpensive surveying method that can be employed even in hard-to-reach areas.

Airborne gravimetry is actually based on the integration of a precise gravimeter and a satellite navigation system (SNS) operating in a differential mode with phase measurements, into a single measuring system. The gravimeter itself may be considered as an almost perfect system that can measure the gravitational acceleration vector at any local point, and may therefore be considered as a very complicated accelerometer.

The gravimeter readings can be presented as a sum of three components: $\tilde{g}^{\mathrm{gr}} = \tilde{g} + \delta g + \ddot{h}$, where $\tilde{g}$ is a gravity anomaly, $\delta g$ is a gravimeter error, and $\ddot{h}$ is a vertical acceleration caused by any vertical motion of the aircraft.

The data from the DSNS are $h^{\mathrm{SNS}} = h + \delta h$, where $h$ is unknown altitude and $\delta h$ is the error in the DSNS measurements.

For the purpose of eliminating indications of an unknown component of vertical acceleration in the carrier vehicle, the data from both sensors are used concurrently. The difference between the second integral of the gravimeter indications and the altitude from the DSNS is formed as shown in Figure 9.13.

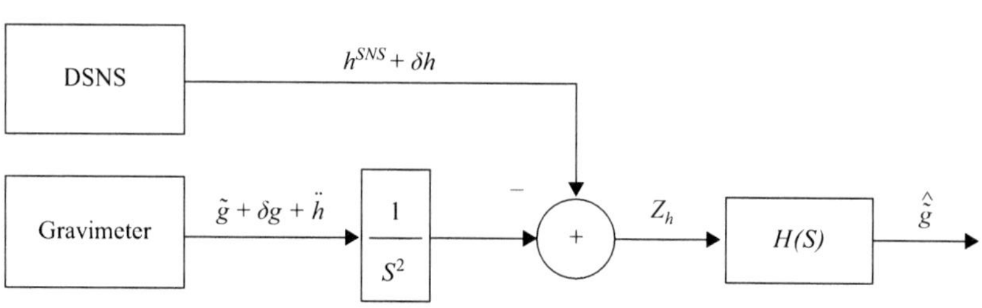


**Figure 9.13.** Block diagram for an AGA system.

The differential measurements may be presented as follows (Kulakova, Nebylov, and Stepanov 2004; Jin, Fathi, and Jekeli 1997):

$$z_h = \frac{\tilde{g}^{\text{gr}}}{s^2} - h^{\text{SNS}} = \frac{\tilde{g} + \delta g}{s^2} - \delta h \,. \tag{9.58}$$

The estimation problem consists of obtaining AGA using differential measurements (9.58), that is, in the design of the filter transfer function $H(s)$. For this purpose, models for $\tilde{g}$, $\delta h$, and $\delta g$ are necessary.

Accepting for such models the data obtained using the gravimeter developed in the Russian CSRI "Elektropribor" and dual-frequency geodetic Canadian *Novatel* receiver, it has been shown (Blazhnov et al. 2002; Stepanov 2002) that the errors in phase measurements are mainly of white-noise character with intensity $R_h$ = (0.005 m)$^2$ s$^{-1}$, and that the gravimeter errors can be described as white noise with intensity $R_{gr}$ = (5 mGal)$^2$ s$^{-1}$.

GA is often described by a process with a spectral density (Jordan 1972) as follows:

$$S_{\tilde{g}}(\omega) = 2a^3 \cdot \sigma_{\tilde{g}}^2 \cdot \frac{5 \cdot \omega^2 + a^2}{(\omega^2 + a^2)^3}, \tag{9.59}$$

where $\sigma_{\tilde{g}}^2$ is the variance, $a = V/\mu$; $V$ being the vehicle speed, and $\mu$ being the correlation distance. If the AGA gradient $\nabla\tilde{g} = \sigma_{\partial\tilde{g}/\partial l}$ (the r.m.s. value of the gravitational increment at a distance $l$, which is assumed to be equal to 1 km) is given, then α can be found from $(\nabla\tilde{g})^2 = 2a^2\sigma_{\tilde{g}}^2$. The AGA corresponding to the model (9.59) obtained by simulation is shown in Figure 9.14.

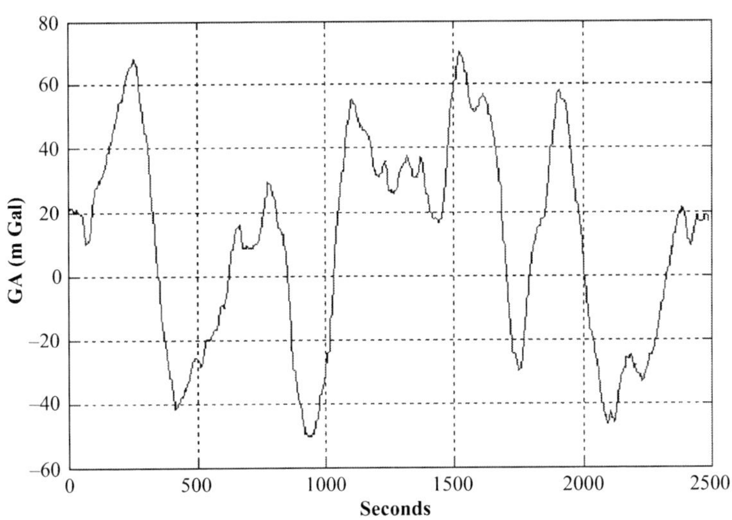


**Figure 9.14.** Pattern of AGA at $\sigma_{\partial\tilde{g}/\partial l}$ = 10 mGal km$^{-1}$, $\sigma_{\tilde{g}}$ = 30 mGal, $V$ = 50 m s$^{-1}$.

It is not infrequent that instead of $a$ (the r.m.s. value of a gravitational increment at a distance, which is assumed equal to 1 km), $\sigma_{\partial\tilde{g}/\partial l}$ is given. In this case the value $a$ can be found from the equation $\sigma^2_{\partial\tilde{g}/\partial l} = 2a^2\sigma^2_{\tilde{g}}$. AGA spectral density is shown in Figure 9.15.

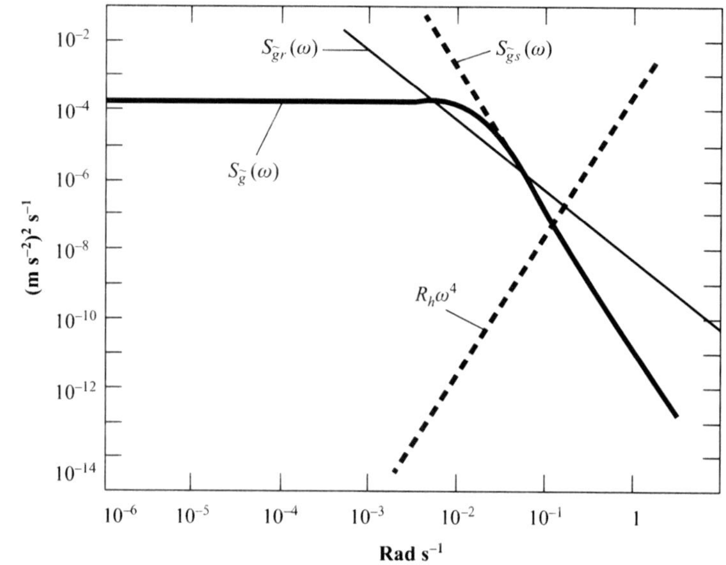


**Figure 9.15.** AGA spectral densities and measurement errors at $\sigma_{\partial\ddot{g}/\partial l} = 10$ mGal km$^{-1}$, $\sigma_{\ddot{g}} = 30$ mGal, and $V = 50$ m s$^{-1}$.

The model for Equation (9.59) is referred to as the Jordan model and is widely used in problems that require stochastic descriptions of AGA. However, it makes the algorithms more complicated in comparison with the case where a simpler model is used to describe AGA. The problem can be simplified by straightening (on a logarithmic scale) the density $S_{\ddot{g}}(\omega)$ in the vicinity of its cross-point with $\omega^4 R_h$ (see Figure 9.15). In this case the Equation (9.59) model is approximated by the second integral of the white noise with intensity $q^2 = 10\alpha^3\sigma_{\ddot{g}}^2$:

$$S_{\ddot{g}_S}(\omega) \approx q_{\ddot{g}} / \omega^4 . \tag{9.60}$$

The optimal filter for AGA description by this Jordan model will be of the fifth order, whereas in the form of the second integral of the white noise it will be of the fourth order.

The considered problem can also be solved by applying a robust approach in which it is necessary to accept as the more authentic model the dispersions of the AGA itself and its first derivative, connected with the particular spectral density by the relations:

$$D_{\ddot{g}_i} = \frac{1}{\pi}\int_0^\infty \omega^{2i} S_{\ddot{g}}(\omega)d\omega, \quad i = 0,1. \tag{9.61}$$

Obviously, numerous spectral densities are in accordance with each particular set of such numerical characteristics. Therefore, while giving information about signals with AGA dispersion $D_{\ddot{g}0} = \sigma_{\ddot{g}}^2$ and first derivative dispersion $D_{\ddot{g}1} = 2a\sigma_{\ddot{g}}^2$, the filter transfer function $H(s)$ determination is defined. However, at such *a priori* indeterminacy, it is impossible to find the error dispersion $D_e$, but as shown in Section 9.2.6, it is possible to estimate its upper bound $\overline{D_e}$. This upper bound must therefore be used as the criterion, minimized during algorithm synthesis, in the considered approach.

The suggested robust approach is intended for filter realization with a preset structure determined according to heuristic considerations. Recalling that the gravimeter readings are integrated twice, this is equivalent to the fact that for the input signal at the synthesized filter, the dispersions of second $D_{h2}$ and third $D_{h3}$ derivatives are given. The order of the older among the restricted derivatives of the useful signal determines the inclination of the spectral density plot in the high frequency domain. Therefore, this order determines the order of the filter used with such a spectral density. Hence, in general, the order of the filter must be not less than the order of the older among the restricted derivatives of the useful signal, which in this case is three. Additionally, because it is necessary to estimate not the original signal, but its second derivative (i.e., the acceleration via altitude measurement), a double differentiation operation with smoothing must be carried out, which is why the filter order should be increased by two. Thus, it is necessary to optimize the transfer function

$$H(s) = \frac{b_2 s^2 + b_3 s^3 + b_4 s^4}{a_0 + a_1 s + a_2 s^2 + a_3 s^3 + a_4 s^4 + a_5 s^5}. \tag{9.62}$$

The transfer function for acceleration $\tilde{H}_r(s) = \frac{1}{s^2} H(s)$ obtained after the optimization of coefficients $\{a_i\}_0^5, \{b_j\}_2^4$, resembles that of a classical low pass filter. It has a flat vertex and an inclination of –60 dB dec$^{-1}$ in the high frequency domain, which validates the supposition that the problem can be solved on the basis of a third order filter. In practice, it has been observed that the precision factor $\overline{D}_e$ increased only a few percent in comparison with the case when a fifth order TF was used. Therefore, given such evidence, it is possible to claim that a robust filter appropriate to the airborne gravimetry problem is of the third order, that is, two orders lower than the filter optimal for the Jordan model.

Table 9.1 presents calculated results as r.m.s. errors for different values of the derivative for AGA equal to 3, 5, and 10 mGal km$^{-1}$. In the calculations, it was assumed that $\sigma_{\ddot{g}} = 30$ mGal and that the velocity was assumed to be $V = 50$ m s$^{-1}$, which is typical for an aerogravimetric survey.

**Table 9.1.** R.m.s. errors in AGA filtering (mGal)

| $\sigma_{\partial\ddot{g}/\partial l}$, **mGal/km** | **3** | **5** | **10** |
|---|---|---|---|
| Optimal algorithm, $\sigma_e$ | 2 | 3.1 | 5.5 |
| Robust algorithm, $\overline{\sigma}_e$ | 2.8 | 4.1 | 6.7 |
| Robust algorithm, $\sigma_{er}$ | 2.8 | 3.9 | 6.4 |

The robust filter is characterized here by two parameters, an upper bound of r.m.s. error, $\overline{\sigma}_e$, and the exact value of the r.m.s. error, $\sigma_{er}$, which characterize the quality of the robust filter in the case of filtering for the process represented as a Jordan model. Comparison by accuracy shows that both filters have similar r.m.s. error values. For the three selected values of AGA derivatives $\sigma_{\partial\ddot{g}/\partial l} = 10$, 5, and 3 mGal km$^{-1}$, the losses in accuracy of the robust filter compared with that of the optimal one are 20, 25, and 30% of the r.m.s. error, respectively. The differences in accuracy between filters for different derivative values $\sigma_{\partial\ddot{g}/\partial l}$ are caused because decreasing these values narrows the spectral density, $S_{\ddot{g}}(\omega)$. This results in a narrowing of the passband of the optimal filter, while the robust filter must take into account all the signals belonging to the preset class.

Note that for the entire class of input signals with preset values $D_{\ddot{g}0}$, $D_{\ddot{g}1}$ accepted for the robust filter synthesis, the filtering error will not exceed the value $\overline{\sigma}_e$.

Before considering filters in terms of their sensitivities to the initial model disagreements, it is expedient to compare their gain plots, shown in Figure 9.16.

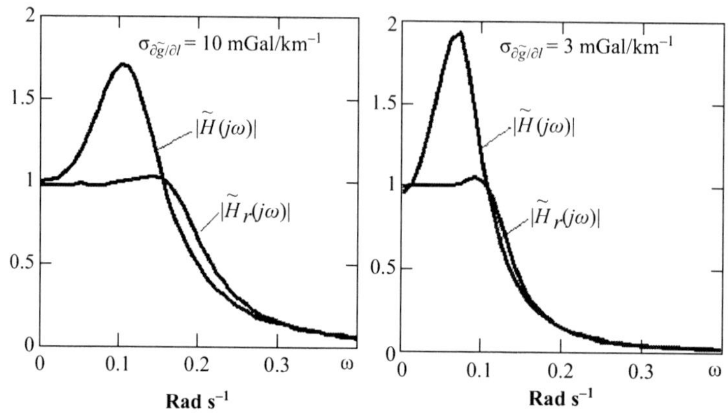


**Figure 9.16.** Gain versus acceleration plots for optimal $|\tilde{H}(j\omega)|$ and robust $|\tilde{H}_r(j\omega)|$ filters.

It was mentioned earlier that the gain plot of a robust filter has a flat vertex. By comparison, the existence of gain peaks in the plot for an optimal filter makes it too sensitive to the presence of any additional noise power in the frequencies where the gain coefficient is high. It is also sensitive to any deviation in the accepted model that causes a decrease in the useful signal energy in this range. Therefore, intuitively, though the robust filter has low sensitivity, no special precautions are needed during its synthesis.

It is interesting to solve the "inverse" problem to the optimal filtering one, that is, to find a useful signal spectrum $S_r(\omega)$, observed with added white noise interference, for which the synthesized robust filter will be optimal. Such investigations (with a slight approximation) have shown that a robust filter can be considered as corresponding to the solution of the optimal filtering problem when the AGA are described by the Wiener process.

The "robust" spectral density, $S_{\ddot{g}r}(\omega)$, is shown in Figure 9.15, and it follows that robust filters can be adjusted for a spectral density that uses a greater frequency bound. Therefore, robust filters will offer low sensitivity to those disagreements in otherwise acceptable models that cause extensions in the spectral density. Hence, a robust filter, in comparison with an optimal filter adjusted for the Jordan model, will exhibit a greater filtering error while narrowing the AGA spectral density.

It could be said that knowledge of the AGA dispersion value $\sigma_{\ddot{g}}^2$ has rather little influence on robust algorithm synthesis. This means that if only the parameter, $D_{\ddot{g}1}$, that contains information about both dispersion and correlation intervals is preset, then the precision factor (upper bound $\overline{D}_e$) will vary only slightly.

Application of the $H_2/H_\infty$ approach in the problem of airborne gravimetry has been developed by Kulakova, Nebylov, and Stepanov 2008).

### 9.3.4 THE ORBITAL VERTICANT

An orbital verticant (for constructing a local vertical) should indicate the direction to the center of the Earth at any moment during satellite motion along a near-Earth orbit. Usually, this is achieved by an integrated measuring system containing positional and inertial sensors (Figure 9.17). An infrared vertical (Chertok 2006 or a radio vertical (RV) installed on a horizontally stabilized platform (SP) may be considered for use as a positional sensor. This can be collocated with an accelerometer (A) perceiving an acceleration in the orbital motion of the satellite, and also the component of an acceleration due to gravity arising from a deviation in the platform (and therefore of an accelerometer axis of sensitivity) from a proper attitude by some small angle $e$. Thus, the accelerometer defines the acceleration as

$$a = eg + Rd^2\gamma / dt^2, \tag{9.63}$$

where $g$ is the gravitational acceleration, $R$ is the Earth's radius and $d^2\gamma / dt^2$ is the orbital angular acceleration (the current orbital angle being $\gamma$ ).

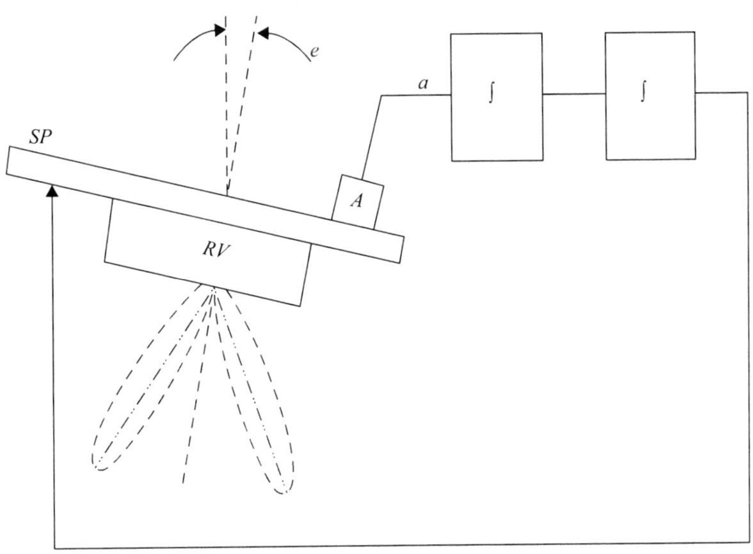


**Figure 9.17.** General arrangement of an orbital verticant.

Naturally, the accelerometer also has an interrupt component $\upsilon_a$ of the output signal (an error with the dimensions of acceleration), which it is admissible to consider in a first approximation as a quasi-constant variable with a maximum possible value $\upsilon_{aM}$.

The positional sensor immediately measures any error in vertical keeping $e$. If either an infrared or a radio vertical sensor is used, its error $\upsilon_p$ can be considered as white noise with a spectral density $S_{\upsilon p}(\omega) = S_{\upsilon p}$.

The measured acceleration $a$ is doubly integrated and acts on the angle setter of the stabilized platform, and the signal from the positional sensor with a scale factor $k_p$ acts at the same point after single-valued integration. Thus the stabilized platform is turned through an angle $\vartheta$ given by

$$\theta = \int k_2(\int k_1 a dt + k_p e)dt, \tag{9.64}$$

where $k_1, k_2$ are transmission factors of the first and second integrators. (The interrupt signals of both sensors are not considered in expression (9.64).)

It is possible to add the feedback path equation to Equations (9.63) and (9.64):

$$e = \gamma - \theta, \tag{9.65}$$

so defining an interconnecting link between a vertical error in $e$, the current vehicle orbital angle in radians $\gamma$, and a stabilized platform angle of rotation $\theta$.

The block diagram in Figure 9.18 corresponds to Equations (9.63) to (9.65).

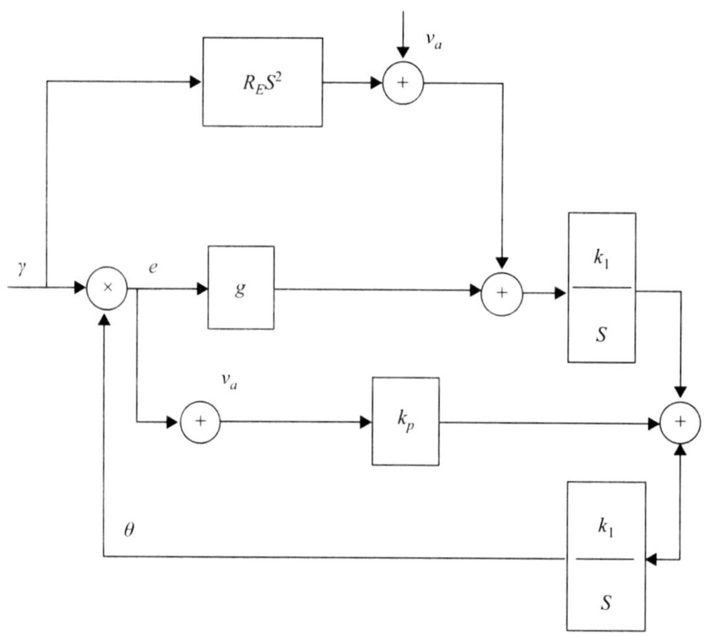


**Figure 9.18.** Block diagram of orbital verticant.

Consolidating Equations (9.63) to (9.65) along with the interrupt components of the sensor signals, and transferring from originals to Laplace transforms, finally results in the Laplace image of the verticant error:

$$E(s) = E_\gamma(s) - E_a(s) - E_p(s), \tag{9.66}$$

where

$$E_\gamma(s) = \frac{(1 - k_1 k_2 R_E)s^2}{s^2 + 2\xi\Omega_0 s + \Omega_0^2}\Gamma(s) \tag{9.67}$$

is an image of the dynamical error,

$$E_a(s) = \frac{k_1 k_2}{s^2 + 2\xi\Omega_0 s + \Omega_0^2} V_a(s) \tag{9.68}$$

is an image of the error due to accelerometer drift, and

$$E_p(s) = \frac{k_1 k_2 s}{s^2 + 2\xi\Omega_0 s + \Omega_0^2} V_p(s) \tag{9.69}$$

is an image of the error due to positional sensor noise.

In equations (9.67) to (9.69), $\Omega_0 = \sqrt{\frac{g}{R_E}}$, this being the frequency of oscillation of a mathematical pendulum with a plumb line length $R_E$ equal to the Earth's radius (the period of such an oscillation $T_0 = 2\pi/\Omega_0 \approx 84.6\,\text{min}$, called the Shuler period, could also be considered); and $\xi = \frac{k_2 k_p}{2\Omega_0}$ being a damping factor.

After returning from the Laplace images of (9.66) to the originals, the total error will appear as

$$e(t) = e_\gamma(t) + e_{va}(t) + e_{vp}(t). \tag{9.70}$$

According to Equation (9.67) the dynamical error $e_\gamma(t)$ will be zero at $k_1 k_2 R_E = 1$, which is why the invariance condition (9.10) in the considered system has the form

$$k_1 k_2 = \frac{1}{R_E}. \tag{9.71}$$

So, in the invariant orbital verticant, the maximal error will have only two components corresponding to the last two summands in (9.70):

$$e_M = e_{vaM} + e_{vpM}. \tag{9.72}$$

These components may be calculated on the basis of the theory given in Section 9.2 as follows:

$$e_{vaM} = \lim_{s\to 0} E_a(s) = \frac{k_1 k_2}{\Omega_0^2} v_{aM} = \frac{v_{aM}}{g}, \tag{9.73}$$

$$e_{vpM} = 3\sigma_{evp} = 3\sqrt{S_{vp}\Delta f_p}, \tag{9.74}$$

where the equivalent bandwidth of the positioning sensor channel is

$$\Delta f_p = \frac{1}{2\pi}\int_{-\infty}^{\infty} \left| \frac{k_1 k_p j\omega}{(j\omega)^2 + 2\xi\, j\omega + \Omega_0^2} \right|^2 d\omega = \frac{(k_2 k_p)^2}{4\xi\Omega_2} = \frac{k_2 k_p}{2}. \tag{9.75}$$

For example, at $v_{aM} = 3\ 10^{-3}$ m s$^{-2}$, $S_{\upsilon p} = 10^{-4}$ rad$^2$ Hz$^{-1}$, $R_E = 6.37\ 10^6$ m, $\Omega_0 = 1.24\ 10^{-3}$ s$^{-1}$, $\xi = 0.5$, and $k_1 = 10^{-4}$, expressions (9.71) to (9.75) permit the following:

$$k_2 = 1\ /10^{-4}\,6.37\ 10^6 = 1.57\ 10^{-3}\ \text{m}^{-1},\ k_p = 2\Omega_0\xi/k_2 = 2\ 1.24\ 10^{-3}\ 0.5/(1.57\ 10^{-3}) = 0.790,$$

$$e_{vaM} = 3\ 10^{-3}/9.81 = 3.06\ 10^{-4} = 1.10\ \text{ang s},\ \Delta f_p = 1.57\ 10^{-3}\ 0.790 = 1.24\ 10^{-3}\ \text{Hz},$$

$$e_{vpM} = 3\sqrt{10^{-4}\ 1.24\ 10^{-3}} = 1.05\ 10^{-3} = 3.78\ \text{ang s},\ e_M = 1.10 + 3.78 = 4.88\ \text{ang s}.$$

## REFERENCES


Andronov, A. A., A. A. Witt, and S. E. Hikin. 1969. *Theory of Oscillations*. Moscow: Fizmatgiz. (In Russian.)

Ahiezer, N. I. 1961. *Classical Problem of Moments*. Moscow: Fizmatgiz. (In Russian.)

Bendat, J. S., and A. G. Piersol. 1966. *Measurement and Analysis of Random Data* (p. 390). New York: Wiley.

Bendat, J. S., and A. G. Piersol. 1971. *Random Data: Analysis and Measurement Procedures* (p. 407). New York: Wiley.

Bendat, J. S., and A. G. Piersol. 1980. *Engineering Applications of Correlation and Spectral Analysis* (p. 320). New York: Wiley.

Besekerskiy, V. A., and A. V. Nebylov. 1993. *Robust Automatic Control Systems* (p. 240). Moscow: Nauka. (In Russian.)

Blazhnov, B.A., L. P. Nesenjuk, V.G. Peshekhonov, A.V.Sokolov, L.S.Elinson et al. 2002. *An integrated mobile gravimetric system. Development and test results.* Proceedings of the 9th Saint-Peterburg International Conference on Integrated Navigation Systems.

Bulgakov, B. V. 1946. "About accumulation of perturbations in linear time invariant oscillating systems." *Journal of Reports of the USSR Academy of Sciences*. 51 (5): 339–42. (In Russian.)

Chertok, B. E. 2006. Rockets and People, V.2 (p. 698). US National Aeronautics and Space Admin.

Duge, D. 1972. *Theoretical and Applied Statistics* (Russian translation). Moscow: Nauka.

Gnoyenski, L. S. 1961. "About accumulation of perturbations in linear systems." *Journal of Applied Mathematics and Mechanics*. 25(6): 317–31. (In Russian.)

Greenlee, T. L., and C. T. Leondes. 1977. "Generalized bounding filters for linear time invariant systems." *Journal of Proceedings of IEEE Conference on Decision Control* 585–90.

Huber, P. J. 1984. *Robust Statistics*. New York: John Wiley and Sons.

Jin, W., D. Fathi, and C. Jekeli. 1997. "INS, GPS, and photogrammetry integration for vector gravimetry." *Proceedings of the International Symposium on Kinematic Systems in Geodesy, Geomatics and Navigation.* Banff, Canada.

Jordan, S. K. 1972. "Self-consistent statistical models for gravity anomaly and undulation of the geoid." Journal of Geophysical Research 77 (20): 3660–70. DOI: 10.1029/JB077i020p03660.

Kalman, R. E. 1960. "A new approach to linear filtering and prediction problems." *Transactions of the ASME - Journal of Basic Engineering* 82: 35–45. DOI: 10.1115/1.3662552.

Karlin, S., and W. J. Studden. 1966. Tchebycheff systems: With applications in analysis and statistics. *Pure and Applied Mathematics*, Vol. XV. New York-London-Sydney: Interscience Publishers John Wiley & Sons.

Kassam, S. A., and Tong Leong Lim. 1977. "Robust Wiener filters." *Journal of the Franklin Institute* 304 (4/5).

Kassam S. A., and H. V. Poor. 1985. "Robust techniques for signal processing: A survey." *Proceedings IEEE* 73: 433–81. DOI: 10.1109/PROC.1985.13167.

Krasovsky, N. N. 1968. *The Theory of Motion Control*. Moscow: Science. (In Russian.)

Krein, M. G., and A. A. Nudelman. 1973. *A Problem of Markov's Moments and Extremal Problems*. Moscow: Nauka. (In Russian.)

Kulakova, V. I., and A. V. Nebylov. 2008. "Ensuring evaluation of signals with bounded dispersions of derivatives." *Automation and Remote Control* 1: 83–96.

Kulakova, V. I., A. V. Nebylov, and O. A. Stepanov. 2004. "Application of the robust approach to the problem of airborne gravimetry." *Proceedings of the 16th IFAC Symposium on Automatic Control in Aerospace*. St. Petersburg, Russia. pp.354–9.

Kulakova, V. I., A. V. Nebylov, and O. A. Stepanov. 2008. "Application of $H_2/H_\infty$ approach in the problem of airborne gravimetry." *Gyroscopy and Navigation* 2: 53–62. (In Russian.)

Looze, P., and H. Poor. 1983. "Minimax control of linear stochastic systems with noise uncertainty." Journal of IEEE (Institute of Electrical and Electronics Engineers)Transactions of Automat. Control. AC-28 (9): 882–8. DOI: 10.1109/TAC.1983.1103353.

Magni, J. F., S. Bennani, and J. Terlouw. (Editors). 1997. *Robust Flight Control: A Design Challenge*. London: Springer-Verlag. DOI: 10.1007/BFb0113842.

Nebylov, A. V. 2004. "Ensuring control accuracy." *Lecture Notes in Control and Information Sciences*, 305, Heidelberg, Germany: Springer-Verlag.

Nebylov, A. V., and P. Wilson. 2002. *Ekranoplane - Controlled Flight Close to Surface*. Southampton, UK: WIT-Press.

Stepanov, O. A. 2002. "Integrated inertial-satellite navigation systems." *Gyroscopy and Navigation*, CSRI "Elektropribor", St. Petersburg, 1: 23–45. (In Russian.)

Tsipkin, Ja. Z., and A. S. Pozniak. 1981. "Optimal and robust algorithms of optimization at presence of correlated parasites." *Journal of the USSR Academy of Sciences Reports*. 258(6): 1330–3. (In Russian.)

Wiener, N. 1949. *Extrapolation, Interpolation and Smoothing of Stationary Time Series*. New York: Wiley.

# Epilogue

The authors and editors hope that the reader has found in this book the knowledge and data required; and to conclude, it is appropriate to offer some explanation of how the included subject matter was chosen.

Initially, it should be noted that technology progresses very quickly in the field of aerospace sensors and significant new results appear every year. Therefore, though the book claims to provide comprehensive and indepth examinations of basic concepts and current problems, it may not reflect the very latest results. However, the content may be complemented by checking the relevant scientific and technical journals and the publications following major conferences such those of the IFAC, AIAA, IEEE, and IET amongst others.

In part, the spacial limitations of the present volume modified the selection of the sensor material presented, which means that it does not purport to include absolutely all aerospace sensors. For example, it does not include star sensors and some other optical devices that are the most accurate gauges of angular orientation in spacecraft and space probes, and are capable of ensuring the accuracy of attitude control to a few arc seconds. However, the main problems encountered in the construction of such sensors lie not with the optical systems, but with the digital image processing algorithms for constellations as obtained by the CCDs.

An even more complex rôle is in the province of algorithms for information processing in some radio navigation and guidance systems for providing the emission, reception, and processing of specialized radio signals. These systems include radio complexes for long and short range navigation, landing systems, satellite navigation systems, and homing systems. Such measuring systems, in terms of integrated navigation and motion control design, are actually sources of information and, as suggested in this book, may be considered as a special group of aerospace sensors. In view of this, it was decided to publish a second, or companion volume with descriptions of the above-mentioned radio aids as aerospace sensors, under the title *Aerospace Navigation and Guidance Systems.* (At the time of writing, this is in process of publication.)

This second book, taken in concert with the present one, form a two-volume duo that addresses most of the problems met in aerospace sensor investigation, design, and construction for motion control systems. The authors and editors hope that readers will not disregard this second book and will consider both books on "Aerospace Sensors," parts one and two, as an integral edition useful for gaining an understanding of modern concepts in motion control and aerospace vehicle navigation.

Finally, it should be remembered that each aerospace vehicle is a very complex machine that results from the efforts of many people, all highly qualified in the various relevant areas of science and technology. It is hoped that this book may also contribute to better mutual understanding between these participants and assist them in their chosen areas of aerospace technology.

A.V. Nebylov
Volume Editor

J. Watson,
Series Editor
2012

# INDEX

## Check Out The Other Mechanical Engineering Titles We Have!

**Automotive Sensors**, *John Turner*
**Centrifugal and Axial Compressor Control**, *Gregory McMillan*
**Virtual Engineering**, *Joe Cecil*
**Chemical Sensors: Fundamentals and Comprehensive Sensor Technologies, Volumes 1 through 6**, as well as **Chemical Sensors: Simulation and Modeling, Volumes 1 through 5**, *Ghenadii Korotcenkov*, Editor
**Biomedical Sensors**, *Deric P. Jones*
**Acoustic High-Frequency Diffraction Theory**, *Federic Molinet*
**Bio-Inspired Engineering**, *Chris Jenkins*
**The Essentials of Finite Element Modeling and Adaptive Refi nement: For Beginning Analysts to Advanced Researchers in Solid Mechanics**, *John O. Dow*

---

## Announcing Digital Content Crafted by Librarians

Momentum Press offers digital content as authoritative treatments of advanced engineering topics, by leaders in their fields. Hosted on ebrary, MP provides practitioners, researchers, faculty and students in engineering, science and industry with innovative electronic content in sensors and controls engineering, advanced energy engineering, manufacturing, and materials science. **Momentum Press offers library-friendly terms:**

- perpetual access for a one-time fee
- no subscriptions or access fees required
- unlimited concurrent usage permitted
- downloadable PDFs provided
- free MARC records included
- free trials

The **Momentum Press** digital library is very affordable, with no obligation to buy in future years.

For more information, please visit **www.momentumpress.net/library** or to set up a trial in the US, please contact **Adam Chesler**, *adam.chesler@momentumpress.net*.

CPSIA information can be obtained at www.ICGtesting.com
Printed in the USA
BVOW062002281112

306759BV00002B/2/P

9 781606 500590